SECOND EDITION

Organic Chemistry

Ralph J. Fessenden
Joan S. Fessenden

University of Montana

Study Guide with Solutions

Willard Grant Press
Boston, Massachusetts

PWS PUBLISHERS
Prindle, Weber & Schmidt · Willard Grant Press · Duxbury Press ·
Statler Office Building · 20 Providence Street · Boston, Massachusetts 02116

 PWS Publishers is a division of Wadsworth, Inc.

Printed in the United States of America
10 9 8 7 6 5 4 3 2 — 86 85 84 83 82

ISBN 0-87150-753-6

Text composition by Rita de Clercq Zubli
Cover design by Martucci Studio

To the Student

This Study Guide is intended to accompany the text Organic Chemistry, 2nd Edition. Included in this guide are some hints to help you study organic chemistry; a chapter-by-chapter discussion of many important features; and answers to the study problems at the end of each chapter in the text. (The answers to the study problems within the chapters are at the end of the text.) In this guide, we have included explanations of the answers where appropriate. These explanations will be especially helpful if you have difficulty answering a particular type of problem.

Sometimes organic nomenclature, bonding, reactions, etc. seem perfectly clear to a student, yet he does poorly on an examination. To avoid being in this situation, use your study time correctly.

What is the correct way to study organic chemistry? It is similar to learning a foreign language: first, understand the material; then, memorize the salient features. Rote memorizing of organic chemistry without first understanding the material is a waste of time. On the other hand, studying only to the point of understanding does not result in good examination grades. Only by coupling understanding and memorizing can you progress to thinking in the language of organic chemistry.

As a check on your understanding, you should work the problems within the chapter as you go along. (Cover the answers to Sample Problems with a sheet of paper, and try to work them as well.)

It is important to <u>write</u> your answers to the problems rather than simply to envision the answers in your mind. There are two reasons for writing out the answers: (1) for practice in the mechanical art of writing organic formulas, and (2) for reinforcing the subject material.

After you have studied the textual material and your lecture notes thoroughly, tackle the chapter-end problems. Some of these problems are drill, while some require thought and ingenuity. See if you can answer the problems yourself, and <u>then</u> check the answers and the text. (If you try to take shortcuts, you are only deceiving yourself.) If your answers are correct, you have a reasonable mastery of the material. If you have several incorrect answers, go back and restudy the chapter, your lecture notes, and the hints in this guide.

As a starting point on what to study, be sure you understand each definition and each reaction in the chapter summary and the summary tables. Be sure you can extrapolate from the generalities of the summaries to the specific reactions mentioned in the chapter. Then, be sure you can extrapolate from the specific reactions in the chapter to reactions of new and different compounds.

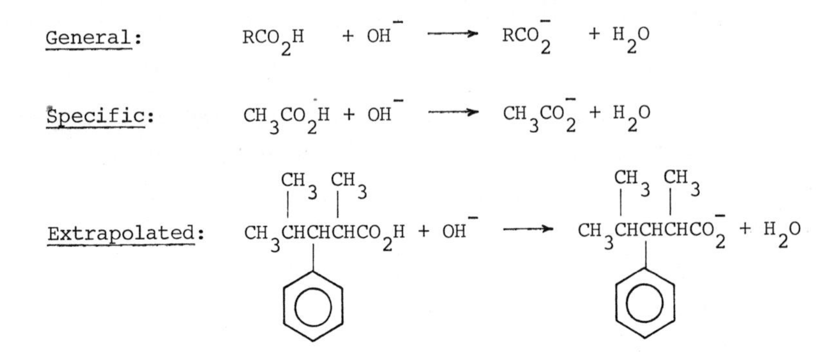

Organic molecules are three-dimensional. Besides a pencil and paper, you will need a molecular model kit. At the start of the course, especially in discussions of stereochemistry, you will find it helpful to construct models of organic molecules and compare them with their representations on paper.

Contents

CHAPTER 1

Atoms and Molecules—A Review

Some Important Features

Electrons are found in shells surrounding the nucleus of an atom. Each shell is composed of one or more atomic orbitals. The first shell contains a 1*s* orbital (spherical); the second shell contains one 2*s* orbital (spherical) and three 2*p* orbitals (dumbbell-shaped and mutually perpendicular). Each orbital can hold zero, one, or two electrons. Electrons are usually contained in the lowest-energy orbitals possible (1*s*, then 2*s*, then 2*p*).

The halogens, oxygen, and nitrogen have fairly high electronegativities (attraction for outer electrons). The metals have low electronegativities, while carbon and hydrogen have intermediate electronegativities.

Chemical bonds are formed by electrons in the outer shell of an atom. Whether ionic or covalent bonds are formed depends on the electronegativity difference between two atoms. Carbon forms covalent bonds with other elements. These covalent bonds may be nonpolar (C—C or C—H) or polar (C—O, C—N, or C—Cl), depending on the electronegativity difference between C and the other element.

Molecules with NH, OH, or HF bonds can form hydrogen bonds with each other or with other molecules containing N, O, or F atoms with unshared electrons.

An acid is a compound that can donate H^+ or accept electrons, while a base is a compound that has unshared electrons that can be donated. Common organic acids are the carboxylic acids, compounds with a $-CO_2H$ group. Common organic bases are amines, compounds with a nitrogen atom bonded to three other atoms.

$$\underset{\text{a carboxylic acid}}{CH_3CO_2H} + \underset{\text{an amine}}{(CH_3)_3N:} \rightleftharpoons CH_3CO_2^- + (CH_3)_3\overset{+}{N}H$$

The strengths of acids and bases are indicated by their pK_a values or pK_b values. A smaller numerical value for pK_a means a stronger acid.

Other important topics covered in this chapter are ionic and covalent bonds, calculation of formal charge, types of chemical formulas, bond lengths and angles, and bond dissociation energies.

Reminders

In stable compounds, carbon forms four bonds, hydrogen forms one bond, and oxygen forms two bonds.

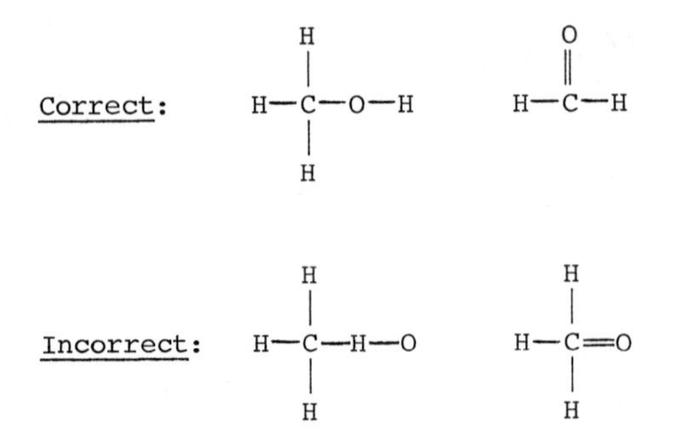

Hydrogen bonds are formed between unshared electrons on O, N, or F and a hydrogen attached to O, N, or F.

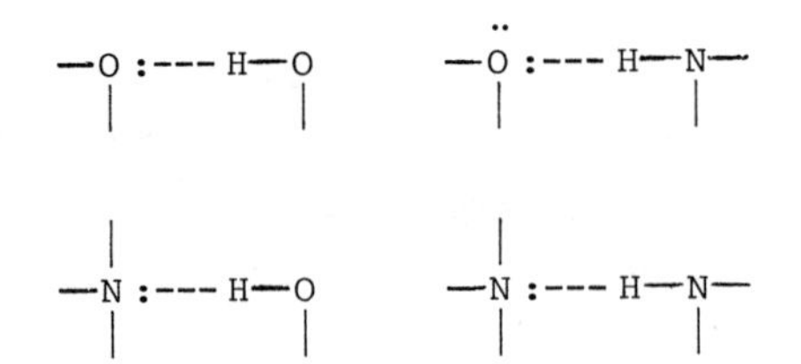

Answers to Problems

1.17 (a) $1s^2\ 2s^2\ 2p^2$ (b) $1s^2\ 2s^2\ 2p^6\ 3s^2\ 3p^2$

(c) $1s^2\ 2s^2\ 2p^6\ 3s^2\ 3p^3$ (d) $1s^2\ 2s^2\ 2p^6\ 3s^2\ 3p^4$

1.18 (a) Na (b) Cl (c) Ne

1.19 (b) and (c), because each has the same number of electrons in its outer shell.

1.20 90°, which is the angle between p orbitals.

1.21 (a) Si (b) B (c) C

Atomic radii increase as we proceed down any column or to the left in any row of the periodic table.

1.22 (a) O (b) N (c) C (d) O

1.23 (a) $CH_3CO_2^-$ Na^+ — $CH_3CO_2^-$: covalent; between $CH_3CO_2^-$ and Na^+: ionic

(b) all covalent

(c) Li^+ ^-OH — between Li^+ and ^-OH: ionic; ^-OH: covalent

(d) CH_3O^- Na^+ — CH_3O^-: covalent; between CH_3O^- and Na^+: ionic

(e) all covalent

(f) Br^- Mg^{2+} ^-OH — between Br^-, Mg^{2+} and ^-OH: ionic; ^-OH: covalent

(g) all covalent (h) all covalent

1.24 Li, 1; Be, 2; B, 3; C, 4; N, 3; O, 2; F, 1; Ne, 0

1.25 Four. Each covalent bond contains two electrons; therefore, a second-period element, with four orbitals in the outer shell, can accommodate a maximum of eight electrons.

1.26

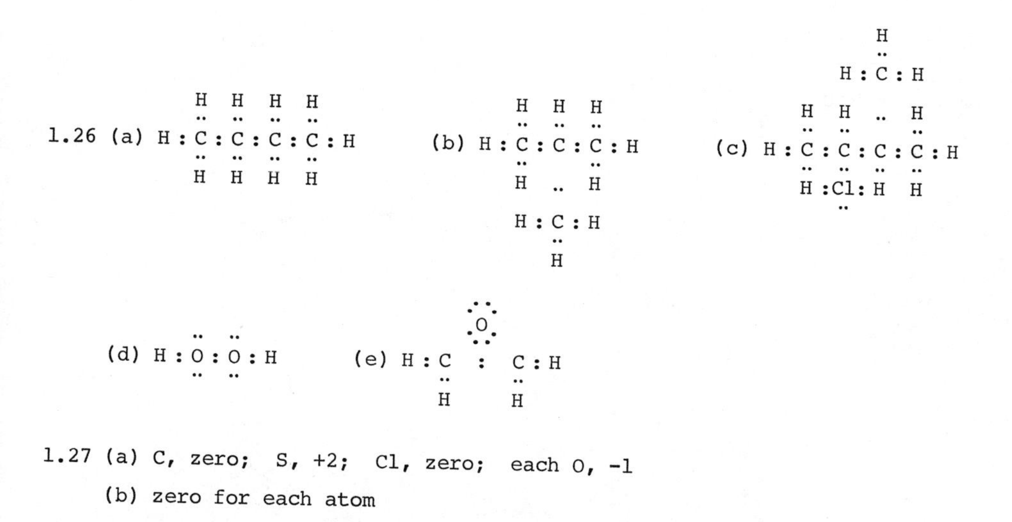

1.27 (a) C, zero; S, +2; Cl, zero; each O, -1

(b) zero for each atom

(c) each C, zero; O, zero; right-hand O, -1

(d) each C, zero; S, +1; O, -1 (See the sample calculations in the text.)

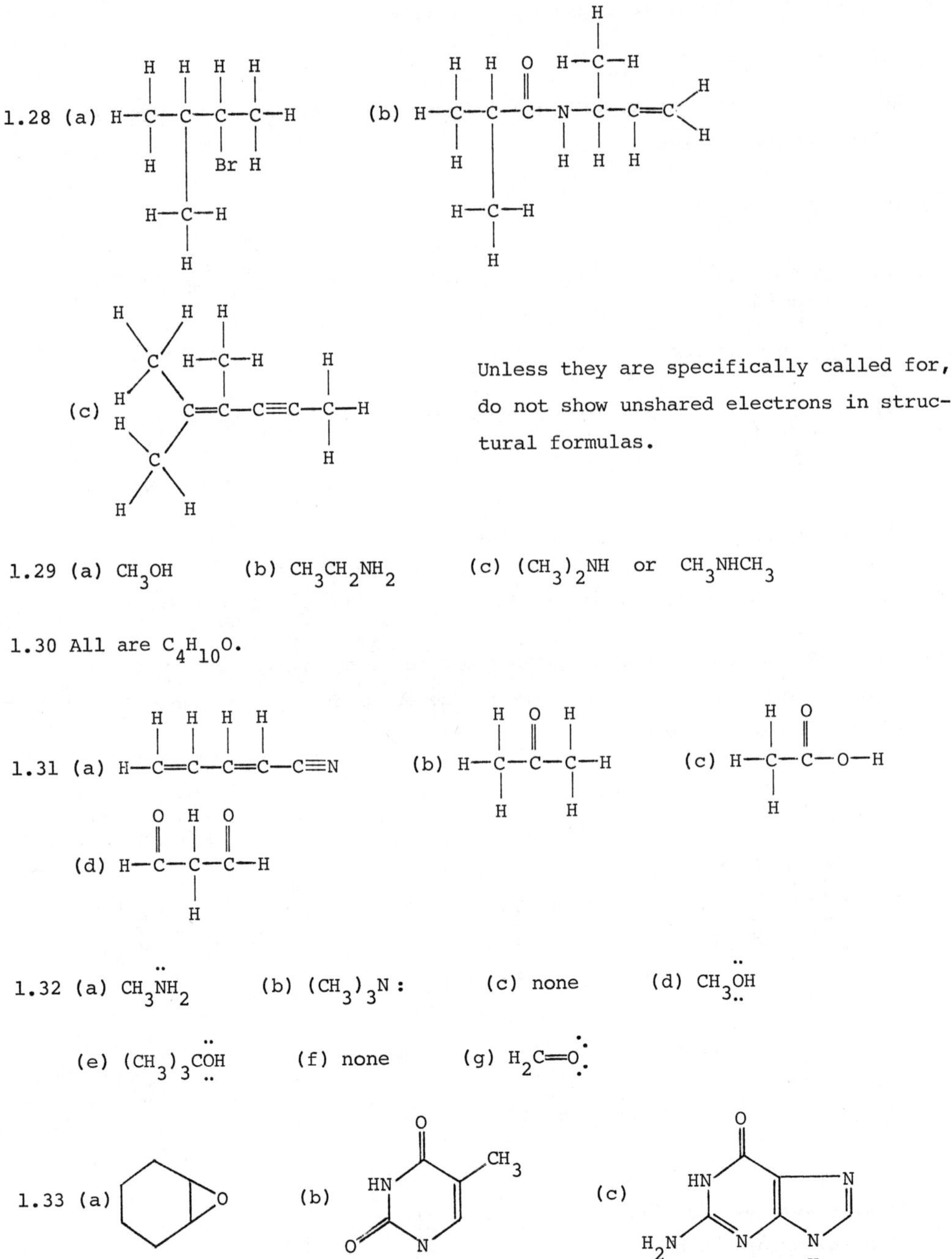

1.28 (a) (b) (c)

Unless they are specifically called for, do not show unshared electrons in structural formulas.

1.29 (a) CH_3OH (b) $CH_3CH_2NH_2$ (c) $(CH_3)_2NH$ or CH_3NHCH_3

1.30 All are $C_4H_{10}O$.

1.31 (a) (b) (c) (d)

1.32 (a) $CH_3\ddot{N}H_2$ (b) $(CH_3)_3N:$ (c) none (d) $CH_3\ddot{O}H$ (with a second unshared pair below the O)

(e) $(CH_3)_3C\ddot{O}H$ (with a second unshared pair below the O) (f) none (g) $H_2C{=}\ddot{O}:$ (two unshared pairs on O)

1.33 (a) (b)

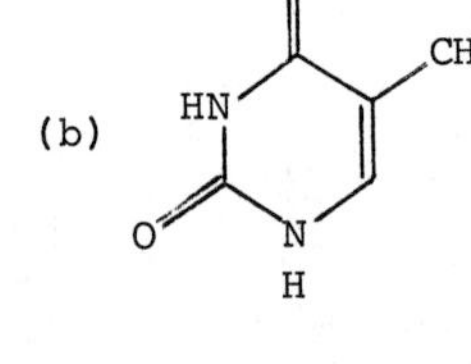

(c)

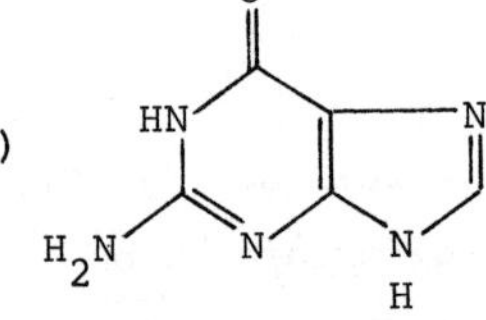

(d)

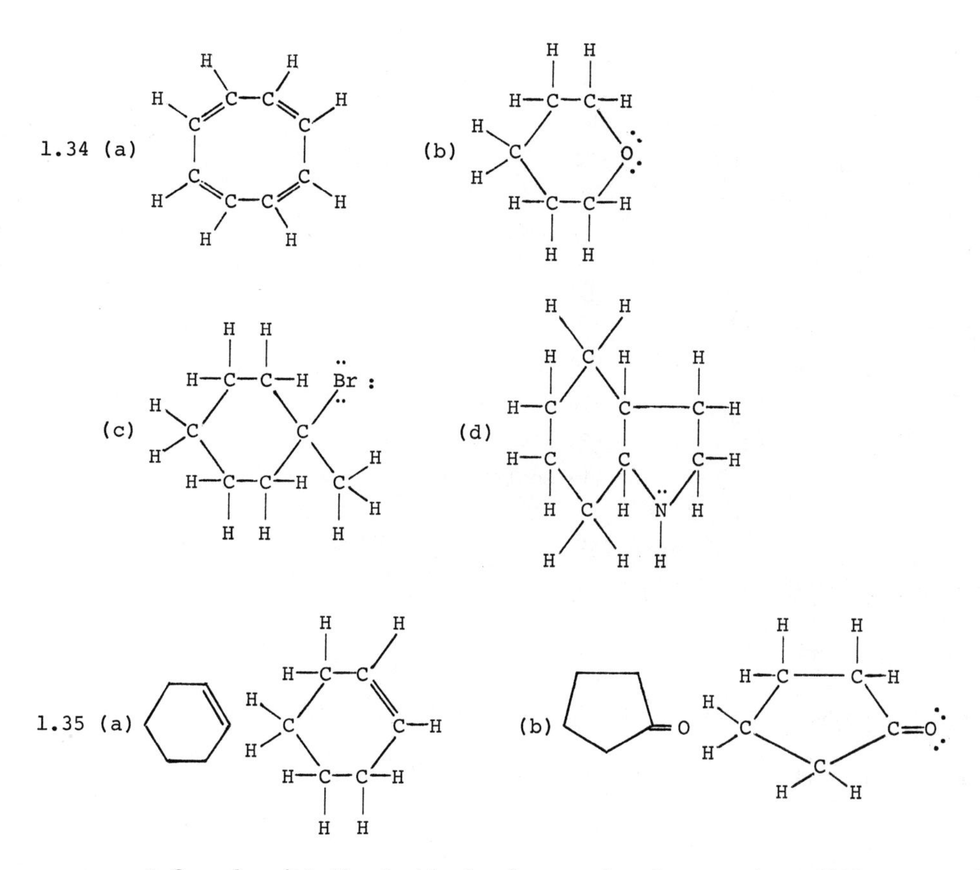

A formula with the double bond or carbonyl group in a different position represents the same compound and is also correct. For example:

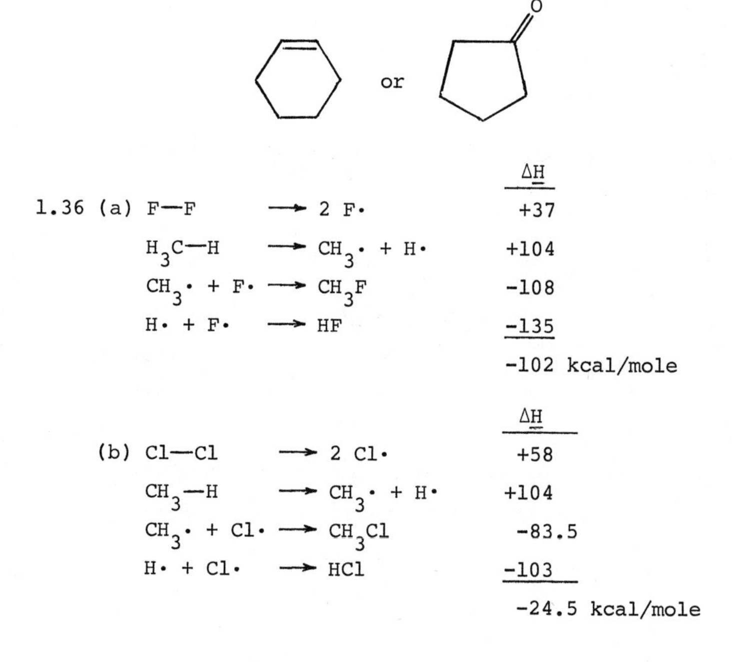

1.36 (a)

		ΔH
F—F	$\longrightarrow$ 2 F·	+37
H_3C—H	$\longrightarrow$ CH_3· + H·	+104
CH_3· + F·	$\longrightarrow$ CH_3F	-108
H· + F·	$\longrightarrow$ HF	-135
		-102 kcal/mole

(b)

		ΔH
Cl—Cl	$\longrightarrow$ 2 Cl·	+58
CH_3—H	$\longrightarrow$ CH_3· + H·	+104
CH_3· + Cl·	$\longrightarrow$ CH_3Cl	-83.5
H· + Cl·	$\longrightarrow$ HCl	-103
		-24.5 kcal/mole

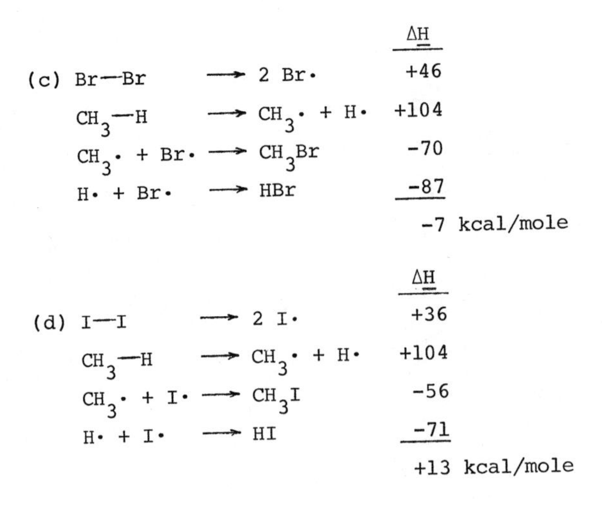

			homolytic		heterolytic
1.37 (a)	$CH_3CH_2—\ddot{Cl}:$	→	$CH_3\dot{C}H_2 + :\ddot{Cl}·$	or	$CH_3\overset{+}{C}H_2 + :\ddot{Cl}:^-$
(b)	$H—\ddot{O}H$	→	$H· + ·\ddot{O}H$	or	$H^+ + {}^-:\ddot{O}H$
(c)	$H—\ddot{N}H_2$	→	$H· + ·\ddot{N}H_2$	or	$H^+ + {}^-:\ddot{N}H_2$
(d)	$CH_3—\ddot{O}H$	→	$CH_3· + ·\ddot{O}H$	or	$CH_3^+ + {}^-:\ddot{O}H$
(e)	$CH_3\ddot{O}—H$	→	$CH_3\ddot{O}· + ·H$	or	$CH_3\ddot{O}:^- + H^+$

In heterolytic cleavage, the bonding electrons always go with the more electronegative atom.

1.38 (a) $\overset{\delta-}{C}—\overset{\delta+}{Mg}$ (b) $\overset{\delta+}{C}—\overset{\delta-}{Br}$ (c) $\overset{\delta+}{C}—\overset{\delta-}{O}$ (d) $\overset{\delta+}{C}—\overset{\delta-}{Cl}$ (e) $\overset{\delta-}{C}—\overset{\delta+}{H}$

(f) $\overset{\delta-}{C}—\overset{\delta+}{B}$

The direction of the dipole is determined by comparison of the electronegativities of the bonded atoms.

1.39 (a) $\overset{\delta+}{CH_3}—\overset{\delta-}{Ⓞ}—\overset{\delta+}{H}$ (b) $CH_3\overset{\delta+}{C}CH_3$ with $=\overset{\delta-}{Ⓞ}$ (c) $\overset{\delta-}{Ⓕ}—\overset{\delta+}{CH_2}CO_2H$

(d) $(CH_3)_2NCH_2\overset{\delta+}{CH_2}—\overset{\delta-}{Ⓞ}—\overset{\delta+}{H}$

In (a) and (d), the most electronegative atom is bonded to two other atoms; therefore, it is necessary to show the polarity of both of its bonds.

1.40 (a) $CH_3CH_2CH_3$, $CH_3CH_2CH_2NH_2$, $CH_3CH_2CH_2OH$

(b) $CH_3CH_2CH_2I$, $CH_3CH_2CH_2Br$, $CH_3CH_2CH_2Cl$

The ranking is based upon differences in electronegativities of atoms in the polar groups. A greater electronegativity difference means a more polar molecule.

1.41 (a) $(CH_3)_2\underset{H}{N}$:--- $H\ddot{N}(CH_3)_2$

(f) [cyclic structure: H—O—CH₂—CH₂—Ö(CH₃) with dashed H-bond from H to O]

$CH_3\ddot{O}CH_2CH_2OH$ ⋮ $HOCH_2CH_2OCH_3$

$CH_3OCH_2CH_2\underset{H}{\ddot{O}}$:--- $HOCH_2CH_2OCH_3$

The compounds in (b) - (e) cannot undergo hydrogen bonding with other molecules of their own kind because they do not contain H bonded to O, N, or F.

1.42 With itself: (a), (b), and (e); with water: (a), (b), (d), (e), and (f). Any compound that can hydrogen bond with itself can also hydrogen bond with water. Compounds (d) and (f) have unshared electrons that can form a hydrogen bond with the partially positive hydrogen atom of water.

1.43 (a) $(CH_3)_2NH$ ---: $\overset{H}{N}(CH_3)_2$ (b) $(CH_3)_2NH$ ---: $\ddot{O}H_2$

(c) $(CH_3)_2\overset{H}{N}$:--- H_2O (d) $H_2\ddot{O}$:--- H_2O

Nitrogen is less electronegative than oxygen; therefore, its unshared electrons are more loosely held and more available for hydrogen bonding. The OH bond is very polar (and the H is more positive) because of the high electronegativity of oxygen. For these two reasons, the N --- HO hydrogen bond in (c) is the strongest hydrogen bond in the solution.

1.44 (a) CH_3OH + ^-OH (b) $CH_3NH_3^+$ + Cl^- (c) $^-O_2CCO_2^-$ + 2 H_2O

(d) [cyclic structure with $\overset{+}{N}H_2$] (e) CH_3CO_2H (f) $CH_3NH_3^+$ + $CH_3CO_2^-$

(g) $CH_3CO_2^-$ + CH_3OH (h) CH_3NH_2 + CH_3OH (i) CO_3^{2-} + 2 CH_3OH

1.45 (a) 4.75 (b) 10 (c) 4.3 (d) ∿ 16 (e) ∿ 43

(e) < (d) < (b) < (a) < (c)

The calculation of pK_a values is discussed in Section 1.10A of the text. These values may also be obtained directly from a calculator with a log function. (Remember, however, to change the sign: $pK_a = -\log K_a$. Also, be careful to watch significant figures.)

1.46 (a) 13 (b) 9.4 (c) 3.2 (d) 5.8 (a) < (b) < (d) < (c)

1.47 (a) $(CH_3)_2\overset{+}{N}H_2$ (b) $CH_3CH_2CH_2CO_2^-$

1.48 (c) < (d) < (b) < (a). Remember, the conjugate base of a stronger acid is a weaker base.

1.49 Either a mixture of NaBr and LiCl or a mixture of NaCl and LiBr in water yields Na^+, Br^-, Li^+, and Cl^-. A solution of CH_3Cl and NaBr yields CH_3Cl, Na^+, and Br^-. This is different from a solution of CH_3Br and NaCl, which yields CH_3Br, Na^+, and Cl^-.

1.50 $PbCl_4$ is a covalent compound, while $PbCl_2$ is ionic.

1.51 (a) ^-CN, because carbon has a formal charge of -1 and N has a formal charge of 0.

(b) $^-C{\equiv}CH$, because the left-hand carbon has a formal charge of -1 and the right-hand carbon has a formal charge of 0.

(c) $^+CH_2OH$, because carbon has a formal charge of +1 and the other atoms have formal charges of 0.

(d) CH_3O^-, because carbon has a formal charge of 0, and oxygen has a formal charge of -1.

(e) $CH_2{=}CH\overset{+}{C}H_2$, because the right-hand carbon has a formal charge of +1 and the other atoms have formal charges of 0.

(f) $(CH_3)_3C^+$, for the same reason as stated in (e).

(g) $^-NH_2$, because N has a formal charge of -1 and the H atoms have formal charges of 0.

(h) $CH_3\overset{O}{\overset{\|}{C}}O^-$, because the right-hand oxygen has a formal charge of -1 and the other atoms have formal charges of 0.

1.52 (a) (b) (c)

Four- and three-membered rings would also be correct for (c). For example,

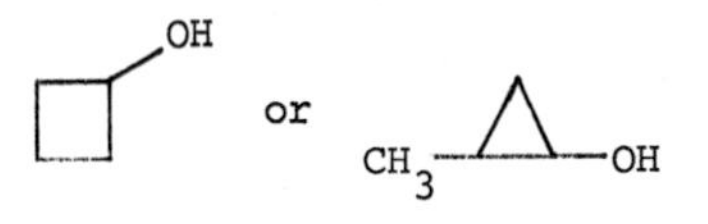

1.53 (a) $CH_3CH{=}CH_2$

(b) $CH_3\overset{O}{\overset{\|}{C}}H$

Although $CH_2{=}CHOH$ must be considered a correct answer to this problem, it is unstable and is rapidly converted to $CH_3\overset{O}{\overset{\|}{C}}H$ (see Section 9.8).

(c) $CH_2{=}CHOCH_2CH_3$ or $CH_3\overset{O}{\overset{\|}{C}}CH_2CH_3$ are both correct answers. There are also other correct answers.

1.54 (a)

			ΔH
Cl_2	$\longrightarrow$	$2\ Cl\cdot$	+58
$CH_3CH_2{-}H$	$\longrightarrow$	$CH_3CH_2\cdot + H\cdot$	+98
$CH_3CH_2\cdot + Cl\cdot$	$\longrightarrow$	CH_3CH_2Cl	-81.5
$H\cdot + Cl\cdot$	$\longrightarrow$	HCl	-103
			-28.5 kcal/mole

(b)

			ΔH
Br_2	$\longrightarrow$	$2\ Br\cdot$	+46
$CH_3CH_2{-}H$	$\longrightarrow$	$CH_3CH_2\cdot + H\cdot$	+98
$CH_3CH_2\cdot + Br\cdot$	$\longrightarrow$	CH_3CH_2Br	-68
$H\cdot + Br\cdot$	$\longrightarrow$	HBr	-87
			-11 kcal/mole

The reaction in (a) liberates more energy.

1.55 (a) $(CH_3)_2CH{-}\ddot{\underset{..}{Br}}: \longrightarrow (CH_3)_2\overset{+}{C}H + :\ddot{\underset{..}{Br}}:^-$

(b) $CH_3CH_2{-}Li \longrightarrow CH_3\ddot{C}H_2^- + Li^+$

(Lithium is less electronegative than carbon.)

(c) $(CH_3)_2CH{-}\ddot{\underset{..}{O}}{-}CH(CH_3)_2 \longrightarrow (CH_3)_2CH\ddot{\underset{..}{O}}:^- + {}^+CH(CH_3)_2$

(Cleavage of the other C—O bond yields the same ions.)

(d) CH_2CH_2 (ring with $:\ddot{O}$) $\longrightarrow$ $:\ddot{\underset{..}{O}}:^-{-}CH_2\overset{+}{C}H_2$

(Because a ring bond is broken, the + and - charges remain in the same structure.)

1.56 a planar molecule: BF_3 (F—B bonds at 120°, 120°, 120°) The three BF bond moments cancel in vector addition.

1.57 (b) < (c) < (a). The order is based upon hydrogen bonding with water. Compound (b) forms no hydrogen bonds. Compound (c) does form hydrogen bonds with water (but not with other ether molecules). Compound (a) has both hydrogen-bonding C═O and OH groups.

1.58 The two compounds are equally soluble in water because they have the same molecular weights and both form hydrogen bonds with water. Pure 1-butanol can form hydrogen bonds with itself and therefore has a higher boiling point than diethyl ether, which does not have a partially positive hydrogen and thus cannot form hydrogen bonds with itself.

1.59 (a) Lewis acid, $AlCl_3$; Lewis base, $(CH_3)_3CCl$

(b) Lewis acid, $(CH_3)_3C^+$; Lewis base, $CH_2{=}CH_2$

(c) Lewis acid, $(CH_3)_3C^+$; Lewis base, H_2O

(d) Lewis acid, Br_2; Lewis base, $CH_2{=}CH_2$

1.60
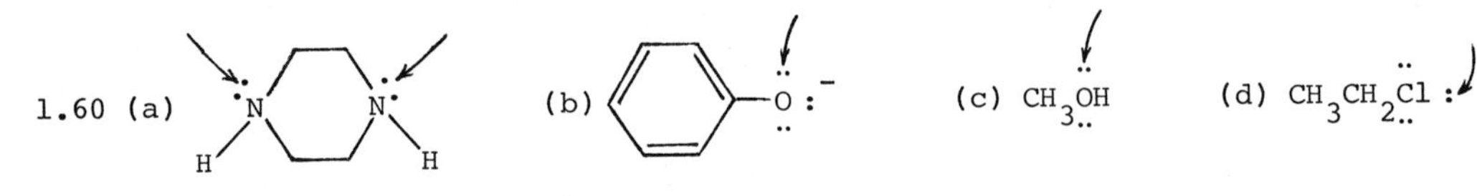

1.61 (a) $CH_3\ddot{O}H + H{-}OSO_2OH \rightleftharpoons CH_3\overset{+}{O}H_2 + {}^-{:}OSO_2OH$

(b) $CH_3\ddot{O}{-}H + {}^-{:}NH_2 \rightleftharpoons CH_3\ddot{O}{:}^- + {:}NH_3$

1.62 (a) Because Mg is electropositive, the bonding electrons are drawn toward the carbon atom. Thus, the carbon atom carries a partial negative charge:

$$\overset{\delta-}{H_3C}:\overset{\delta+}{MgI}$$

Carbon is not very electronegative; therefore, a species containing a partially negative carbon can act as an electron donor and is a strong base. CH_3MgI is a stronger base than ^-OH because oxygen is more electronegative than carbon and can carry a negative charge more readily.

(b) When treated with water, $CH_3{-}MgI$ removes a proton from the water:

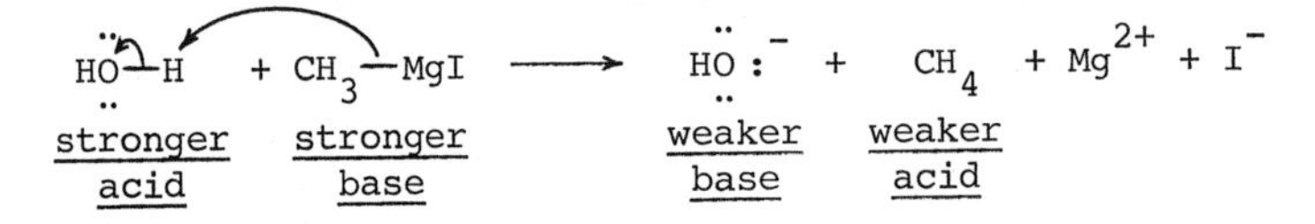

CHAPTER 2

Orbitals and Their Role in Covalent Bonding

Some Important Features

A bonding molecular orbital is formed by the overlap of two atomic or hybrid orbitals of the same phase. A bonding molecular orbital is of lower energy (more stable) than the two original orbitals. An antibonding molecular orbital is formed by interference of two atomic or hybrid orbitals of opposite phase and is of higher energy (less stable) than the two original orbitals.

A sigma (σ) molecular orbital results from the end-to-end overlap of atomic or hybrid orbitals. A pi (π) molecular orbital results from side-to-side overlap of p orbitals. A pi orbital is more exposed and is usually of higher energy (more reactive) than a sigma orbital.

A carbon atom can hybridize its atomic orbitals in one of three ways:

sp^3: four tetrahedral single bonds

sp^2: two single bonds and one double bond

sp : one single bond and one triple bond

The bond length from a carbon atom decreases with increasing s character — that is, sp bonds are the shortest and sp^3 bonds are the longest.

Some important functional groups are listed in Table 2.2 in the text (p.60). Note the different ways of writing each functional group.

A nitrogen atom with three bonds to other atoms (as in NH_2 or RNH_2) has a pair of unshared electrons that can be donated to an electron-deficient species:

$\ddot{N}H_3 + H^+ \longrightarrow NH_4^+$. An oxygen atom in compounds has two bonds to other atoms (as in H_2O, ROH, or $R_2C{=}O$) and has two pairs of unshared electrons. A carbon-oxygen or a carbon-nitrogen bond is polar because oxygen and nitrogen are more electronegative than carbon.

Conjugated double bonds can occur only if two *p* orbitals involved in two different pi bonds can partially overlap their sides. For this reason, the pi bonds must join adjacent atoms for conjugation to occur. Benzene (C_6H_6) is a symmetrical cyclic molecule in which six *p* electrons are completely delocalized (see Figure 2.23, p. 67).

Resonance structures are used to show complete or partial delocalization of electronic charge, and differ only in the positions of electrons. The major contributors are the resonance structures of lowest energy (greatest stabilization). Reread the "rules" in Section 2.9C.

Reminders

Remember the valences: C = 4, O = 2, H = 1.

To determine the approximate bond angles around a carbon atom, ask yourself "What is the hybridization of that carbon?" If the answer is *sp*^{3}, then the bond angles are approximately 109°; if the answer is *sp*^{2}, 120°; if *sp*, 180°.

A doubly bonded carbon atom is *sp*^{2} hybridized and has three planar, trigonal sigma bonds plus a pi bond.

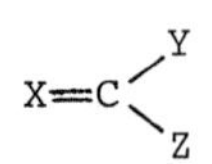

The bonds to X, Y, and Z are planar.

Answers to Problems

2.17 All are the same because C is tetrahedral.

2.18 Only (b) and (c) could be real compounds. Structure (a) has too many hydrogens for two carbons. In (a) and (d), there are odd numbers of hydrogens. (Because carbon forms four bonds, there must be an even number of hydrogens in a compound containing only C and H.) One way to solve a problem of this type is to write the symbols for the carbon atoms, then insert the hydrogens by trial and error. (A more sophisticated method of prediction of formulas is presented in Section 3.1B.)

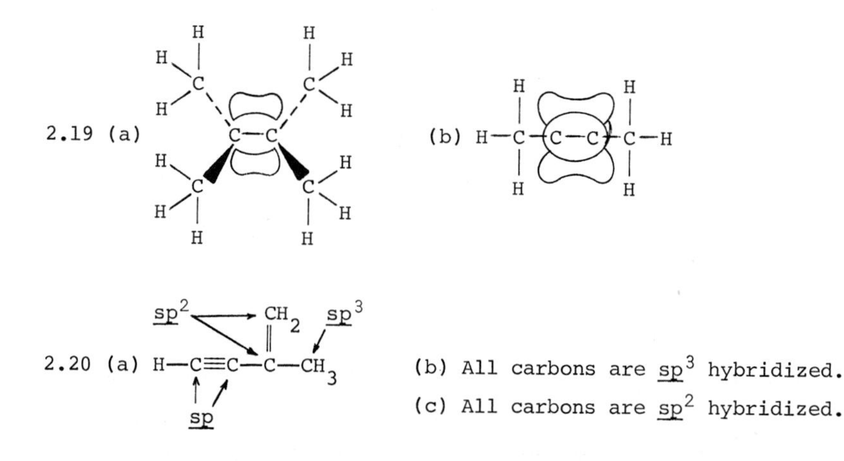

(b) All carbons are sp^3 hybridized.

(c) All carbons are sp^2 hybridized.

2.21 In (a) and (b), the indicated carbon atoms are bonded to sp^2 carbons and therefore lie in the same plane as the sp^2 carbons. In (c), the CH_3 group bonded directly to the sp^2 carbons lies in the plane of the sp^2 carbons, but the other CH_3 group can rotate around its sigma bond. This CH_3 group sometimes lies in the same plane as the sp^2 carbons, but is not restricted to this position.

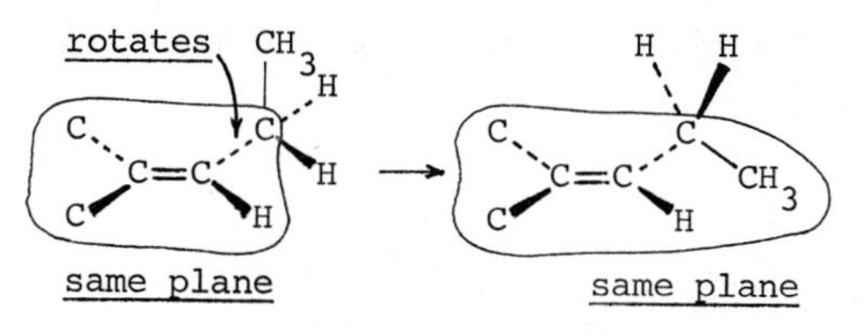

2.22 (a) 120° around sp^2 carbon. (b) 180° around sp carbon.
(c) 120° around sp^2 carbon. (d) 120° around sp^2 carbon.
(e) 109° around sp^3 carbon. (f) 109° around sp^3 carbon.

The actual bond angles might differ slightly from those predicted because of dipole-dipole repulsions and attractions.

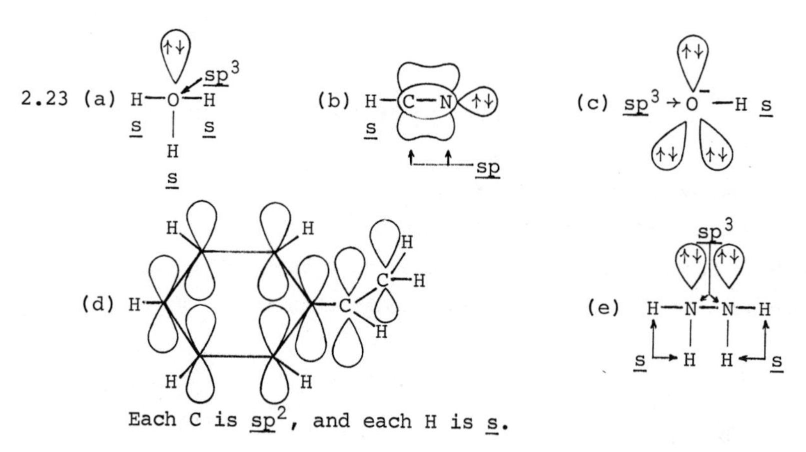

Each C is sp^2, and each H is s.

(f) 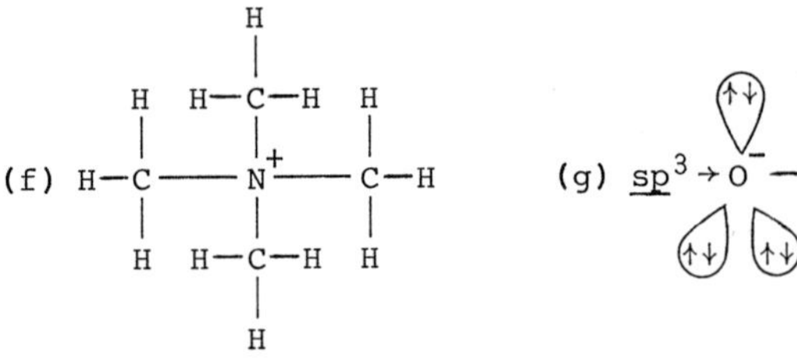

Each C and N is $\underline{sp}^3$ and each H is $\underline{s}$.

(g) $\underline{sp}^3 \rightarrow \bar{O}$—$CH_3$ (C is $\underline{sp}^3$; each H is $\underline{s}$)

2.24 (a) σ* —, π_2^* —, π_1 ↑↓, σ ↑↓ (b) σ* —, π_2^* ↓, π_1 ↑, σ ↑↓

In (a), the ground state, the electrons are paired and fill the two lowest-energy orbitals. Upon excitation, an electron is promoted from the highest occupied molecular orbital (π_1) to the lowest unoccupied molecular orbital (π_2^*). Any other transition would require a greater amount of energy.

2.25 (a) 2 (b) 2 (c) 2 (d) 2

The answers are based upon the relative amounts of $\underline{s}$ character in the bonds. The greater the amount of $\underline{s}$ character, the shorter is the bond length.

2.26 (a)

Each atom in the ring is $\underline{sp}^3$.

(b) 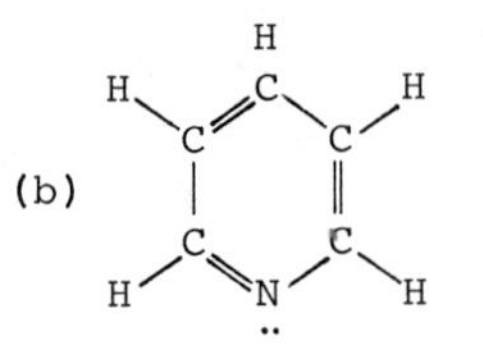

Each atom in the ring is $\underline{sp}^2$.

2.27 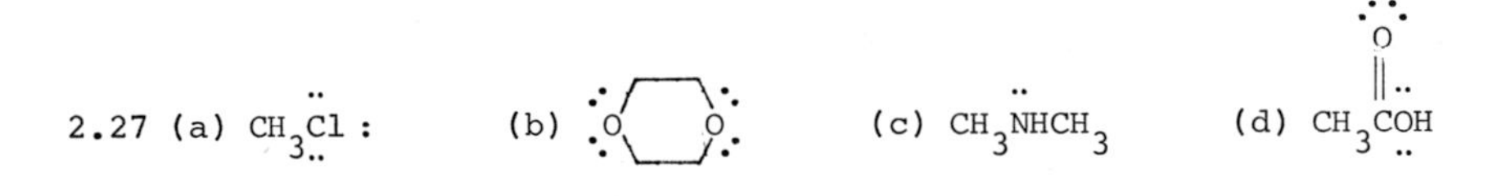

(a) $CH_3\ddot{C}l:$ (b) (c) $CH_3\ddot{N}HCH_3$ (d) CH_3COH

2.28 (a) CH_3CH_2OH, because an O—H bond is more polar (has a greater difference in electronegativity) than a C—O bond.

(b) $(CH_3)_2C{=}O$, because the electrons of the C—O pi bond are more easily polarized than those of the C—O sigma bond.

(c) CH_3NH_2, because the N—H bond is more polar than the C—N bond.

(d) CH_3CH_2Cl, because the C—Cl bond is more polar than the C—H bond.

2.29 (a) CH_3CH (with C=O circled) aldehyde

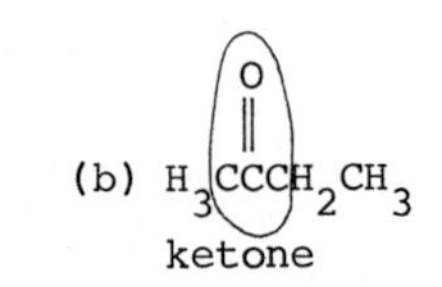

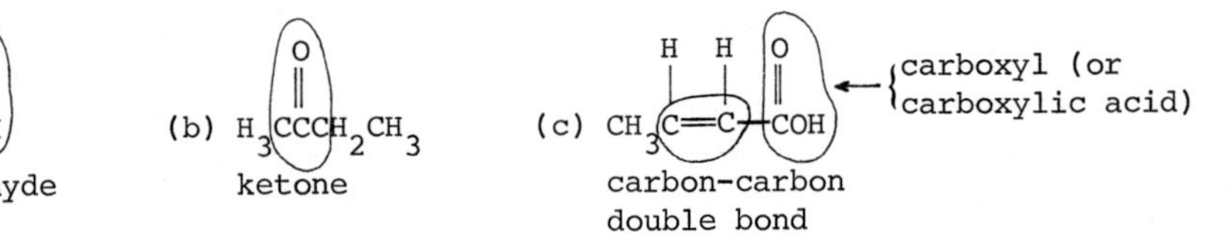

(b) $H_3CCCH_2CH_3$ ketone

(c) $CH_3C{=}C{-}COH$ carbon-carbon double bond; carboxyl (or carboxylic acid)

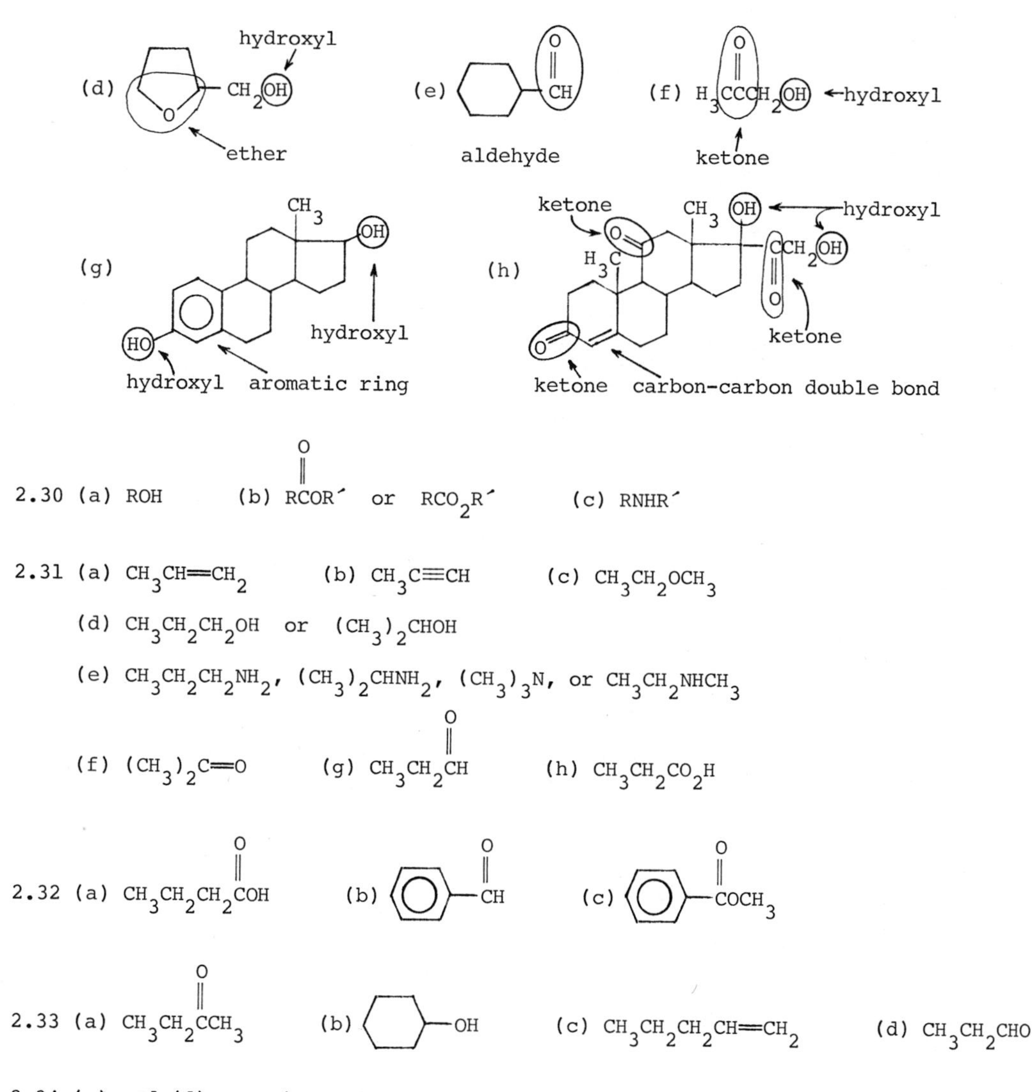

2.30 (a) ROH (b) $RCOR'$ (C=O) or RCO_2R' (c) $RNHR'$

2.31 (a) $CH_3CH{=}CH_2$ (b) $CH_3C{\equiv}CH$ (c) $CH_3CH_2OCH_3$

(d) $CH_3CH_2CH_2OH$ or $(CH_3)_2CHOH$

(e) $CH_3CH_2CH_2NH_2$, $(CH_3)_2CHNH_2$, $(CH_3)_3N$, or $CH_3CH_2NHCH_3$

(f) $(CH_3)_2C{=}O$ (g) CH_3CH_2CH (C=O) (h) $CH_3CH_2CO_2H$

2.32 (a) $CH_3CH_2CH_2COH$ (C=O) (b) C_6H_5–CH (C=O) (c) C_6H_5–$COCH_3$ (C=O)

2.33 (a) $CH_3CH_2CCH_3$ (C=O) (b) cyclohexyl–OH (c) $CH_3CH_2CH_2CH{=}CH_2$ (d) CH_3CH_2CHO

2.34 (a) and (d) contain conjugated double bonds and thus have delocalized pi electrons. In (e), the C=O group is conjugated with one pi bond of the triple bond.

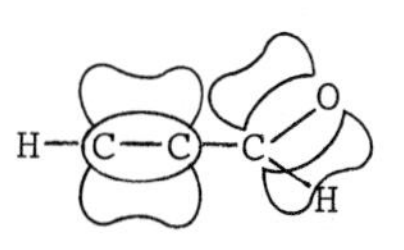

(b), (c), and (f) do not contain conjugation; thus, the pi electrons are localized.

2.35 (a) $CH_2{=}CHCH{=}CHCH_3$ (b) $CH_2{=}CHCH_2CH{=}CH_2$

2.36 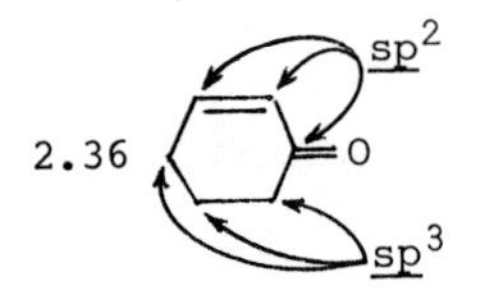

2.37 (a) equilibrium (b) equilibrium (c) resonance (d) resonance
The structures in (a) and (b) differ in the positions of atoms, while the structures in (c) and (d) differ only in the positions of electrons.

2.38 (a), (b), (d), and (f) show resonance structures (structures that differ only in the positions of electrons), while (c) and (e) show structures in equilibrium (structures with different arrangements of atoms).

2.39 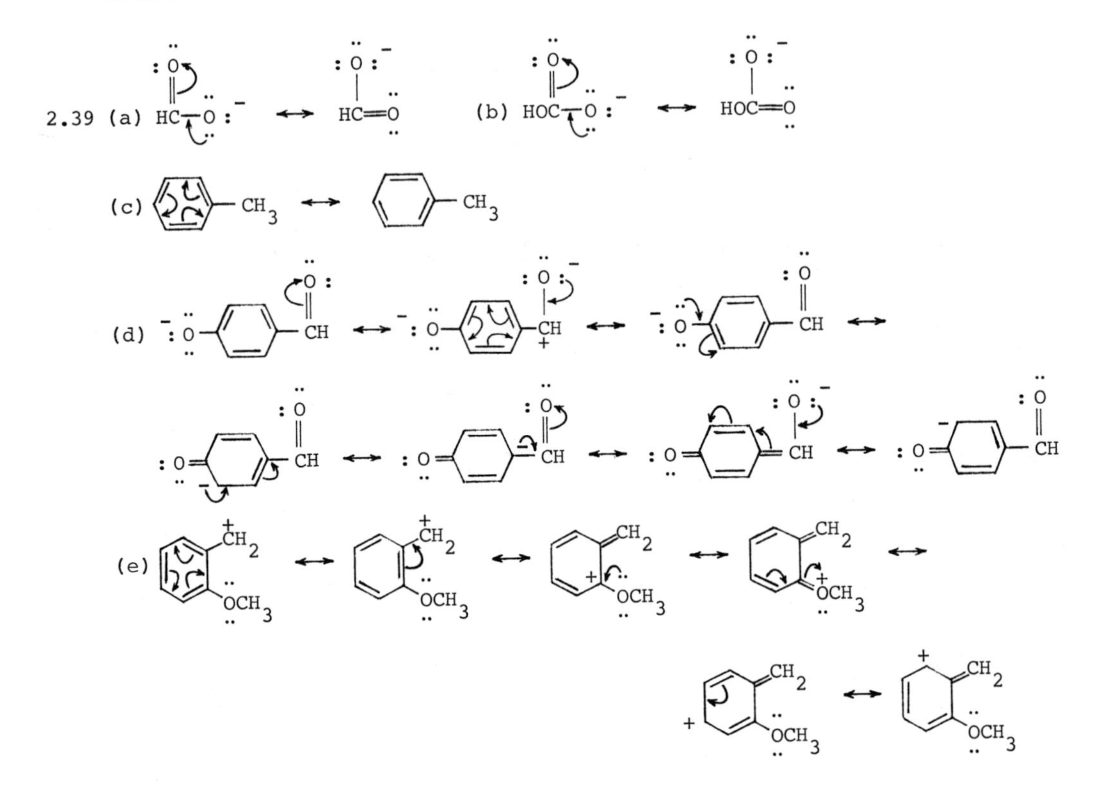

2.40 (a) $CH_3CH_2\overset{O}{\overset{\|}{C}}NH_2$ because each C and N has an octet and there is no charge separation.

(b) $CH_3\overset{O}{\overset{\|}{C}}OCH_3$ because each carbon and oxygen has an octet and there is no charge separation.

(c) $C_6H_5O^-$ because the electronegative oxygen carries the negative charge and the aromatic pi cloud is intact.

2.41 (a) (b) (c)

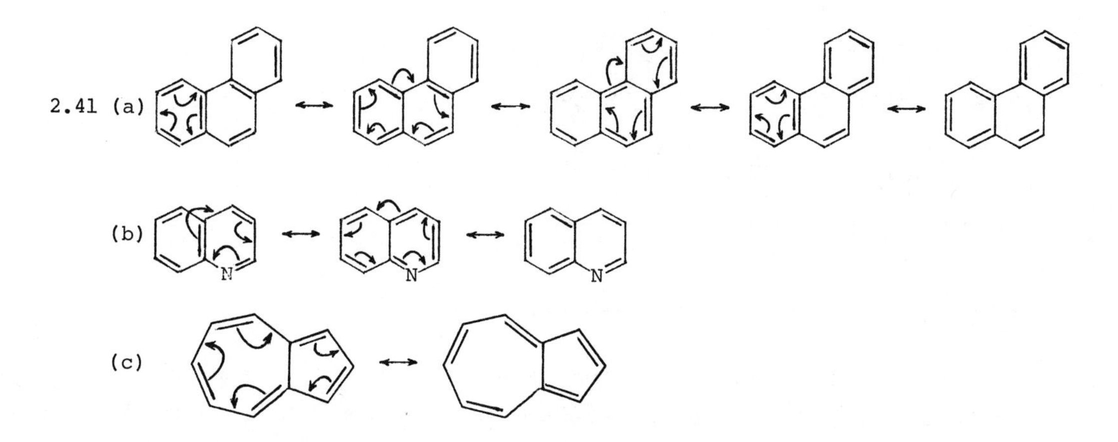

2.42 (a) The first cation is more stabilized, because its positive charge can be delocalized by the ring.

$\overset{+}{C}HCH_3$ ⟷ $=CHCH_3$ ⟷ $+$ $=CHCH_3$ ⟷ $=CHCH_3$

(b) The second cation because its positive charge can be delocalized by the double bond.

$CH_2{=}CH{-}\overset{+}{C}HCH_3$ ⟷ $\overset{+}{C}H_2CH{=}CHCH_3$

(c) The second anion because the nitro group helps share the negative charge.

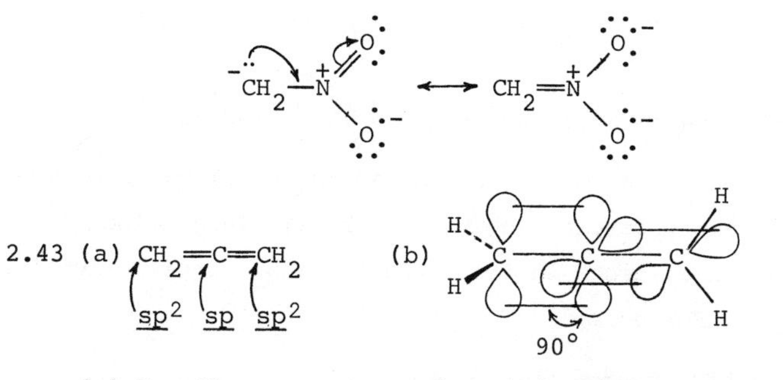

(c) No, they cannot overlap their sides.

(d) No, two pi bonds cannot delocalize electronic charge without overlap.

2.44 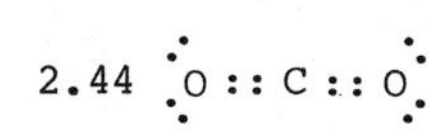

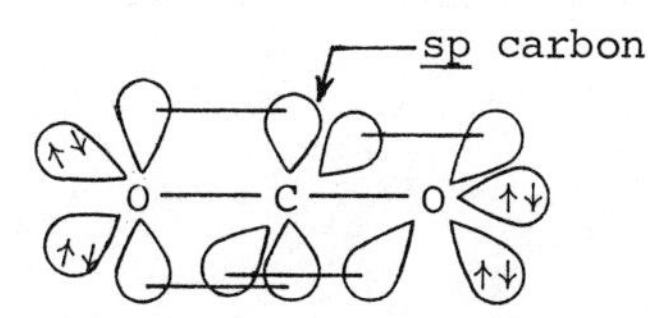

With an O—C—O bond angle of 180° and two pi bonds, the carbon atom must be sp hybridized.

2.45

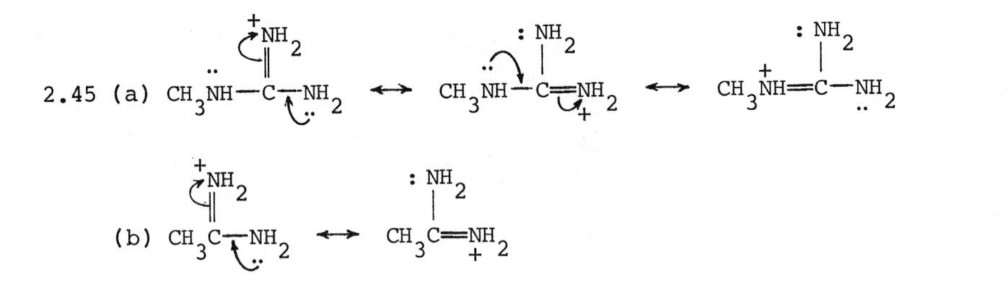

2.46 H_2^+, or $H \cdot H^+$, σ^* ___ σ ↑ H_2^-, or $H : H\cdot^-$, σ^* ↑ σ ↑↓

The H_2^+ ion contains one less electron than H_2 — that is, the group contains one electron. The H_2^- ion contains one more electron than H_2. Since the sigma bonding orbital in H_2^- is filled, the third electron must be in the antibonding orbital.

2.47 Carbon is in the sp state and has two sp-s bonds with the hydrogen atoms. The carbon atom has two p orbitals with one electron in each p orbital.

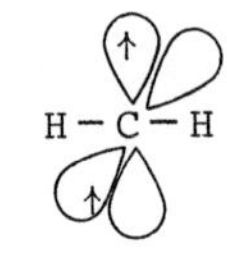

2.48

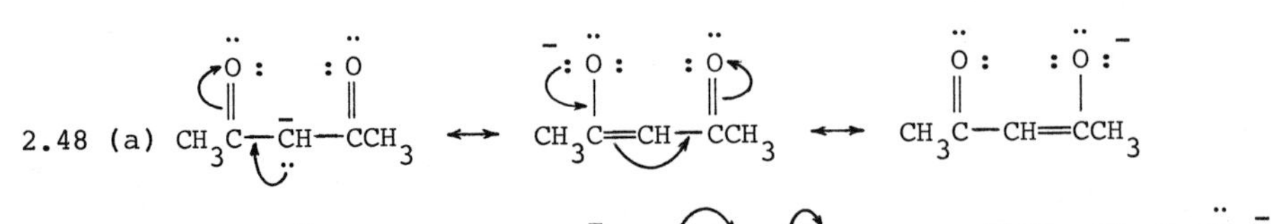

(b) : N≡C—$\bar{C}$H—C≡N : ↔ : $\bar{N}$=C=CH—C≡N : ↔ : N≡C—CH=C=N :$^-$

In each case, the negative charge of the resultant anion is delocalized. The result is that three atoms carry partial negative charges. This delocalization stabilizes the anion and allows a shift of the acid-base equilibrium in favor of the organic anion. For example:

δ− δ− δ− negative charge delocalized
N≡C==CH==C≡N

2.49 In the benzyl cation, the positive charge is delocalized by the benzene ring:

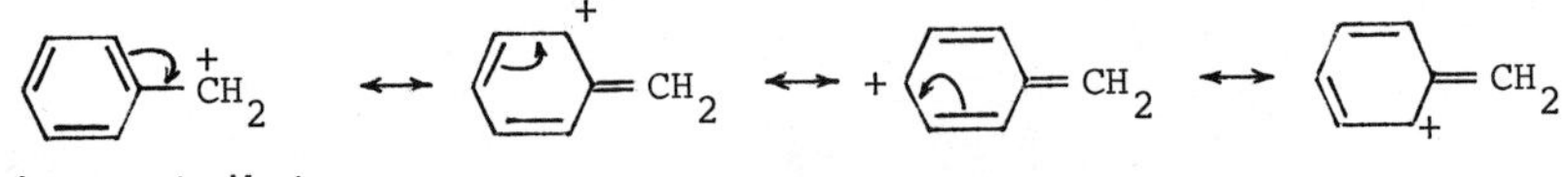

major contributor

In the triphenylmethyl cation, the positive charge is delocalized by three benzene rings. Ten atoms, instead of four, help share the positive charge:

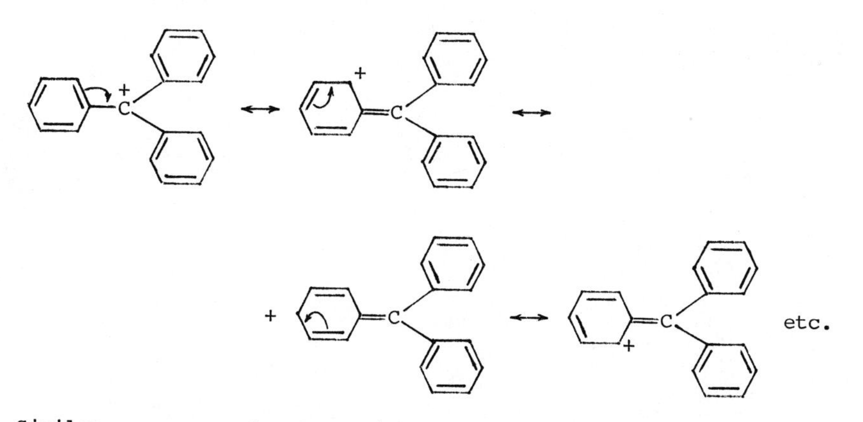

Similar resonance structures involving each of the other two benzene rings can be written.

CHAPTER 3

Structural Isomerism, Nomenclature, and Alkanes

Some Important Features

Structural isomers are compounds with the same molecular formula but different structures (orders of attachment of atoms).

A saturated open-chain alkane (branched or not) has the general formula C_nH_{2n+2}. We subtract two hydrogen atoms from the general formula for each double bond or for each ring.

We will not reiterate the nomenclature of organic compounds here, but refer you to Chapter 3 in the text. (For quick reference, see the appendix in the text.) Remember to: (1) number a chain or ring to give the lowest number to the principal functional group, and (2) group like substituents together.

Alkanes are generally nonreactive, but they do undergo combustion and reaction with halogens. The heat of combustion for alkanes increases with increasing molecular weight and also with increasing energy contained in the bonds.

Reminders

The same molecule may be drawn in a number of ways:

Not isomers: [cyclohexane]—OH ≡ [cyclohexane with OH] ≡ [cyclohexane with HO]

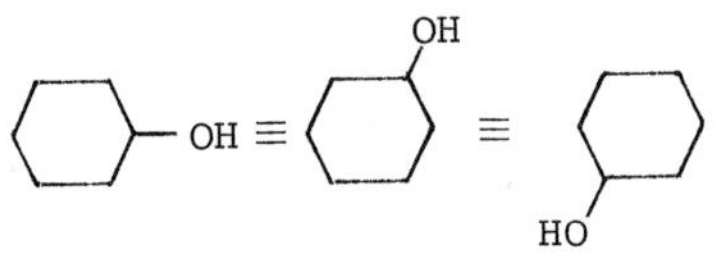

To be structural isomers, two molecules must have the atoms attached in different orders:

Isomers: (cyclohexyl)–OH CH_3–(cyclopentyl)–OH

Some instructors suggest naming potential isomers; if the names are the same, the structures represent the same compound. (This is a valid approach only if you name both structures correctly.)

When drawing isomers for a particular molecular formula, use the general formula. If a compound contains a ring, a structural isomer can contain a double bond instead.

△ and $CH_2{=}CHCH_3$: both $C_{\underline{n}}H_{2\underline{n}}$

General formulas may be extrapolated to compounds other than hydrocarbons:

$CH_3CH_2CH_2OH = C_{\underline{n}}H_{2\underline{n}+2}O$
↑ no double bond or ring

$CH_3CH_2CH(=O) = C_{\underline{n}}H_{2\underline{n}}O$
↑ one double bond or ring

Because nitrogen has a valence of 3, take special care in using general formulas with nitrogen compounds.

$CH_3CH_2NH_2$	$CH_2{=}CHNH_2$ or $CH_3CH{=}NH$	$HC{\equiv}CNH_2$ or $CH_3C{\equiv}N$
C_2H_7N	C_2H_5N	C_2H_3N
$C_{\underline{n}}H_{2\underline{n}+3}N$	$C_{\underline{n}}H_{2\underline{n}+1}N$	$C_{\underline{n}}H_{2\underline{n}-1}N$

Answers to Problems

3.12 (a), (b), (c), (d), (f)

3.13 (c), (d), (e), and (f) have no structural isomers. A typical structural isomer for (a) would be $CH_2{=}CHCH_2OH$, and for (b), CH_3CHBr_2.

3.14 (c) and (f) are isomers. In (a), the formulas represent a five-carbon and a six-carbon alcohol. In (b), (d), and (e), the formulas represent the same compounds.

3.15 (a) $CH_3(CH_2)_4CH_3$, $(CH_3)_2CHCH_2CH_2CH_3$, $(CH_3)_3CCH_2CH_3$, $(CH_3)_2CHCH(CH_3)_2$,

$CH_3CH_2CH(CH_3)CH_2CH_3$

(b) $CH_3CH_2CH_2CH_2OH$, $(CH_3)_2CHCH_2OH$, $CH_3CH_2CH(CH_3)OH$, $(CH_3)_3COH$

(c) $CH_3CH_2CH_2CH_2NH_2$, $(CH_3)_2CHCH_2NH_2$, $CH_3CH_2CH(CH_3)NH_2$, $(CH_3)_3CNH_2$,

$CH_3CH_2CH_2NHCH_3$, $(CH_3)_2CHNHCH_3$, $(CH_3CH_2)_2NH$, $CH_3CH_2N(CH_3)_2$

(d) $CH_3CH_2CHBrCl$, $CH_3CHBrCH_2Cl$, $BrCH_2CH_2CH_2Cl$, $CH_3CHClCH_2Br$, $(CH_3)_2CBrCl$

(e) $CH_3CH_2C{\equiv}CH$, $CH_3C{\equiv}CCH_3$, $CH_2{=}CHCH{=}CH_2$, $CH_2{=}C{=}CHCH_3$,

[cyclobutene], CH_3–[cyclopropenyl], CH_3–[cyclopropenyl], $CH_2{=}$[cyclopropane]

3.16 (a) and (f) are isomers; (b) and (e) are the same compound.

3.17 (a) C_6H_{12}, $C_{\underline{n}}H_{2\underline{n}}$ (b) C_8H_{14}, $C_{\underline{n}}H_{2\underline{n}-2}$ (c) C_8H_{14}, $C_{\underline{n}}H_{2\underline{n}-2}$

(d) $C_{14}H_{24}$, $C_{\underline{n}}H_{2\underline{n}-4}$

3.18 (a) $CH_3CH{=}CH_2$ (b) [bicyclic ketone structure] (c) [cyclic keto ether structure with O] (d) [cyclohexadiene structure]

(There are other correct answers.)

3.19 (a) $CH_3CH_2C{\equiv}CCH_2CH_3$ (b) $CH_3\overset{O}{\overset{\|}{C}}OH$ (c) $CH_3\overset{O}{\overset{\|}{C}}CH_3$ (There may be other correct answers.)

3.20 CH_3OH, CH_3CH_2OH, $CH_3CH_2CH_2OH$, $CH_3CH_2CH_2CH_2OH$, $CH_3(CH_2)_4OH$

3.21 $CH_3(CH_2)_4Br$, $CH_3(CH_2)_5Br$, $CH_3(CH_2)_6Br$, $CH_3(CH_2)_7Br$, $CH_3(CH_2)_8Br$, $CH_3(CH_2)_9Br$

3.22 (a) $(CH_3)_3C(CH_2)_5CH_3$ (b) $(CH_3CH_2)_2CHCH(CH_2CH_3)CH_2CH_2CH_3$

(c) $(CH_3)_2CHCH_2C(CH_3)(CH_2CH_3)(CH_2)_4CH_3$ (d) [cyclohexane with two $CH(CH_3)_2$ groups] (e) $CH_3CH_2CH(CH_3)$–[cyclopentyl]

(f) $(CH_3)_3C$–[phenyl] (g) $(CH_3)_2CHCH_2$–[cycloheptyl] (h) $CH_3(CH_2)_4$–[cyclohexyl with CH_3]

(i) $(CH_3CH_2CH_2)_2CHCH(CH_3)_2$

3.23 (a) 3-ethylpentane (b) 2,3-dimethylbutane

(c) 5-ethyl-1,2,3-trimethylcyclohexane (d) t-butylcyclopentane

(e) 2,2,4,4-tetramethylheptane (f) 2,2,3,4,4-pentamethylpentane

Be sure that the parent carbon chain is numbered from the end that will give the smallest numbers. In (d), no number is necessary for a monosubstituted cycloalkane.

3.24 (a) 3-ethylhexane; the longest chain was not chosen as the parent.

(b) 3-methyloctane; the parent was not numbered to give the smallest numbers.

(c) 2,2-dimethylpropane; same as (a).

(d) 3-ethyl-3,4-dimethyldecane; the methyl groups were not grouped together.

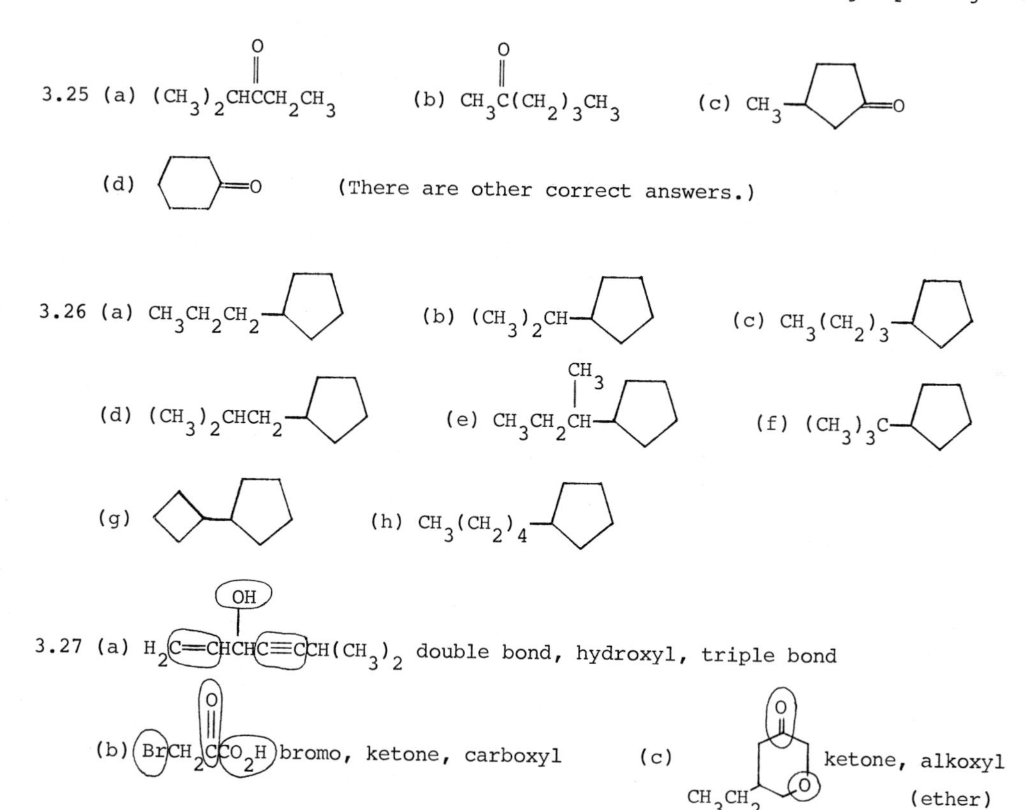

(d) (cyclohexane ring)=O (There are other correct answers.)

3.26 (a) $CH_3CH_2CH_2$—(cyclopentane ring) (b) $(CH_3)_2CH$—(cyclopentane ring) (c) $CH_3(CH_2)_3$—(cyclopentane ring)

(g) (cyclopropane–cyclopentane) (h) $CH_3(CH_2)_4$—(cyclopentane ring)

3.27 (a) H_2C=$CHCHC$≡$CCH(CH_3)_2$ with OH on third carbon; double bond, hydroxyl, triple bond

Note that (c) is not an ester because the oxygen is not bonded directly to the carbonyl carbon.

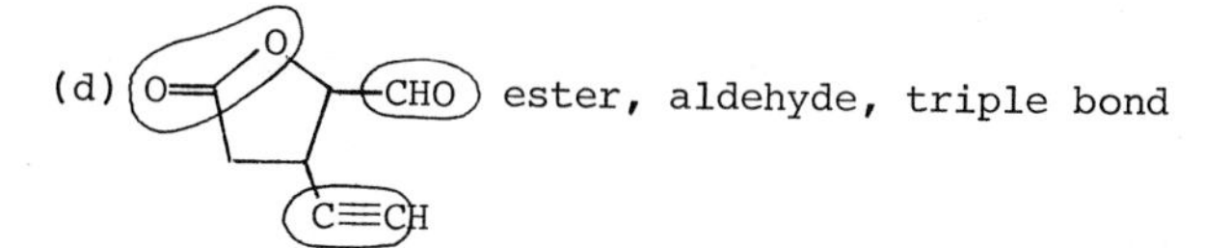

3.28 (a) $CH_3(CH_2)_4Cl$, $CH_3CHCl(CH_2)_2CH_3$, $CH_3CH_2CHClCH_2CH_3$ (b) [cyclopentyl]–Cl

(c) $ClCH_2C(CH_3)_2CH_2CH_3$, $(CH_3)_3CCHClCH_3$, $(CH_3)_3CCH_2CH_2Cl$ (d) $(CH_3)_3CCH_2Cl$

3.29 (a) 1,2,3,4,5,6-hexachlorocyclohexane (b) 1,1,2-trichloroethane
(c) 2-bromo-3-chloro-2,3-dimethylbutane (d) nitromethane

3.30 (a) $BrCH(C_6H_5)CH(C_6H_5)CH_3$ (b) CCl_3CCl_3 (c) $HOCH_2CHI(CH_2)_5CH_3$ (d) [cyclohexane ring with CH_3 and two Cl]

3.31 (a) 2-pentene (b) 2-methyl-2-butene (c) 1,3-cyclopentadiene
(d) 3-methyl-1,3,5-hexatriene (e) 2-butyne (f) 1,1,2-trichloroethene
(g) 3-bromo-1-cyclohexene (h) 1,6-dicyclohexyl-1,3,5-hexatriyne

3.32 (a) $CH_3(CH_2)_5CHO$, heptanal (b) Cl_2CHCO_2H, dichloroethanoic acid

(c) $CH_3(CH_2)_3C(=O)(CH_2)_3CH_3$, 5-nonanone

3.33 (a) 5,5,5-trichloro-4-methyl-2-pentene (The chain is numbered to give the principal functional group the lowest number.)
(b) 3-nitro-1-propene (c) 2-phenyl-1-ethylamine
(d) ethyl 2-chloropropanoate

3.34 (a) $CH_3CH_2C(=O)OCH(CH_3)_2$ (b) $CH_3(CH_2)_3C(CH_2CH_3)(CH_3CH_2)$–$CH(CH_3)CH_2CH(=O)$ (c) [cyclohexyl]–$CH(OH)CH_3$

3.35 (a) $-\Delta \underline{H}$ for one —CH_2— group can be calculated by subtracting the $-\Delta \underline{H}$ for $CH_3CH_2CH_3$ from that for $CH_3CH_2CH_2CH_3$:

$-\Delta \underline{H}$ for —CH_2— = 688 kcal/mole - 531 kcal/mole = 157 kcal/mole

$-\Delta \underline{H}$ for $CH_3(CH_2)_6CH_3$ is the ($-\Delta \underline{H}$ for $CH_3CH_2CH_2CH_3$) + ($-\Delta \underline{H}$ for <u>four</u> CH_2 groups)

$-\Delta \underline{H}$ for octane = 688 kcal/mole + 4(157) kcal/mole
= approx. 1316 kcal/mole

(b) No, the heat of combustion of 2-methylheptane could not be predicted with any accuracy because this compound is branched, not continuous-chain.

3.36 (a) hexane (b) 2-butene (c) 1-pentanol

Branching decreases the boiling point by interfering with van der Waals attractions.

3.37 (a) Small droplets of water (insoluble in gasoline) freeze and clog the carburetor or other portions of the gas line.

(b) Methanol, which is soluble in both gasoline and water, increases the solubility of the water in the gasoline. Methanol also depresses the freezing point of water.

3.38 $CH_3CH_2\overset{O}{\overset{\|}{C}}CH_3$

Because the general formula is $C_{\underline{n}}H_{2\underline{n}}O$, there must be one site of unsaturation or one ring. Since the compound is a ketone, there can be no carbon-carbon double bond nor a ring. Only one four-carbon, open-chain ketone fits the formula.

3.39 (cyclobutyl)–OH; cyclopropane with OH and CH_3; cyclopropane with CH_2OH; cyclopropane with H_3C and OH on the same carbon

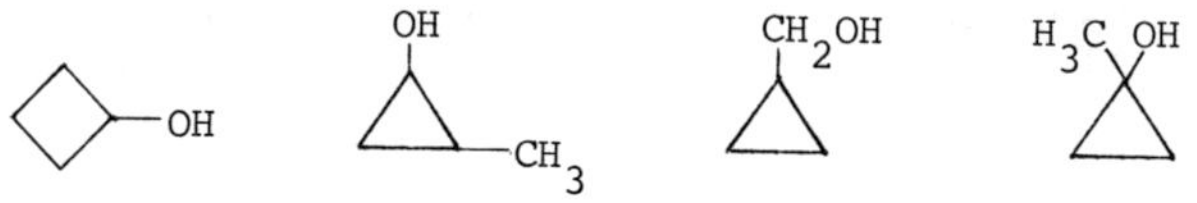

The general formula is $C_{\underline{n}}H_{2\underline{n}}O$; therefore, there must be either one site of unsaturation or one ring. Since the problem states that the compound is an alcohol and contains no carbon-carbon double bonds, the structure must contain either a three- or four-membered ring.

3.40 $CH_3CH_2CH_2CO_2H$, $(CH_3)_2CHCO_2H$

Again, the general formula is $C_{\underline{n}}H_{2\underline{n}}O$. The unsaturation must occur in the C=O of the carboxyl group; therefore, the carboxylic acid must be open-chain.

3.41 $H\overset{O}{\overset{\|}{C}}CH_2CH_2CO_2H$, $H\overset{O}{\overset{\|}{C}}\underset{CH_3}{\underset{|}{C}H}CO_2H$

Both the aldehyde and the carboxyl groups must be at the ends of a chain. Thus, the other two carbon atoms must be positioned between these two functional groups.

3.42 $CH_3\overset{O}{\overset{\|}{C}}CH_2\overset{O}{\overset{\|}{C}}H$, $CH_3CH_2\underset{\underset{O}{\|}}{\overset{O}{\overset{\|}{C}}}CH$, $H\overset{O}{\overset{\|}{C}}CH_2CH_2\overset{O}{\overset{\|}{C}}H$, $H\overset{O}{\overset{\|}{C}}\underset{\underset{CH_3}{|}}{C}H\overset{O}{\overset{\|}{C}}H$, $CH_3\overset{O}{\overset{\|}{C}}—\overset{O}{\overset{\|}{C}}CH_3$

One oxygen is in a carbonyl group, leaving one oxygen to be determined. Since the second oxygen is not in a hydroxyl, ether, ester, or carboxyl grouping, we conclude that the second oxygen is also part of a carbonyl group and that both carbonyl groups are aldehyde or keto groups. The general formula $C_{\underline{n}}H_{2\underline{n}-2}O_2$ allows for two sites of unsaturation (C═O and C═O); therefore, no ring can be present.

3.43 (a) $CH_3CH_2CH_3 + Cl_2 \xrightarrow{h\nu} CH_3CH_2CH_2Cl + CH_3CHClCH_3 + HCl$

(b) $ClCH_2CH_2CH_2Cl$, $CH_3CHClCH_2Cl$, $CH_3CH_2CHCl_2$, $CH_3CCl_2CH_3$

(c) There are six CH_3 hydrogens and two CH_2 hydrogens in propane. Therefore, a completely random substitution would result in 6 parts $CH_3CH_2CH_2Cl$ and 2 parts $CH_3CHClCH_3$. The ratio of 1-chloroethane to 2-chloroethane would be 6 : 2, or 3 : 1.

3.44 $CH_3CH_2CH_2CH_2CH_2CH_2CH_3 \xrightarrow[\text{catalyst}]{\text{heat}} CH_4 + CH_2{=}CHCH_2CH_2CH_2CH_3$

or $CH_3CH_3 + CH_2{=}CHCH_2CH_2CH_3$

or $CH_3CH{=}CH_2 + CH_3CH_2CH_2CH_3$

or $CH_3CH_2CH_3 + CH_2{=}CHCH_2CH_3$

besides the example shown on page 102 of the text. Cleavage could occur at any position in the chain. In addition, some of these cracking products could be cracked further. For example,

$$CH_3CH_2CH_2CH_3 \longrightarrow CH_2{=}CH_2 + CH_3CH_3$$

3.45 (a) $CH_3(CH_2)_4CH_3 \xrightarrow[\text{catalyst}]{\text{heat}}$ (benzene) $+ 3\ H_2$

(b) $CH_3(CH_2)_5CH_3 \xrightarrow[\text{catalyst}]{\text{heat}}$ (benzene)$—CH_3 + 3\ H_2$

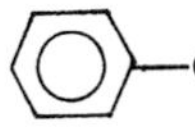

(c) $CH_3(CH_2)_6CH_3 \xrightarrow[\text{catalyst}]{\text{heat}}$
- (benzene)$—CH_2CH_3 + 3\ H_2$
- (benzene)$—CH{=}CH_2 + 4\ H_2$
- (benzene)$—C{\equiv}CH + 5\ H_2$

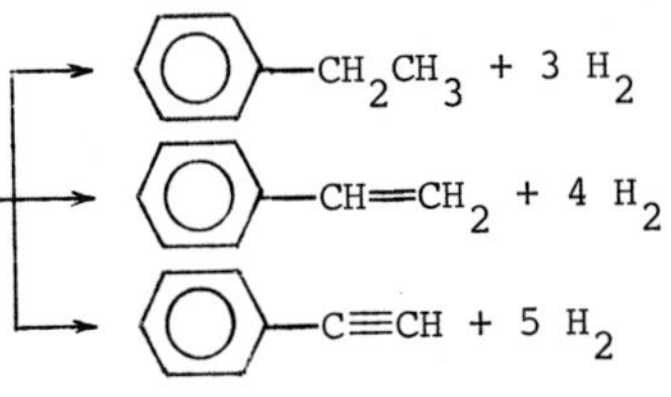

CHAPTER 4

Stereochemistry

Some Important Features

Geometric (or cis, trans) isomerism arises from attachments being on the same side or on opposite sides of a double bond or ring.

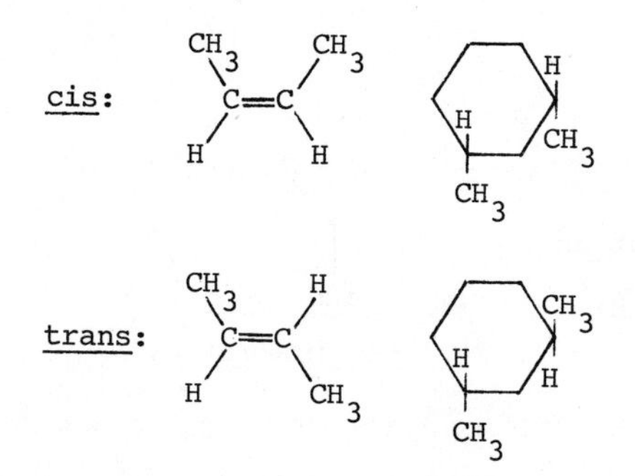

Geometric isomerism is not possible around a double bond if two groups on one carbon atom are the same (two CH_3 groups in the following example).

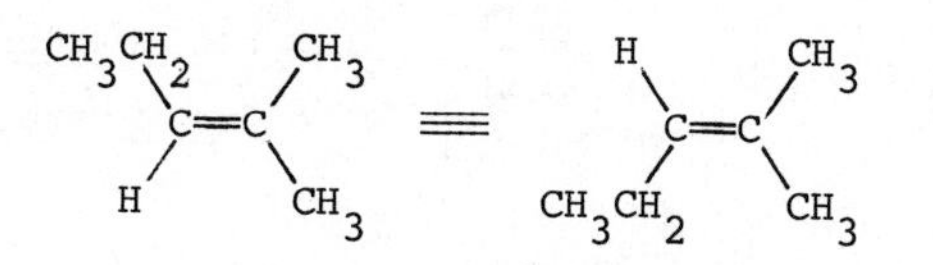

The above two formulas do not represent isomers because they are superimposable. Any superimposable molecules represent the same compound, not isomers. If this feature is not clear to you, make molecular models of the above two formulas and prove to yourself that they are the same.

The priority rules for the (E) and (Z) system are listed in Section 4.1B. The most important feature is that higher atomic number means higher priority. (In the case of isotopes, higher atomic mass means higher priority.) If the atoms attached to the sp^2 carbons are the same, proceed along the chain to the first point of difference.

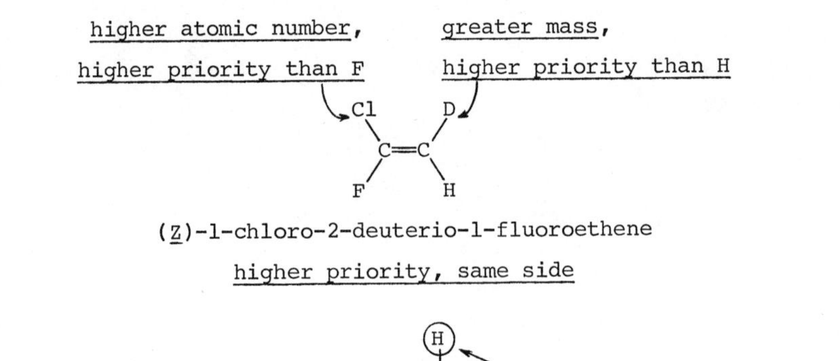

(<u>Z</u>)-1-chloro-2-deuterio-1-fluoroethene
<u>higher priority, same side</u>

(<u>Z</u>)-5-chloro-3-ethyl-2-pentene

A common student error is to sum atomic numbers to arrive at priorities of groups. This is incorrect. Look at individual atoms at the first point of difference, not at groups of atoms.

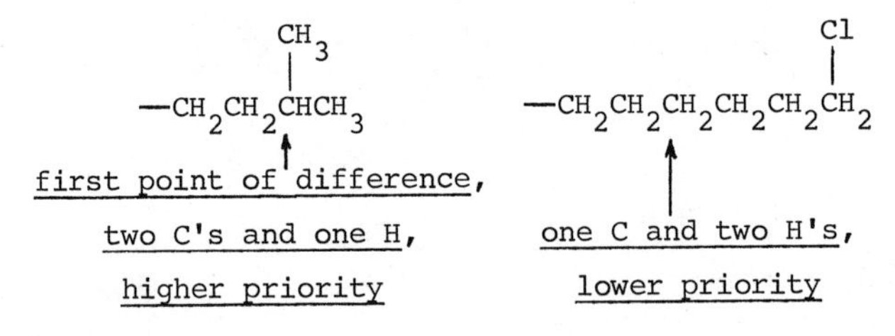

Conformations of molecules are the different shapes they can assume. Anti conformations are generally more stable than eclipsed conformations. (Compare the following Newman projections with molecular models.)

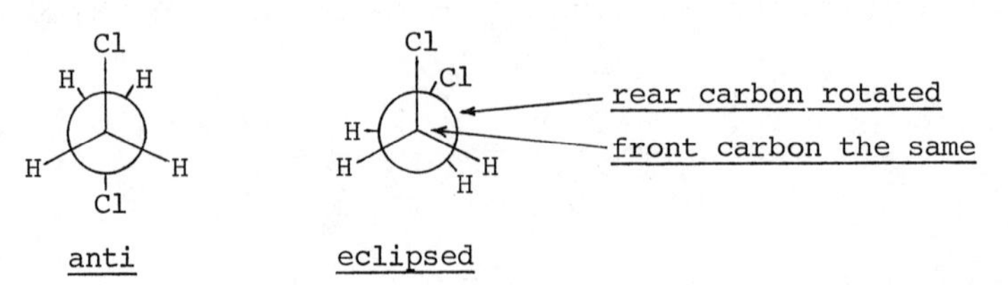

Conformations of cyclic compounds (six-membered rings particularly) should be studied carefully. A molecular model of cyclohexane shows ring conformation much better than a "paper formula" does.

Attachments to the cyclohexane ring in the chair form can be equatorial or axial. (The axial substituents are especially easy to see with models.) Although the following conformers are interconvertible, the conformer with the bulkier substituent equatorial is more stable (that is, of lower energy).

Different conformers of the same molecule:

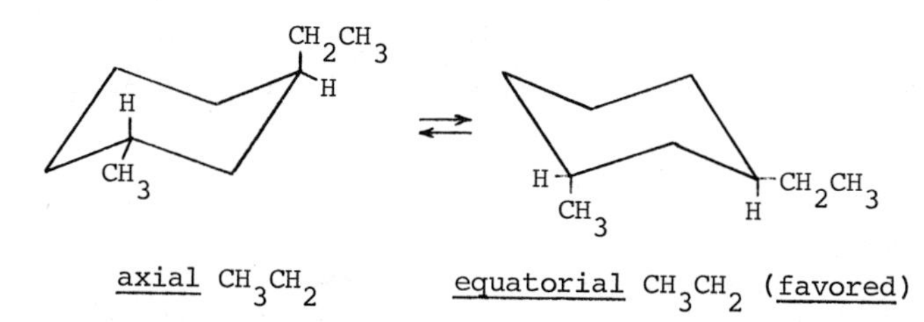

Molecular models are also indispensable in a study of chirality of molecules. A chiral carbon atom is generally one with four different attachments. The presence of one chiral carbon means that a molecule is chiral, or not superimposable on its mirror image, and is capable of rotating the plane of polarization of plane-polarized light.

A pair of nonsuperimposable mirror images are enantiomers of each other. The following formulas show two different ways of representing the enantiomers of 2-chlorobutane.

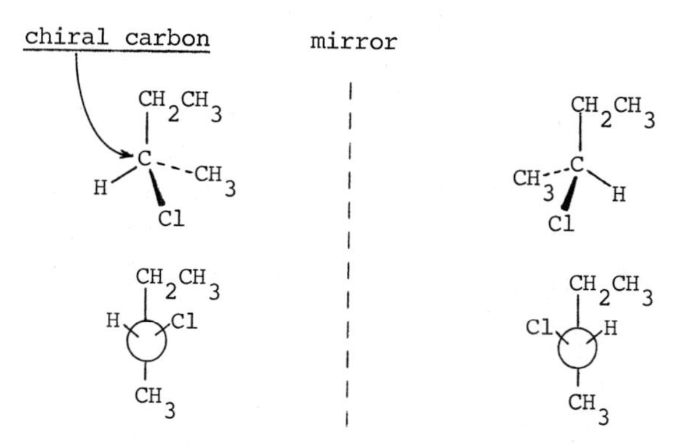

If a carbon atom is sp^2-hybridized, or if it has two identical attachments, the carbon atom is achiral (not chiral). (The chirality of a molecule as a whole, however, is determined by the lack of superimposability on the mirror image.) The following two examples are superimposable on their mirror images. (Try it with models.)

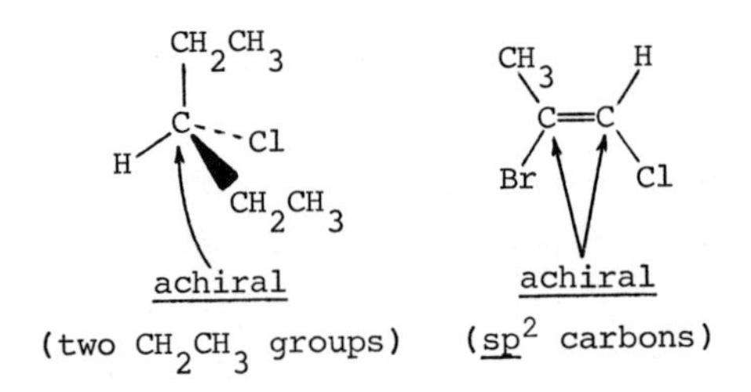

If a molecule contains more than one chiral carbon, there is a possibility of an internal plane of symmetry. A molecule with chiral carbons but with an internal plane of symmetry is the meso form of the compound and cannot rotate the plane of polarization of plane-polarized light.

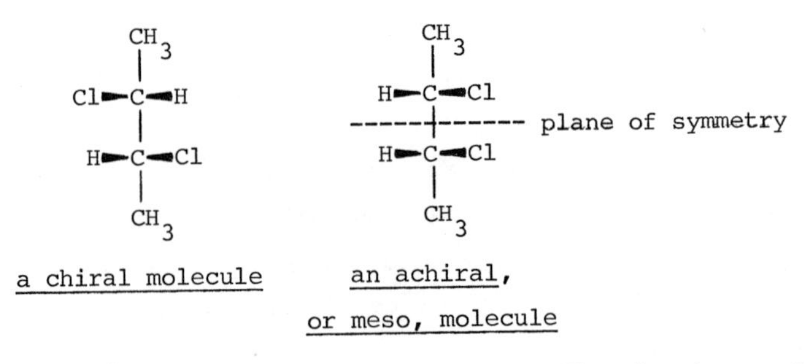

A meso molecule has a "top half" that is the mirror-reflective image of the "bottom half." Because groups can rotate around their bonds, it is not always easy to see at a glance if a structure is a meso form. Molecular models are useful here. Construct the models of the preceding two compounds and their mirror-reflective images, and verify the chirality or achirality of the two. Then, rotate groups around the bonds to see the different conformations each molecule can assume.

The use of Fischer projections is discussed in Section 4.6C. Remember that these projections are simply a shorthand way of representing ball-and-stick or dimensional formulas. For this reason, you can check any formula in a Fischer projection simply by converting it to a dimensional formula. If you do this operation, you are less likely to make errors.

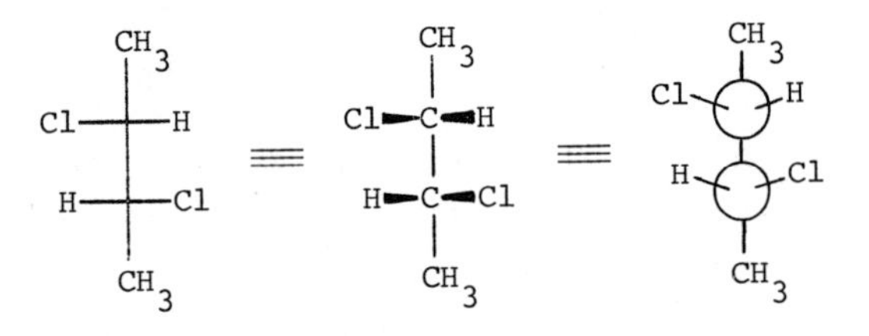

The ($\underline{R}$) and ($\underline{S}$) system of assigning the absolute configuration around chiral carbon atoms is discussed in Section 4.8A. You will probably find assignment of configuration easiest to do with models.

Reminders

To draw a chair form of cyclohexane quickly, follow the steps below (or purchase a template).

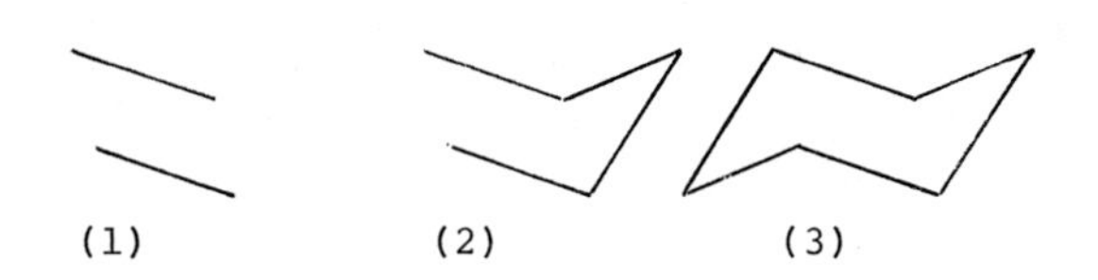

In drawing equatorial and axial substituents, first draw the axial bonds, which are vertical.

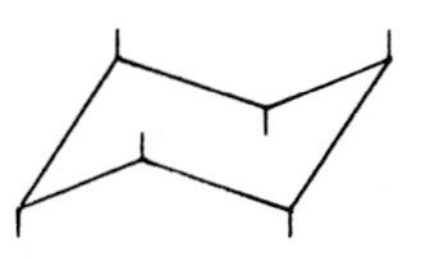

Now draw the equatorial bonds, keeping the bond angles at approximately 109°.

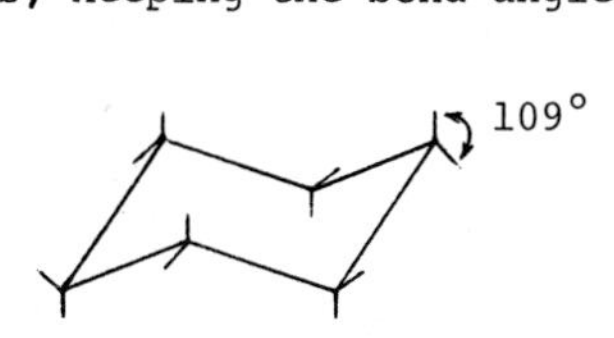

To determine <u>cis</u> and <u>trans</u> on a ring, decide if the groups are on the same side or on opposite sides.

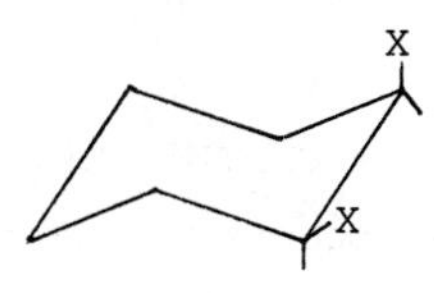

<u>cis</u> (both on "top")

To draw the enantiomer of a dimensional formula, simply reverse two groups, usually those attached to the "front" bonds (B, below).

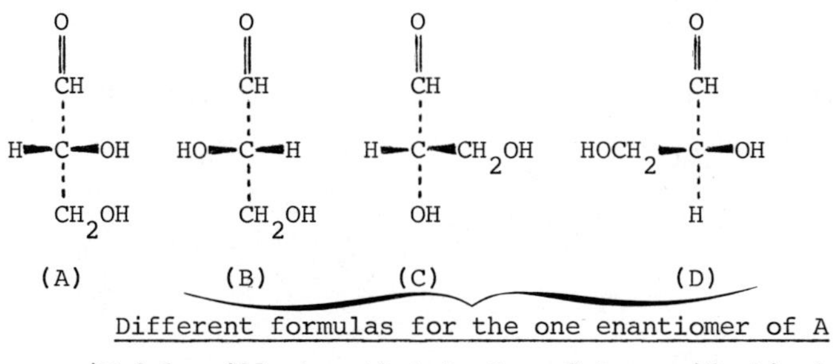

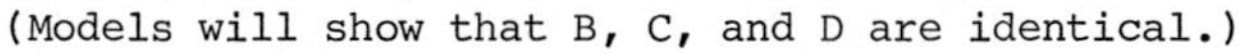

<u>Different formulas for the one enantiomer of A</u>

(Models will show that B, C, and D are identical.)

Answers to Problems

4.16 (a) $CH_3CH_2CH_2CH_2CH{=}CH_2$, no geometric isomer because there are two identical groups (H) on one carbon of the double bond.

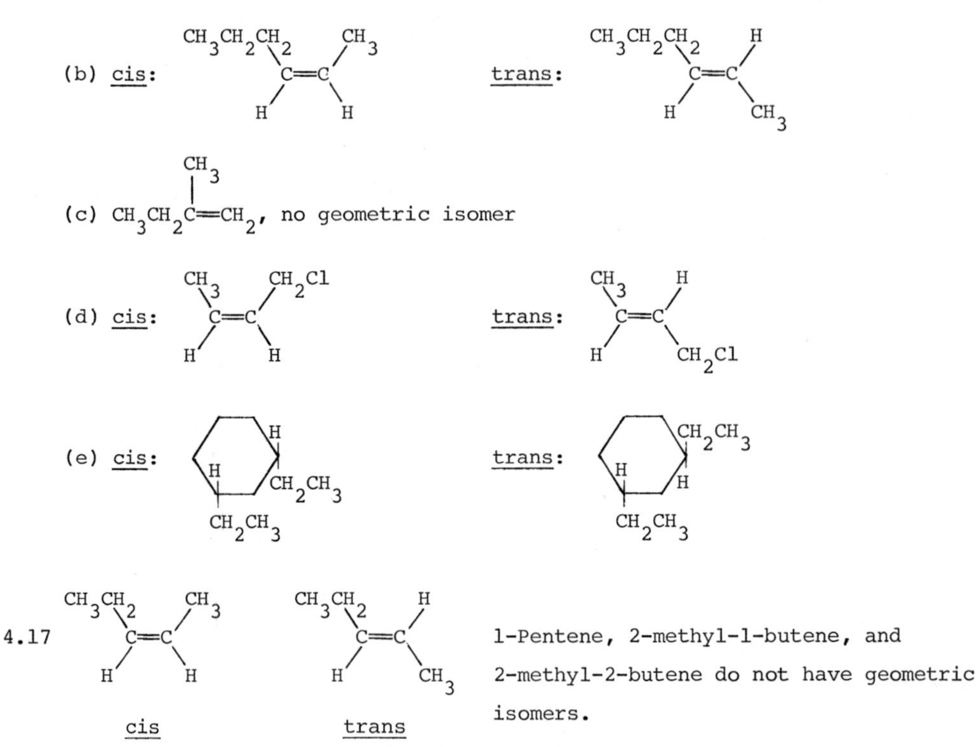

1-Pentene, 2-methyl-1-butene, and 2-methyl-2-butene do not have geometric isomers.

4.18 Compounds (a) $C_6H_5CH{=}CHC_6H_5$, (c) $CH_3CH{=}CHC{\equiv}CH$, and (e) have geometric isomers. In (c), the isomerism occurs around the double bond, not around the triple bond, which is linear. Compounds (b) $CH_2{=}CHC{\equiv}CH$ and (d) $(CH_3)_2C{=}C(CH_3)CH_2CH_3$ do not have geometric isomers because, in each case, one carbon has two identical groups.

4.19 In each case, the higher-priority atoms or groups are circled.

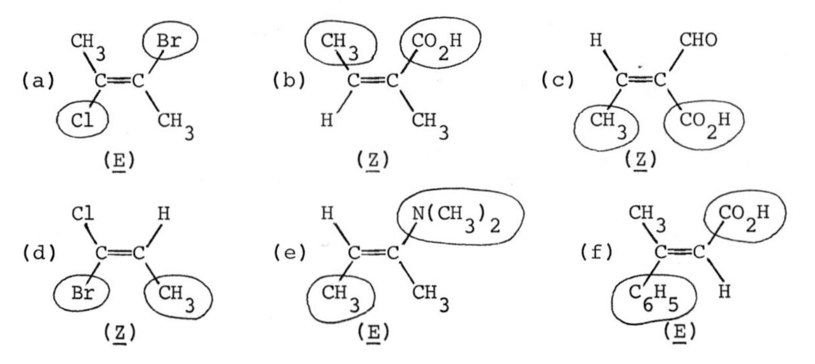

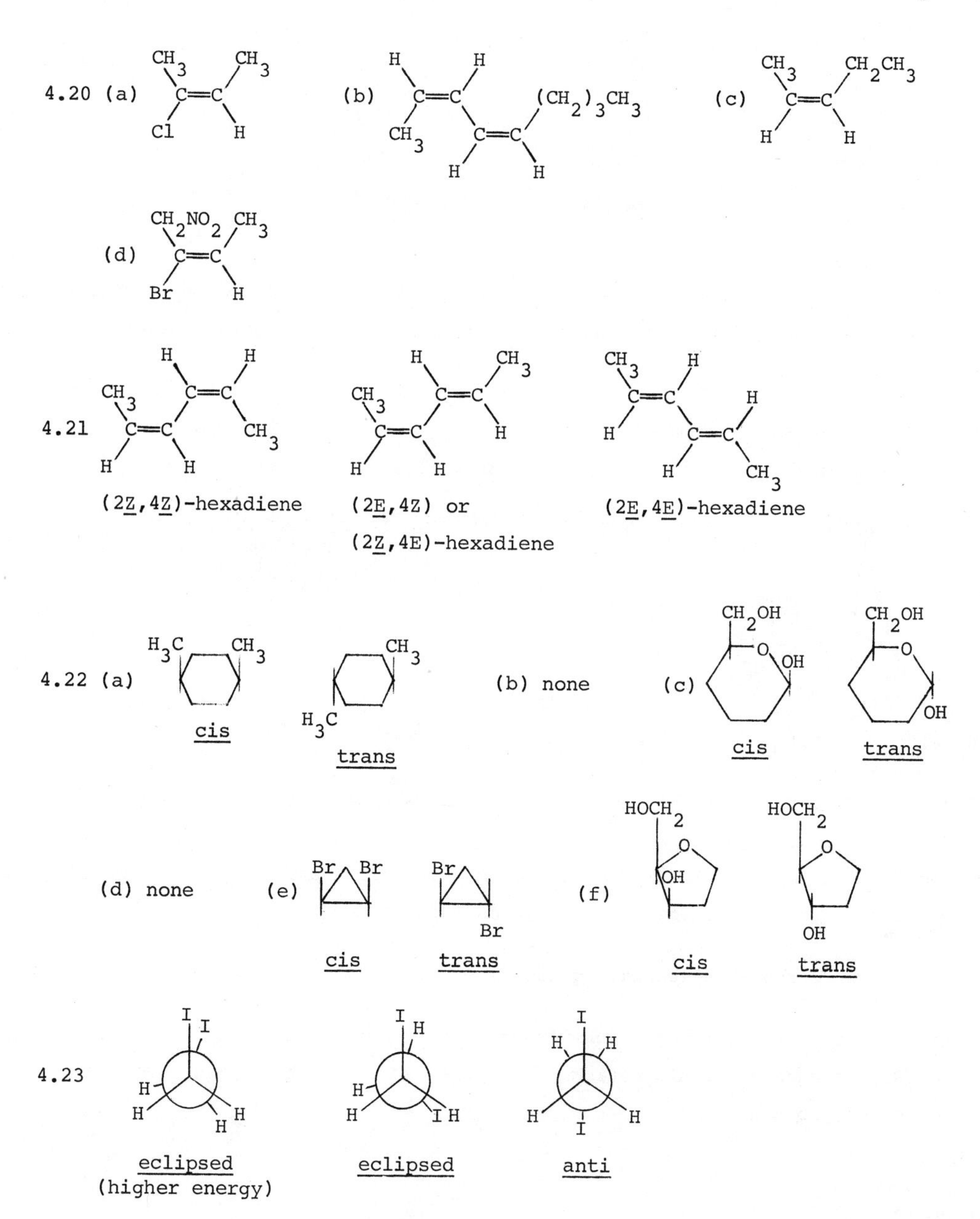

In drawing different Newman projections for a compound, rotate only one of the two carbons of the projection. In our answer, the front carbon is fixed and the rear carbon is rotated. (Note that we have shown the eclipsed atoms not completely eclipsed so that all atoms can be seen.)

4.24

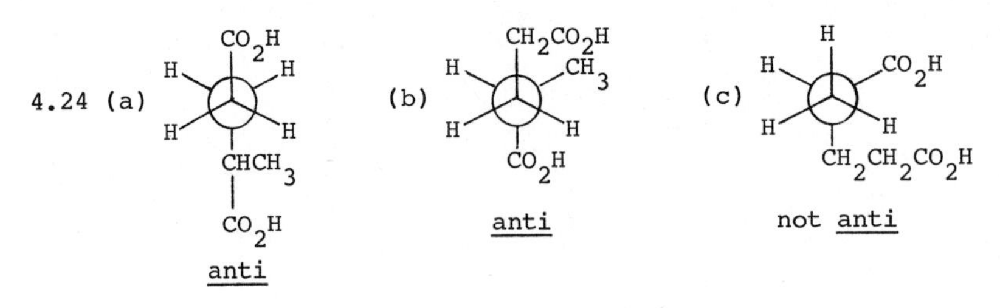

In (a) and (b), the two largest groups are anti. There is no anti-conformation for (c) around the two carbons shown because both large groups are on the same carbon.

4.25 (a) strained three-membered ring (b) strained four-membered ring
(c) unstrained (d) strained three-membered oxygen ring

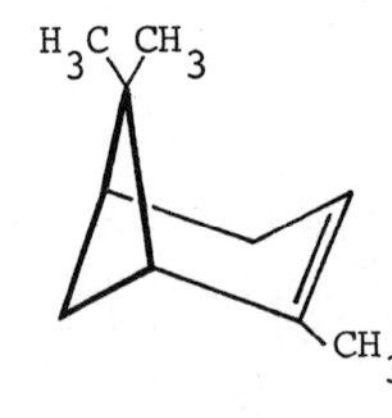

For (b), the four-membered ring is emphasized.

4.26 (a) e (b) a (c) e (d) a

4.27 A boat conformation is of higher energy than a chair form; therefore, (a) and (b) are of higher energy than (c) or (d). The conformation in (d) has two axial groups, while the one in (c) has one axial and one equatorial. Therefore, (c) is the lowest-energy, most-stable conformation.

4.28 (a), because the large t-butyl group is equatorial.

4.29 (a) In the cis-isomer, both substituents can be equatorial.
(b) No, because any conformation of the cis-isomer has one group e and the other group a.

4.30

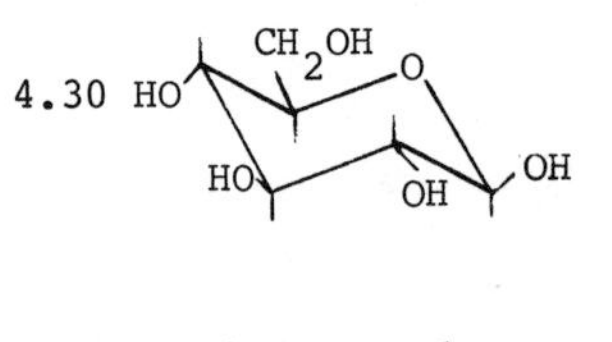

4.31

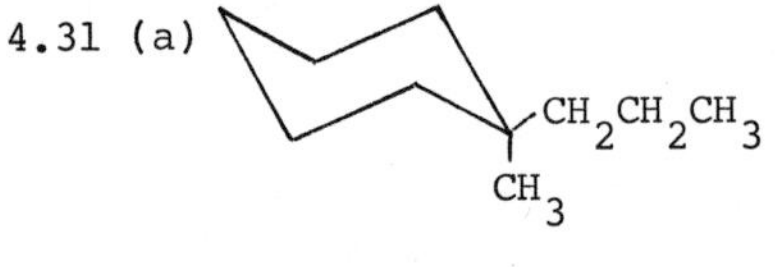

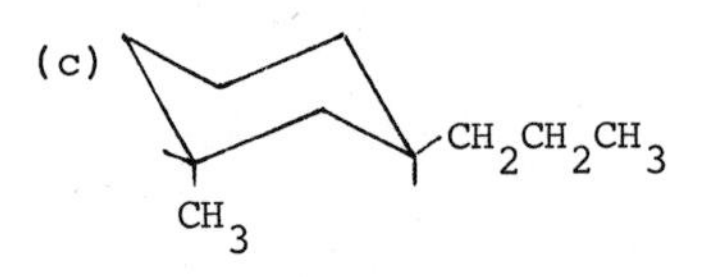

The most stable conformer is the one in which the largest group (propyl) is in an equatorial position.

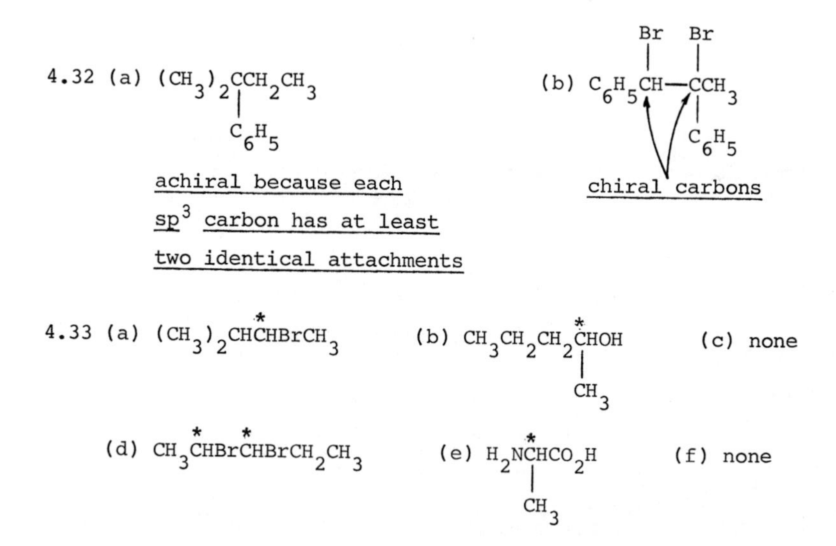

4.32 (a) $(CH_3)_2CCH_2CH_3$
 |
 C_6H_5

achiral because each
sp^3 carbon has at least
two identical attachments

chiral carbons

4.33 (a) $(CH_3)_2CHCHBrCH_3$
 *

(b) $CH_3CH_2CH_2\overset{*}{C}HOH$
 |
 CH_3

(c) none

(d) $CH_3\overset{*}{C}HBr\overset{*}{C}HBrCH_2CH_3$

(e) $H_2N\overset{*}{C}HCO_2H$
 |
 CH_3

(f) none

4.34 None are chiral because each has an internal plane of symmetry in at least one conformation. [Also note that in (a), (b), and (d) each potential chiral carbon has two identical groups attached.] In (c), the plane is perpendicular to the ring at the dashed line.

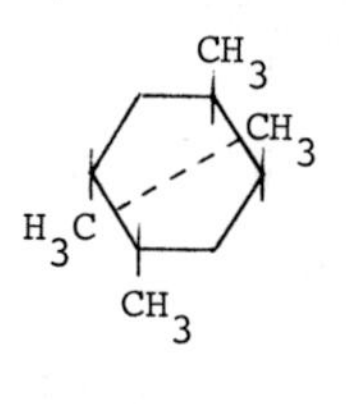

4.35 (a)

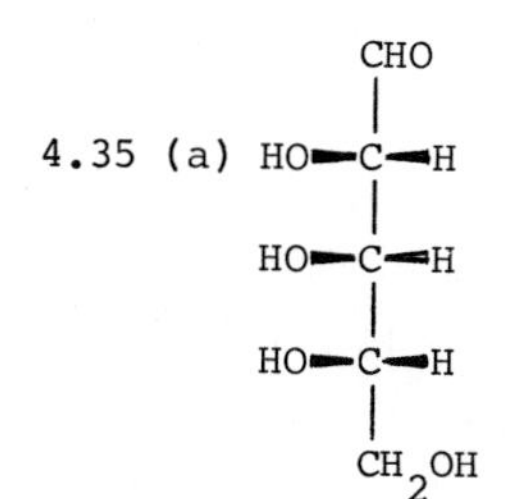

(b)

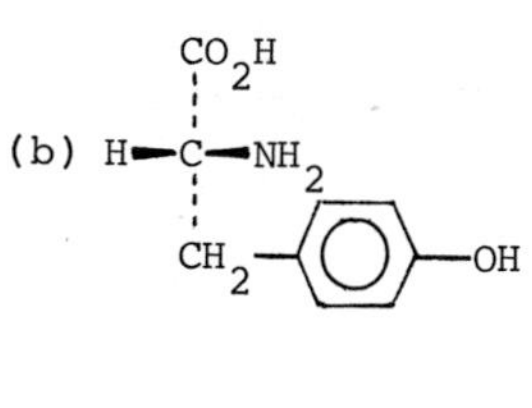

The easiest way to solve this type of problem is to draw a line representing a mirror and then draw the mirror reflection at each chiral carbon. The result is that each group originally on the left is switched to the right and vice versa. (We need not reverse groups on achiral carbon atoms because the configuration is the same regardless of how we draw these groups.)

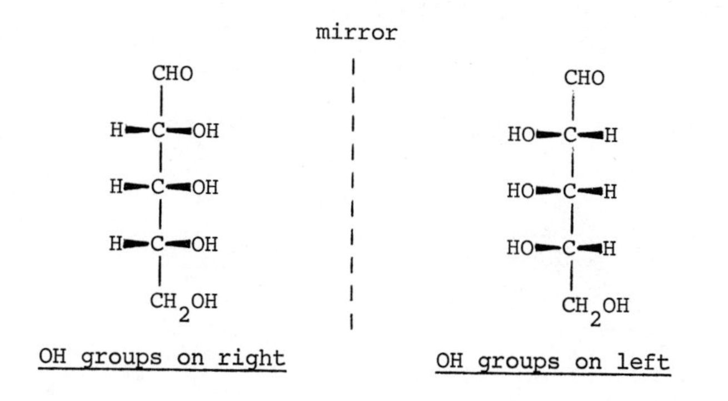

4.36

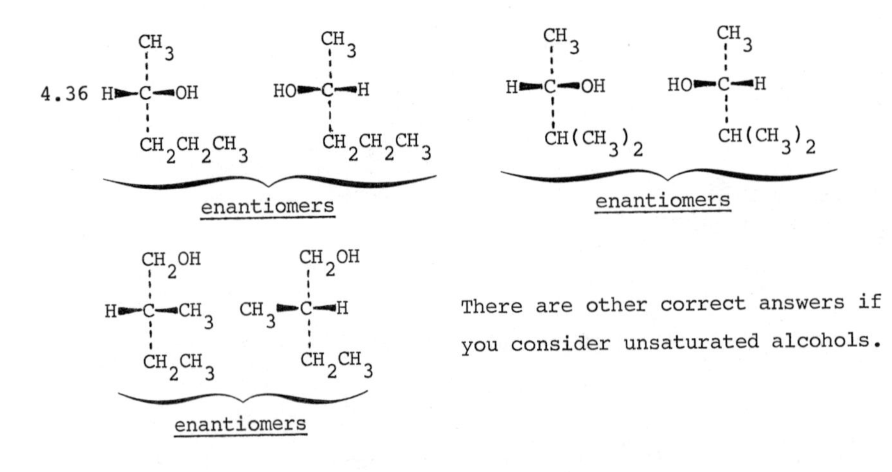

There are other correct answers if you consider unsaturated alcohols.

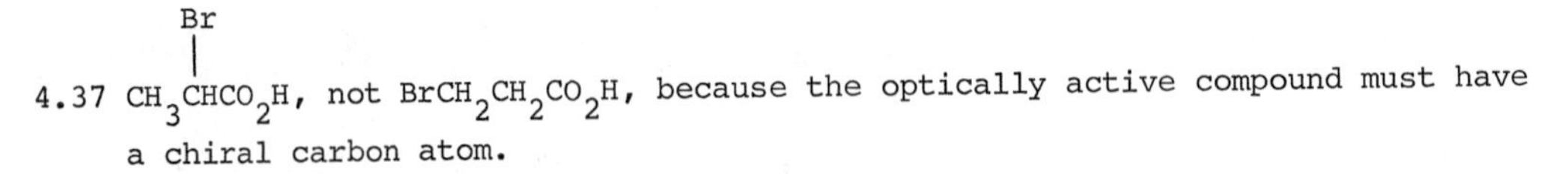

4.37 $CH_3CH(Br)CO_2H$, not $BrCH_2CH_2CO_2H$, because the optically active compound must have a chiral carbon atom.

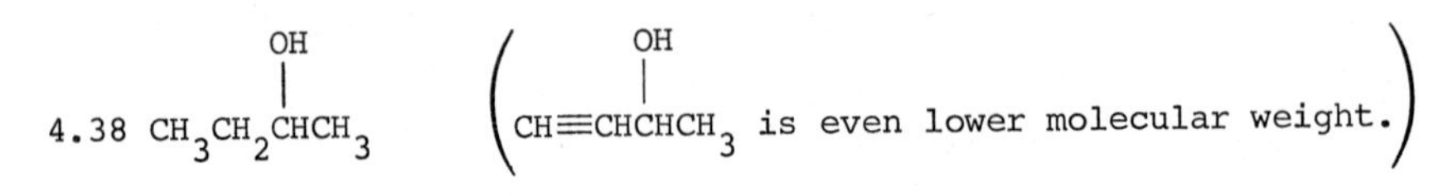

4.38 $CH_3CH_2CH(OH)CH_3$ ($CH{\equiv}CCH(OH)CH_3$ is even lower molecular weight.)

4.39

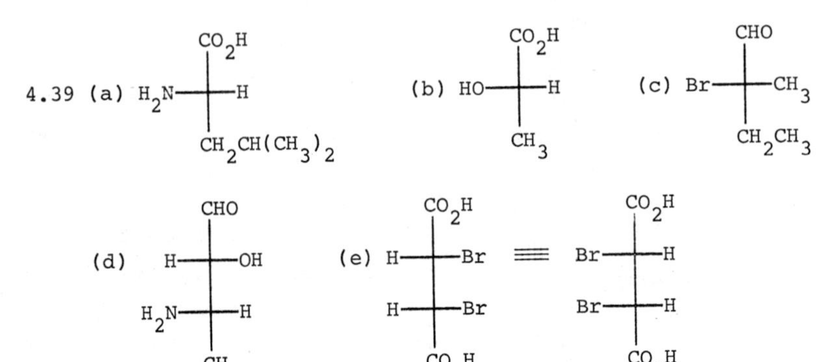

In (e), construct models to verify that the two Fischer projections represent the same compound and not enantiomers.

4.40

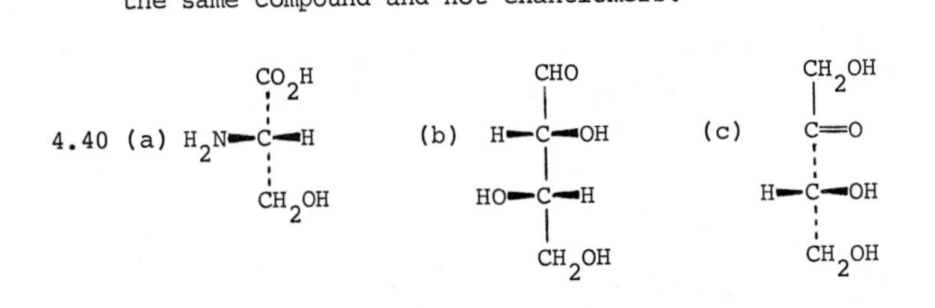

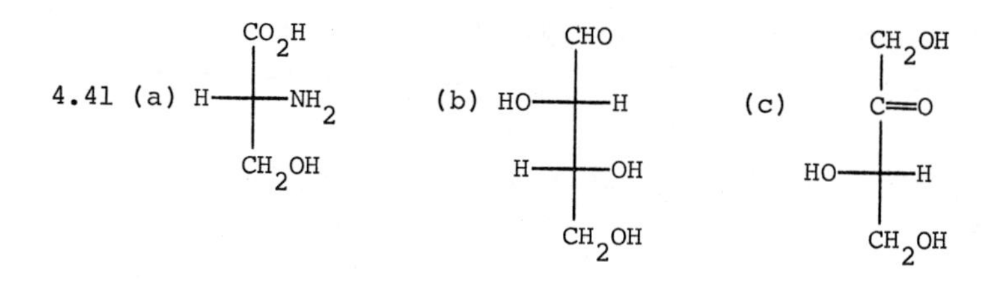

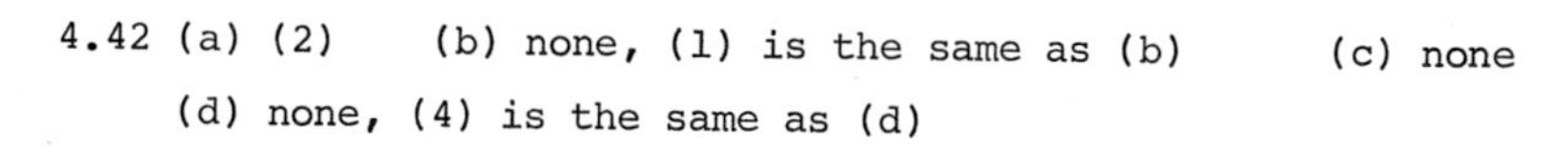
4.42 (a) (2) (b) none, (1) is the same as (b) (c) none
(d) none, (4) is the same as (d)

4.43 (b) and (d). The structures in (a) differ in the projection of only one chiral carbon. The structures in (c) are identical.

4.44 -13.5°

4.45 (b) and (c); (a) does not have a chiral carbon atom.

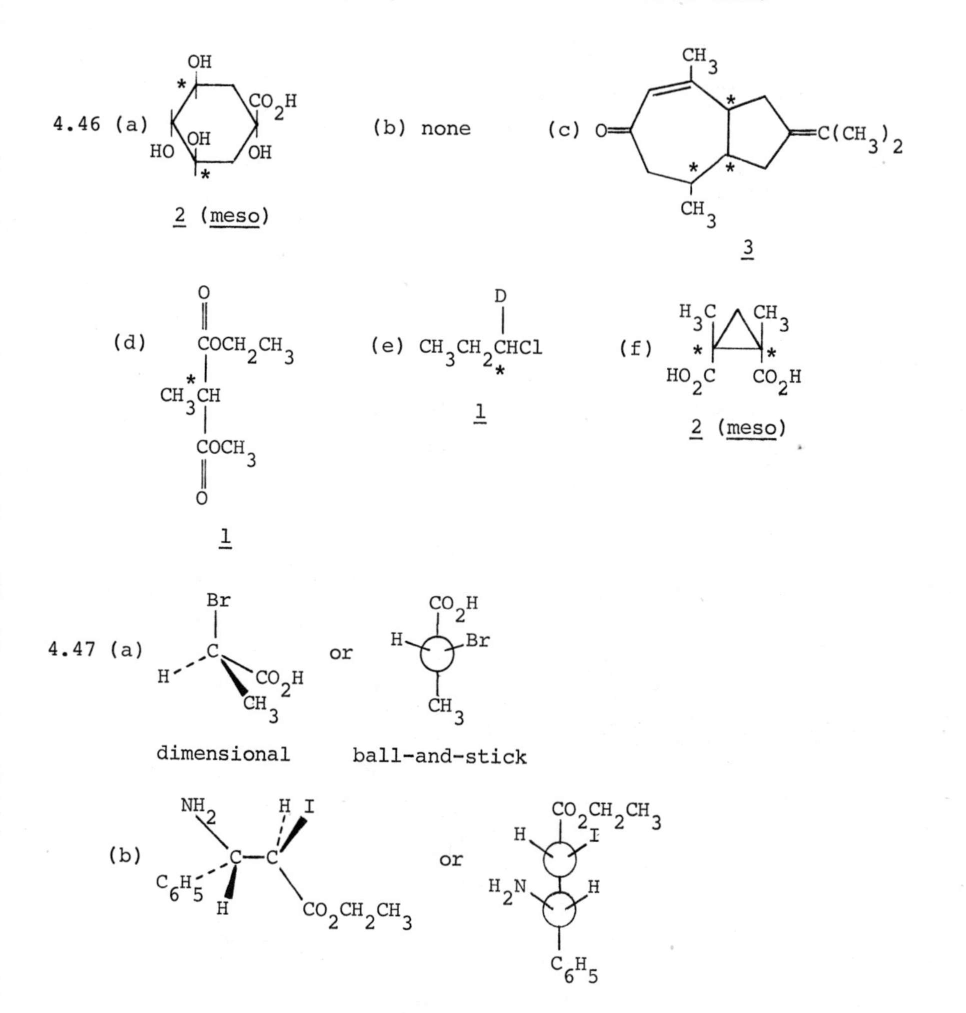

4.48 (a) four (two chiral carbons) (b) two (one chiral carbon)

4.49 (a) I and III are enantiomers (nonsuperimposable mirror images).

(b) I and II are diastereomers, as are II and III (stereoisomers that are not enantiomers).

(c) II is a meso compound; a horizontal plane of symmetry may be drawn through carbon 3.

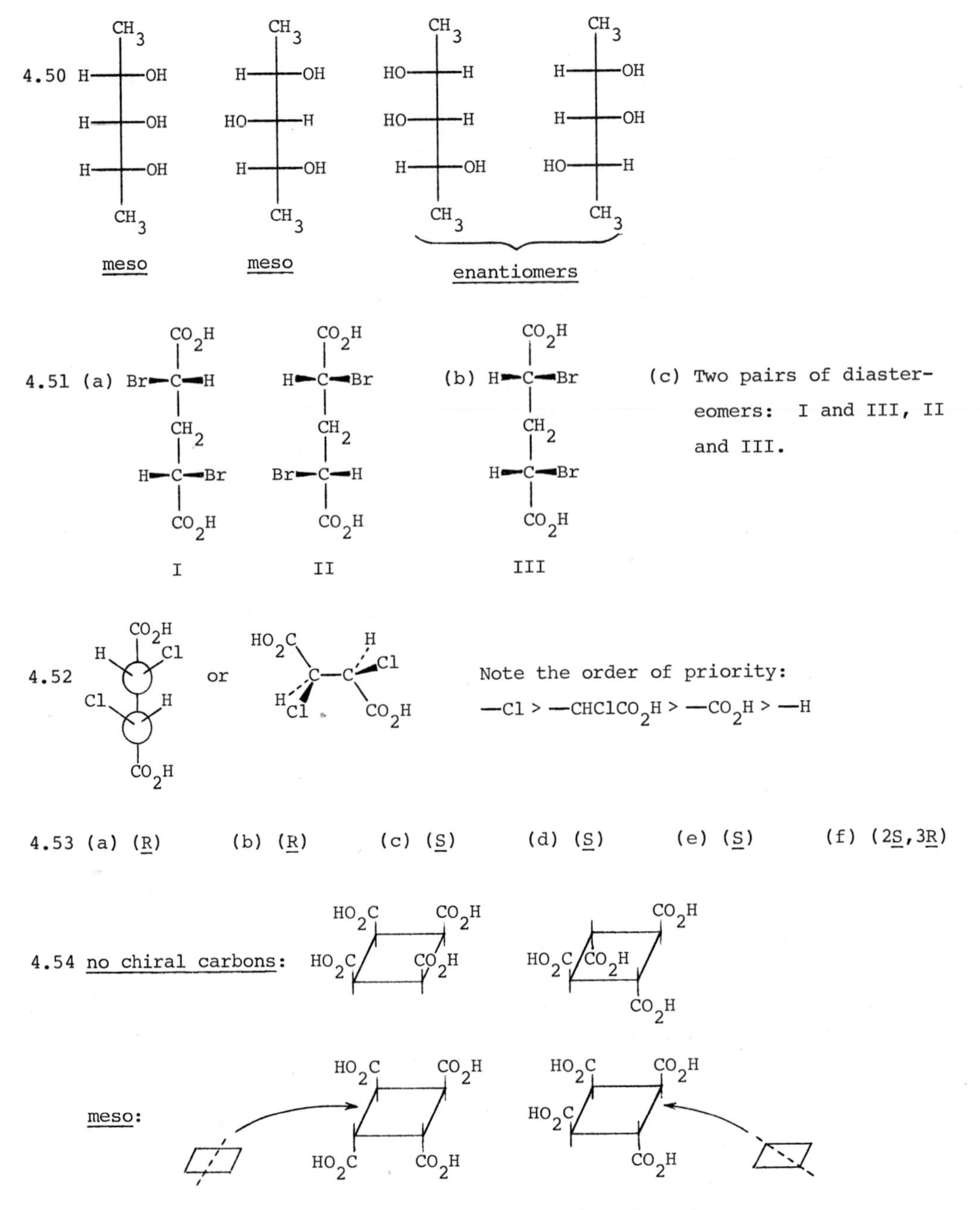

4.55 (a) Three stereoisomers, no enantiomers:

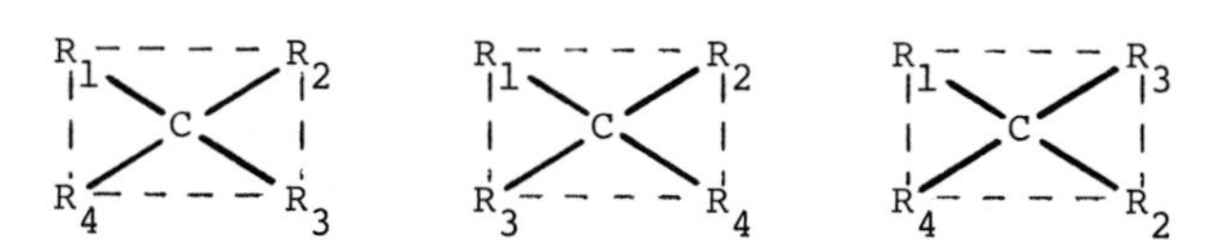

(b) Two stereoisomers, which are a pair of enantiomers:

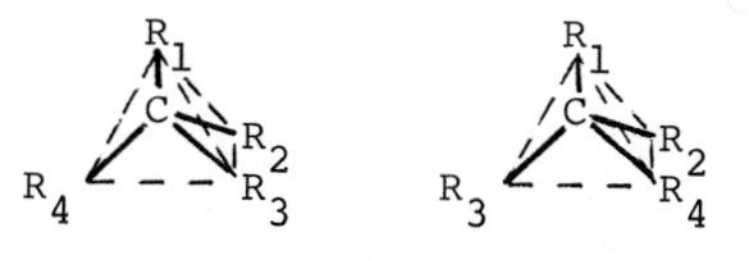

(c) six stereoisomers, which represent three pairs of enantiomers. The members of the enantiomeric pairs are shown one above the other.

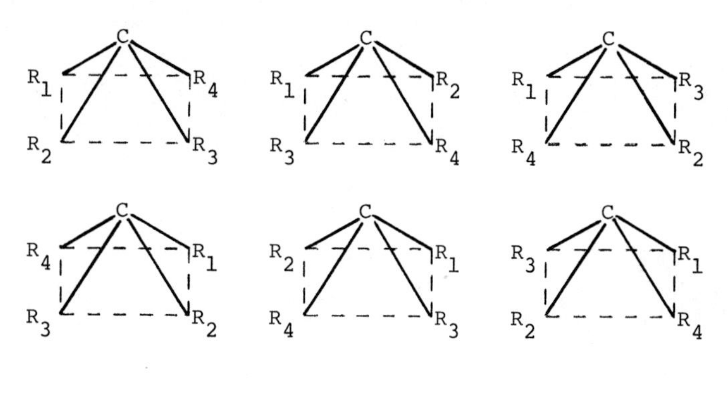

4.56 (a)

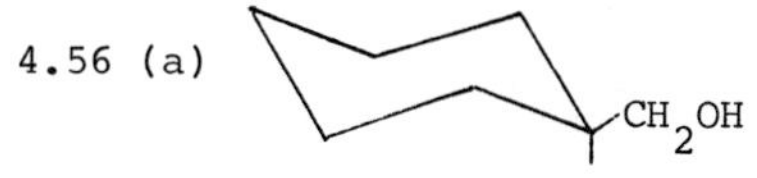

(b)

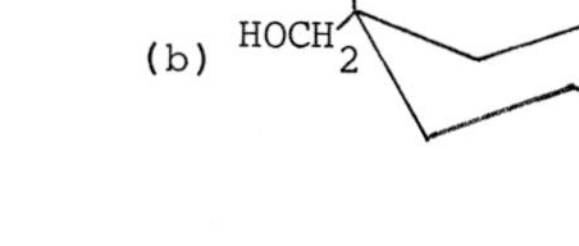

(c) Cl Cl Cl

In (b), the bulkier group is equatorial in the most stable conformation.

4.57 (a)

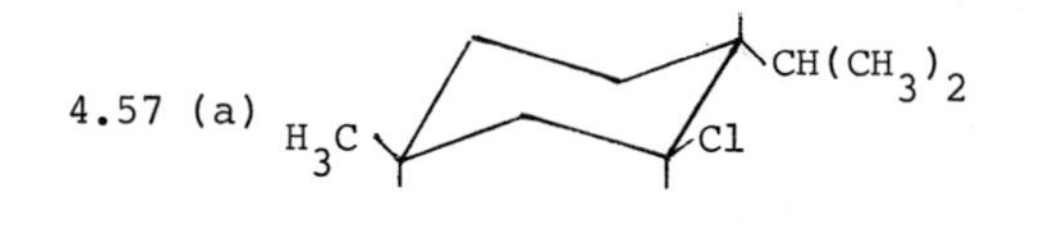

(b)

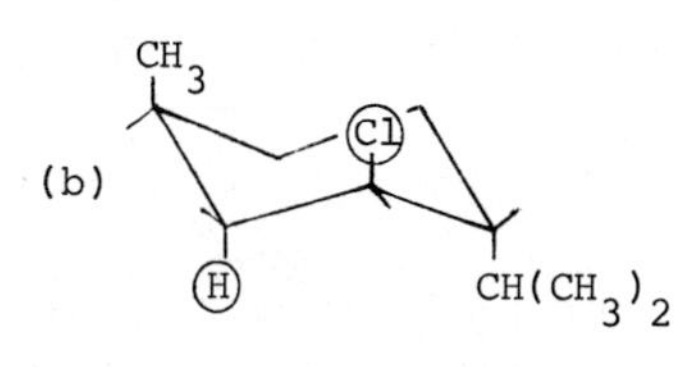

not favored

4.58

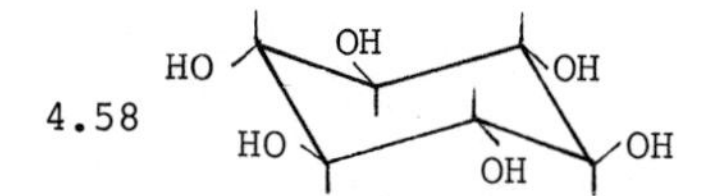

4.59 (a) Using the left-hand chiral carbon as the front carbon:

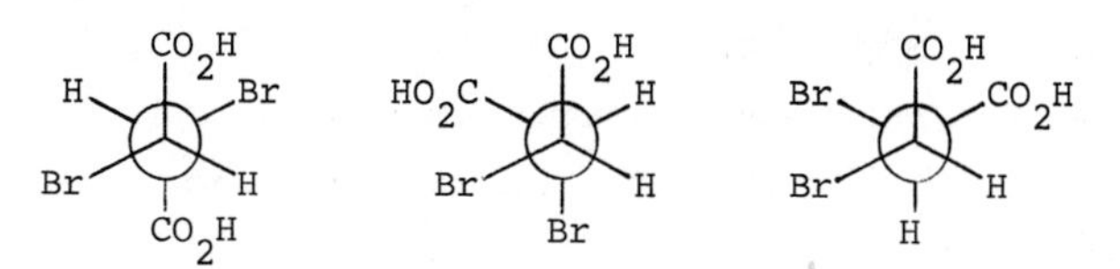

(b)

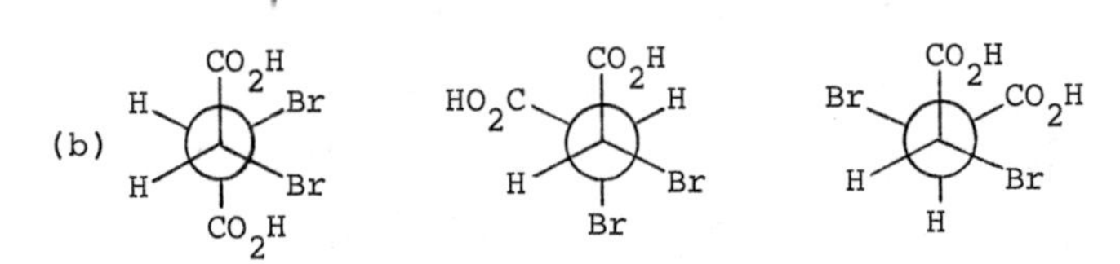

4.60 Using the (1S) carbon as the front carbon:

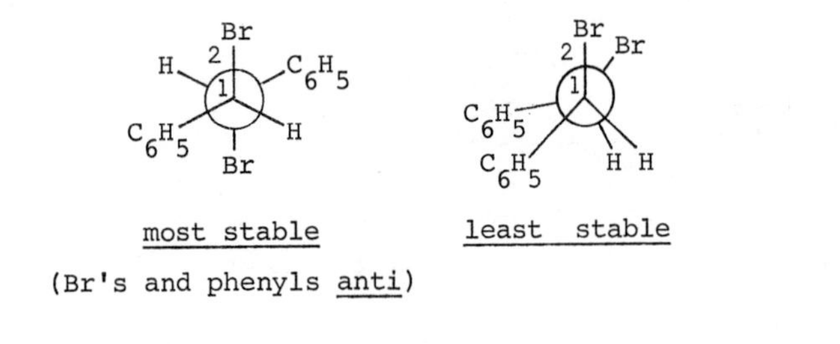

(Br's and phenyls anti)

4.61

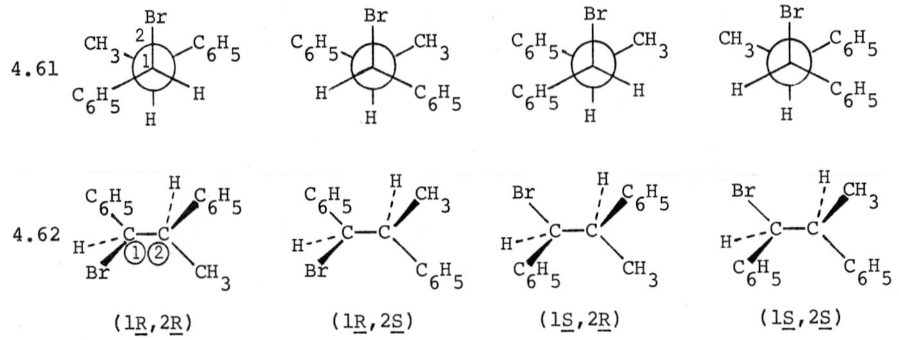

4.62

C6H5, H, C6H5, H---C—C, Br, 1, 2, CH3 (1R,2R)

C6H5, H, CH3, H---C—C, Br, C6H5 (1R,2S)

Br, H, C6H5, H---C—C, C6H5, CH3 (1S,2R)

Br, H, CH3, H---C—C, C6H5, C6H5 (1S,2S)

Note the order of priority:

$$-Br > -\underset{\displaystyle C_6H_5}{CHBr} > -C_6H_5 > -\underset{\displaystyle CH_3}{CHC_6H_5} > CH_3 > H$$

4.63 (a) zero, because it is a racemic mixture.

(b) The solution is, in effect, half racemic and half (S)-enantiomer. The observed rotation is therefore half that of the pure (S)-enantiomer, or +8.0°.

4.64 (a) $[\alpha] = \frac{\alpha}{lc} = \frac{+0.45°}{0.10\ \text{dm} \times 0.200\ \text{g/mL}} = +22.5°$

(b) $[\alpha] = \frac{-3.2°}{1.0\ \text{dm} \times 0.10\ \text{g/mL}} = -32°$

4.65 (b), (c), and (e) would cause rotation. Compound (a) is meso and mixture (d) is racemic: neither of these would cause rotation.

4.66 In (a), the mixture is 50% optically active compound plus 50% optically inactive compound; therefore, the specific rotation is half that of the optically active compound, or -6°. In (b), the mixture is racemic; the specific rotation is therefore zero.

4.67

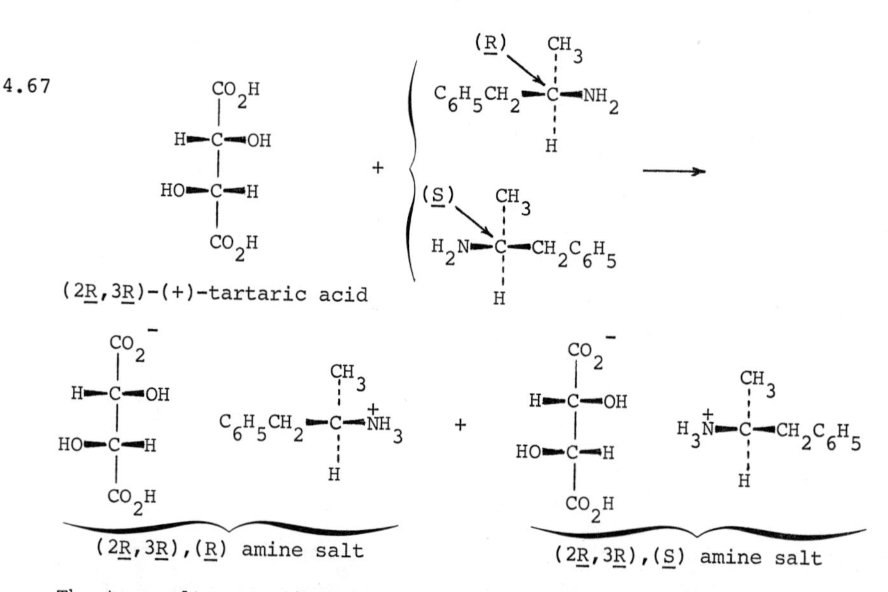

The two salts are diastereomers, not enantiomers, and therefore can be separated by physical means, such as fractional crystallization. Once separated, each enantiomeric amine can be regenerated by treatment of its now-pure salt with aqueous NaOH and extraction of the amine with an organic solvent. (The tartrate anion remains in the water layer.)

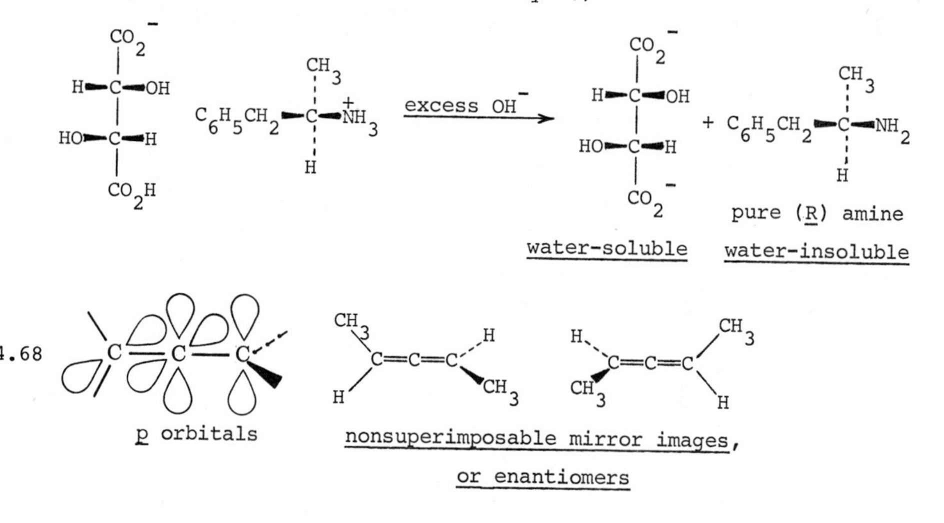

4.68

4.69 (a), (c), (d), (e).

4.70 (a)

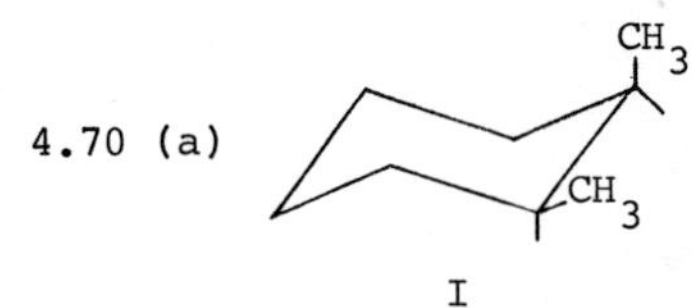

I

cis-1,2-dimethyl-
cyclohexane in one
conformation

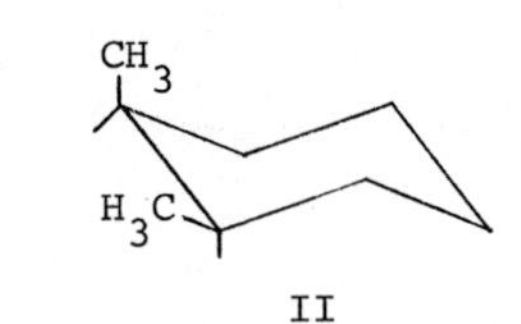

II

nonsuperimposable
mirror image of I

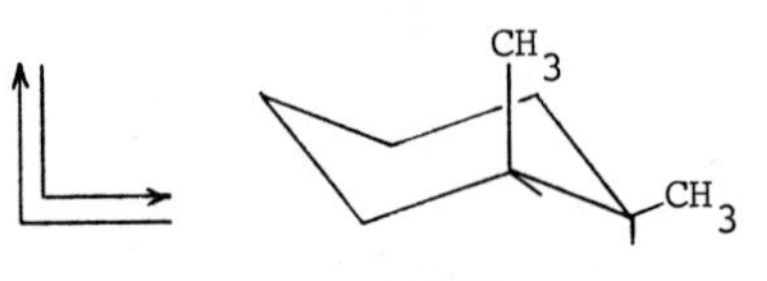

III

superimposable on II

Structure I undergoes rapid interconversion among its conformers. From the conformational formulas, we can see that the mirror image of I is simply another conformer of I. Thus, the two cannot be resolved. (We suggest that you make models to verify the above statements and the mirror-image relationships.)

(b) Yes, the trans compound exists as a pair of nonsuperimposable mirror images that cannot be interconverted. (Try it with models.)

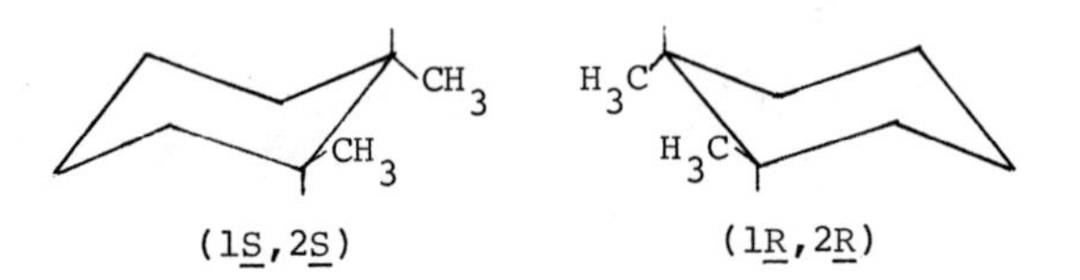

(1S,2S) (1R,2R)

CHAPTER 5

Alkyl Halides; Substitution and Elimination Reactions

Some Important Features

Alkyl halides, but not most aryl halides, can undergo substitution or elimination when treated with a nucleophile or base. Primary alkyl halides and, to an extent, secondary alkyl halides generally react by an S_N2 path. (Tertiary alkyl halides do not undergo S_N2 reaction.) Inversion of configuration at the functional carbon is observed in a typical S_N2 reaction if that carbon is chiral.

$$S_N2\text{: } Nu^- + \underset{\substack{1^\circ \\ (\text{and } 2^\circ)}}{R{-}X} \longrightarrow Nu{-}R + X^-$$

When treated with a very weak nucleophile, such as H_2O or ROH, 2° and 3° alkyl halides can react by an S_N1 path. Primary alkyl halides (unless allylic or benzylic) do not undergo reaction by this pathway.

$$S_N1\text{: } \underset{\substack{3^\circ \\ (\text{and } 2^\circ)}}{R{-}X} \xrightarrow{-X^-} [R^+] \xrightarrow{H_2O} [\overset{+}{R}OH_2] \xrightarrow{-H^+} ROH$$

Carbocation reactions have some disadvantages:

1. Racemization usually occurs because a carbocation is planar and therefore achiral.
2. Rearrangement may occur if a more stable carbocation can be formed by a 1,2-shift.

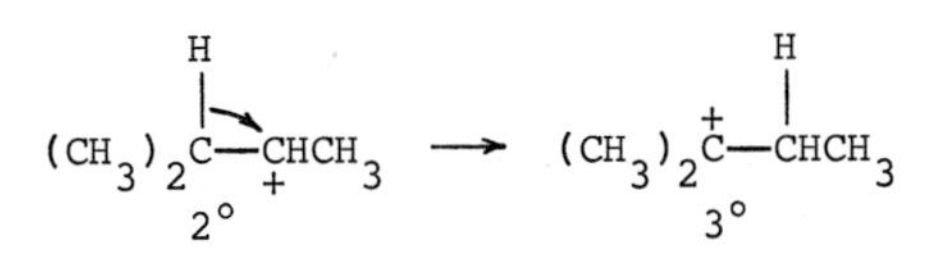

3. S_N1 reactions are accompanied by elimination (by an E1 path); therefore, mixtures of products are observed.

$$\text{E1: } (CH_3)_2\overset{+}{C}-CHCH_3 \xrightarrow{H_2O:} (CH_3)_2C{=}CHCH_3 + H_3O:^+$$

a carbocation

Tertiary (and 2°) alkyl halides undergo elimination by an E2 path when they are heated with a strong base. The principal alkene product of either an E1 or E2 reaction is usually the more substituted, trans alkene.

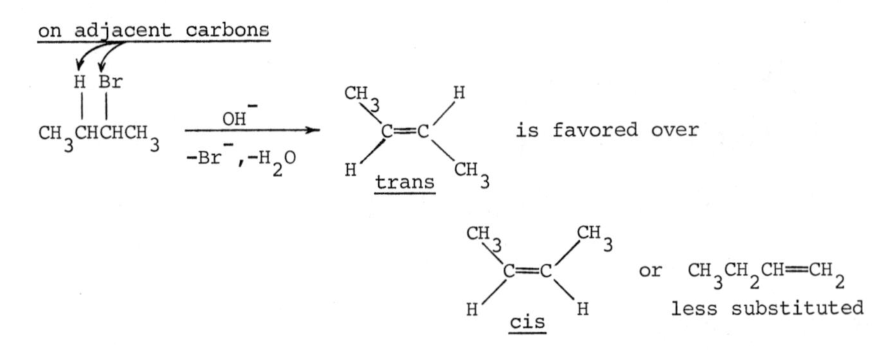

An E2 reaction proceeds by anti-elimination of H^+ and X^- on adjacent carbons. If the halogen atom is attached to a ring carbon, the H and X must be trans and diaxial in order to be eliminated readily.

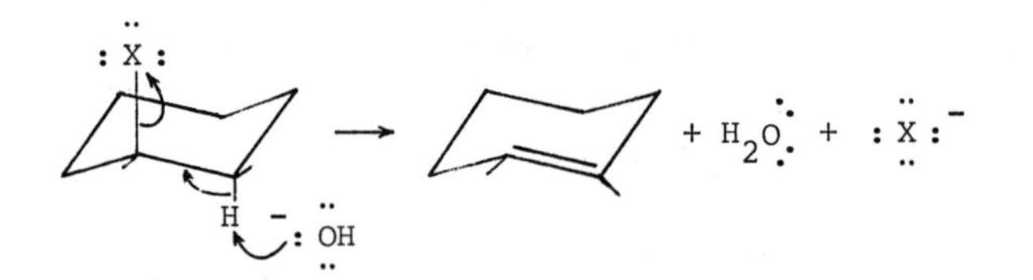

A properly selected alkyl halide yields a single geometric isomer in an E2 reaction.

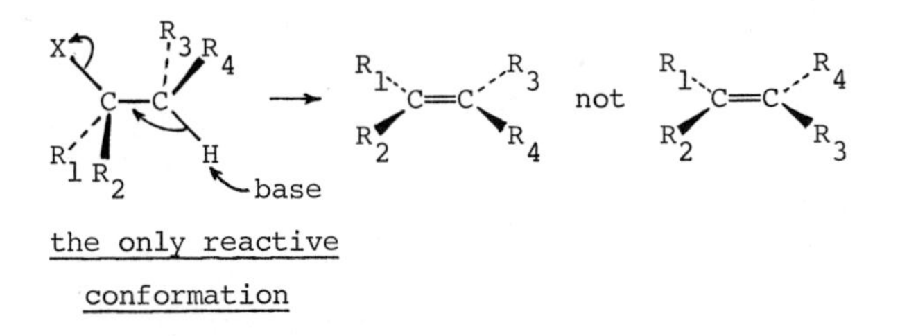

Every reaction step proceeds through a transition state. A lower-energy (more stable) transition state means a lower $\underline{E}_{act}$ and therefore a faster rate of reaction.

In a typical S_N1 or E1 reaction, the rate-determining (slower) step is the formation of the carbocation. Any feature that stabilizes the carbocation also stabilizes the transition state and leads to a faster reaction.

$$RX \xrightarrow{-X^-} [R^+] \xrightarrow{Nu^-} RNu \text{ or alkene}$$

↑ The rate is increased if R^+ is stabilized by the inductive effect, resonance, or a highly polar solvent.

Other important topics discussed in this chapter are how a carbocation is stabilized; first- and second-order rates; the kinetic isotope effect; and the effects of steric hindrance on the reactivity of alkyl halides and on the products of E2 reactions (Saytseff versus Hofmann products). You may also want to review the rules for writing resonance structures in Chapter 2.

Reminders

Although S_N2 reactions can yield optically active products from optically active reactants, S_N1, E1, and E2 reactions all lead to racemization or loss of chirality at the functional carbon.

$RX + Nu^-$ $\xrightarrow{S_N2}$ Nu—R (inverted)

$RX + Nu^-$ $\xrightarrow{S_N1}$ $[R^+]$ (achiral) ⟶ RNu (racemized)

$RX + Nu^-$ $\xrightarrow{\text{E1 or E2}}$ alkenes

($\underline{sp}^2$ carbons are achiral)

Remember the following general rules of reactivity:

S_N2:	$CH_3X > 1^\circ RX > 2^\circ RX$
S_N1 (or E1):	$3^\circ RX > 2^\circ RX$
E2:	$3^\circ RX > 2^\circ RX$

Strong Nu^-, high concentration: S_N2 for 1° and 2° RX

Weak Nu^- (H_2O, etc.): S_N1 for 2° and 3° RX

Strong base: E2 for 2° and 3° RX

Answers to Problems

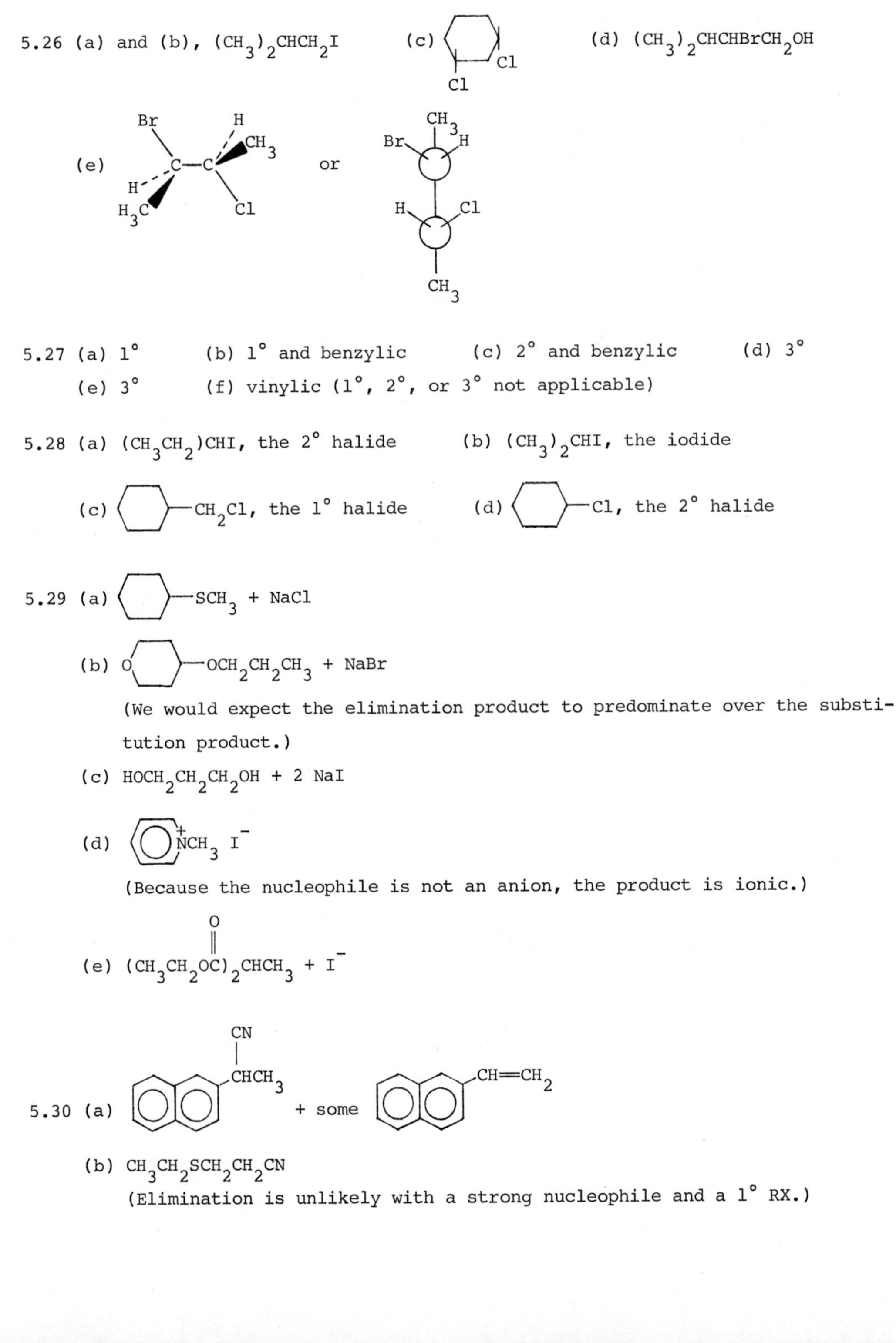

5.25 (a) 3,3,3-trichloro-1-propene (b) 1,3-dibromobutane
(c) 1-bromo-1-methylcyclohexane (d) trans-2-chloro-1-cyclopentanol
(e) (R)-2-iodopropanoic acid

5.26 (a) and (b), $(CH_3)_2CHCH_2I$ (c) Cl Cl (d) $(CH_3)_2CHCHBrCH_2OH$

(e) Br H CH_3 C—C H H_3C Cl or CH_3 Br H H Cl CH_3

5.27 (a) 1° (b) 1° and benzylic (c) 2° and benzylic (d) 3°
(e) 3° (f) vinylic (1°, 2°, or 3° not applicable)

5.28 (a) $(CH_3CH_2)CHI$, the 2° halide (b) $(CH_3)_2CHI$, the iodide
(c) —CH_2Cl, the 1° halide (d) —Cl, the 2° halide

5.29 (a) —SCH_3 + NaCl
(b) O —$OCH_2CH_2CH_3$ + NaBr
(We would expect the elimination product to predominate over the substitution product.)
(c) $HOCH_2CH_2CH_2OH$ + 2 NaI
(d) $\overset{+}{N}CH_3$ I^-
(Because the nucleophile is not an anion, the product is ionic.)
(e) $(CH_3CH_2OC)_2CHCH_3$ + I^-

5.30 (a) CN $CHCH_3$ + some $CH{=}CH_2$
(b) $CH_3CH_2SCH_2CH_2CN$
(Elimination is unlikely with a strong nucleophile and a 1° RX.)

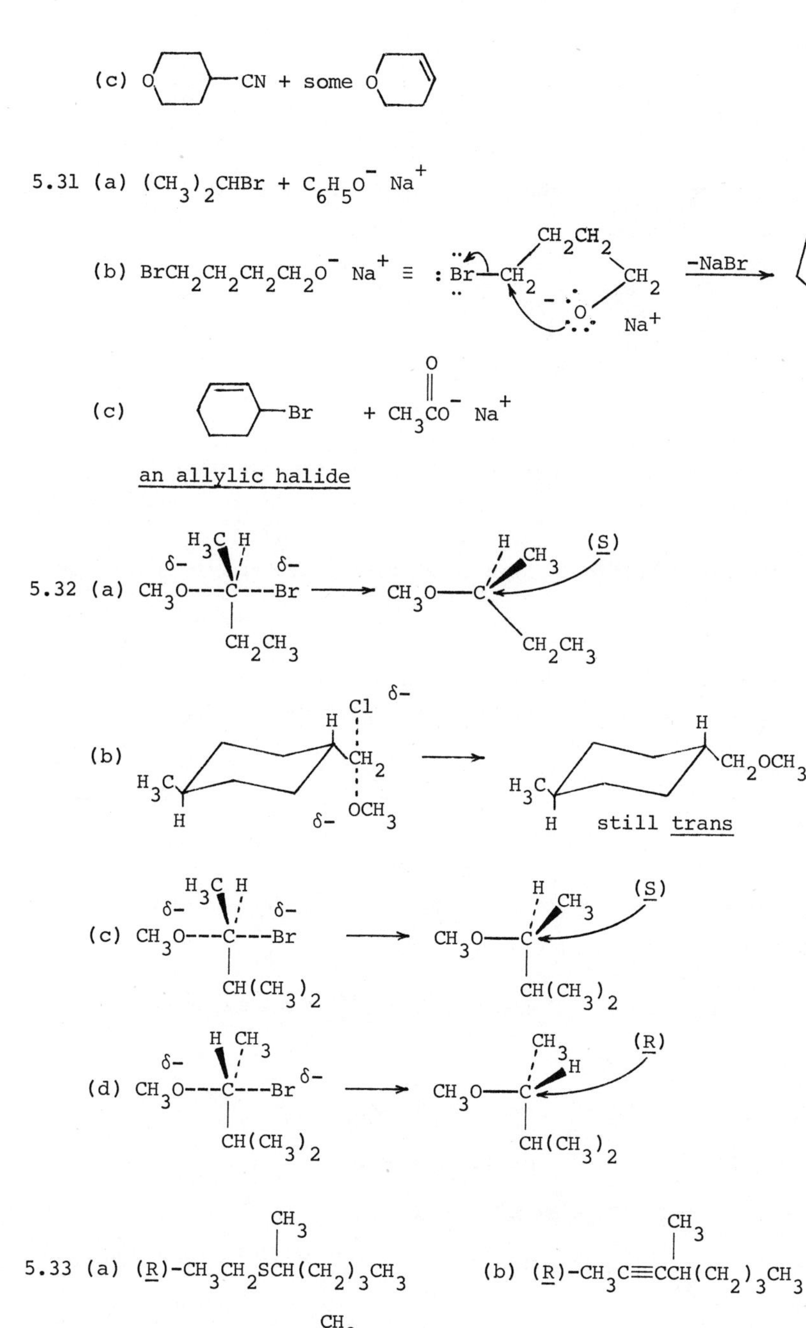

5.31 (a) $(CH_3)_2CHBr + C_6H_5O^- Na^+$

(b) $BrCH_2CH_2CH_2CH_2O^- Na^+ \equiv$... $\xrightarrow{-NaBr}$

(c) —Br + $CH_3\overset{\overset{\displaystyle O}{\|}}{C}O^- Na^+$

__an allylic halide__

5.32 (a) $CH_3O---\overset{\overset{\displaystyle H_3C \; H}{}}{\underset{\underset{\displaystyle CH_2CH_3}{}}{C}}---Br \longrightarrow CH_3O-\overset{\overset{\displaystyle H \; CH_3}{}}{\underset{\underset{\displaystyle CH_2CH_3}{}}{C}}$ (S)

(b) [figure] \longrightarrow [figure] still __trans__

(c) $CH_3O---\overset{\overset{\displaystyle H_3C \; H}{}}{\underset{\underset{\displaystyle CH(CH_3)_2}{}}{C}}---Br \longrightarrow CH_3O-\overset{\overset{\displaystyle H \; CH_3}{}}{\underset{\underset{\displaystyle CH(CH_3)_2}{}}{C}}$ (S)

(d) $CH_3O---\overset{\overset{\displaystyle H \; CH_3}{}}{\underset{\underset{\displaystyle CH(CH_3)_2}{}}{C}}---Br \longrightarrow CH_3O-\overset{\overset{\displaystyle CH_3 \; H}{}}{\underset{\underset{\displaystyle CH(CH_3)_2}{}}{C}}$ (R)

5.33 (a) $(\underline{R})-CH_3CH_2\overset{\overset{\displaystyle CH_3}{|}}{S}CH(CH_2)_3CH_3$ (b) $(\underline{R})-CH_3\overset{\overset{\displaystyle CH_3}{|}}{C}\equiv CCH(CH_2)_3CH_3$

(c) $(\underline{R},\underline{R})\text{-}CH_3(CH_2)_3CH(CH_3)OCH(CH_2CH_3)(CH_2)_2CH_3$

In each case, the configuration around the chiral carbon atom of ($\underline{S}$)-2-iodohexane is inverted. In (d), the configuration of the attacking nucleophile is not changed in the reaction.

5.34 (a) OH

H_3C

Since inversion occurs, only the <u>trans</u> isomer is formed in this case. The transition state would be:

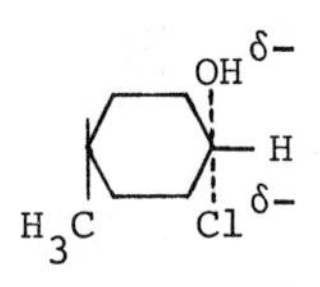

(b) HO OH An intermediate as well as a by-product would be OH Cl

5.35 (a) is preferred because (b) would lead to an alkene instead of the desired ether.

5.36 (a) Reaction 1 (b) Reaction 2 because its E_{act} is lower.

(c) Because Reaction 2 is faster, its products will predominate.

5.37 rate = $k[CH_3I][OH^-]$

(a) The rate is multiplied by 3×2, or 6.

(b) The rate is halved. (c) The rate increases.

(d) The concentration of each reactant is halved; therefore, the rate is divided by four: $(1/2) \times (1/2) = 1/4$.

5.38 The most stable carbocation is (c) because it is 3°. The least stable is (a) because it is 1°.

5.39 (a) $(CH_3)_2CHOH$ + $(CH_3)_2CHOCH_2CH_3$ + some $CH_3CH{=}CH_2$

(b) $(CH_3)_3COH$ + $(CH_3)_3COCH_2CH_3$ + some $(CH_3)_2C{=}CH_2$

(c) $(CH_3)_3COH$ + $(CH_3)_3COCH_2CH_3$ + some $(CH_3)_2C{=}CH_2$

Reaction (c) would be the fastest because RX is 3° and an iodide (a better leaving group).

5.40 (a) CH_3 —I $\xrightarrow[-HI]{H_2O}$ CH_3 —OH

<u>cis</u> <u>cis</u> and <u>trans</u>

(b) (R)-$CH_3CH(I)(CH_2)_5CH_3$ $\xrightarrow[-HI]{H_2O}$ (R)(S)-$CH_3CH(OH)(CH_2)_5CH_3$

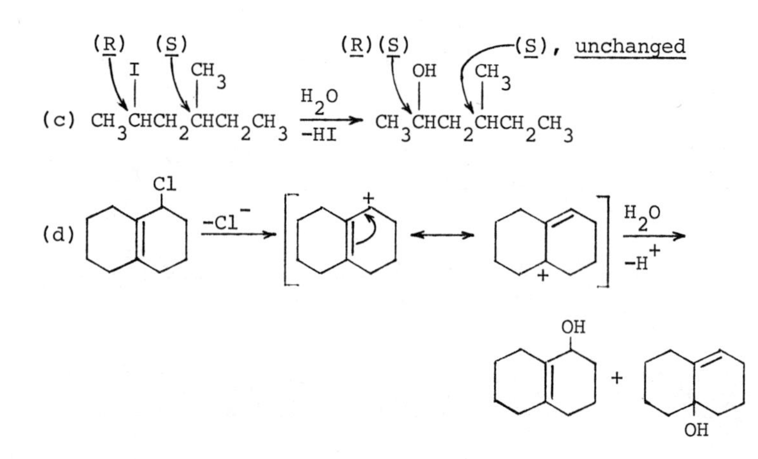

5.41 (a) $(CH_3)_2\overset{+}{C}CH(CH_3)CH_2CH_3$ (methyl shift, 2° to 3° carbocation)

(b) $CH_2{=}CH\overset{+}{C}HCH_2CH_3$ (hydride shift, 2° to allylic carbocation)

(c) $(CH_3)_2\overset{+}{C}CH_2CH_2CH(CH_3)_2$ (hydride shift, 2° to 3°)

(d) 1-ethylcyclohexyl cation (+ on ring carbon bearing $-CH_2CH_3$) (hydride shift, 2° to 3°)

5.42 (a) $[(CH_3)_3C\overset{+}{C}HCH_3] \rightarrow [(CH_3)_2\overset{+}{C}CH(CH_3)_2] \xrightarrow{H_2\ddot{O}:}$

$[(CH_3)_2C(\overset{+}{:}OH_2)CH(CH_3)_2] \underset{}{\overset{-H^+}{\rightleftharpoons}} (CH_3)_2C(\ddot{O}H)CH(CH_3)_2$

(b) $[(CH_3)_2CH\overset{+}{C}HCH_2CH_3] \rightarrow [(CH_3)_2\overset{+}{C}CH_2CH_2CH_3] \xrightarrow{CH_3CH_2\ddot{O}H}$

$[(CH_3)_2C(H\overset{+}{\ddot{O}}CH_2CH_3)CH_2CH_2CH_3] \overset{-H^+}{\rightleftharpoons} (CH_3)_2C(:\ddot{O}CH_2CH_3)CH_2CH_2CH_3$

5.43 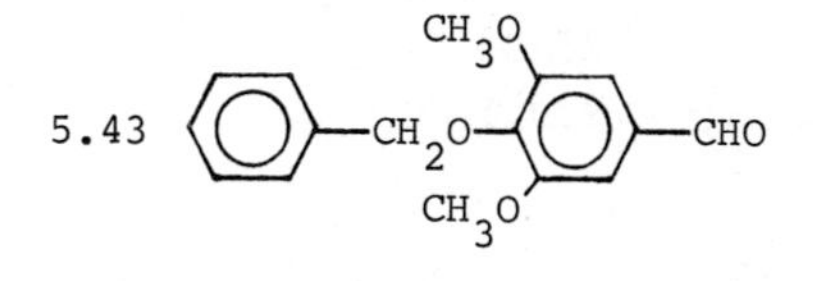

The attacking nucleophile is the product of reaction (a).

5.44 (a) 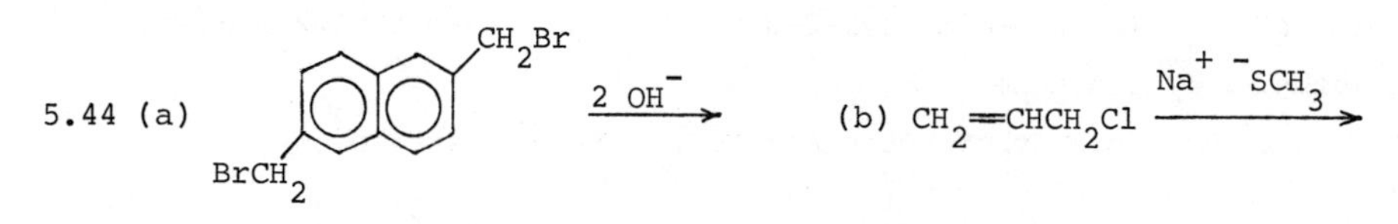

5.45 (a) $CH_2=CH-CH=CH-\overset{+}{C}H_2 \leftrightarrow CH_2=CH-\overset{+}{C}H-CH=CH_2 \leftrightarrow \overset{+}{C}H_2-CH=CH-CH=CH_2$

(b) 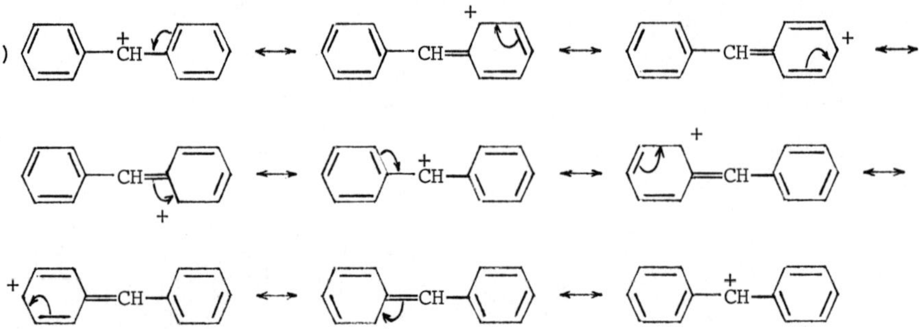 [resonance structures of the diphenylmethyl cation]

plus other benzenoid resonance structures.

(c) 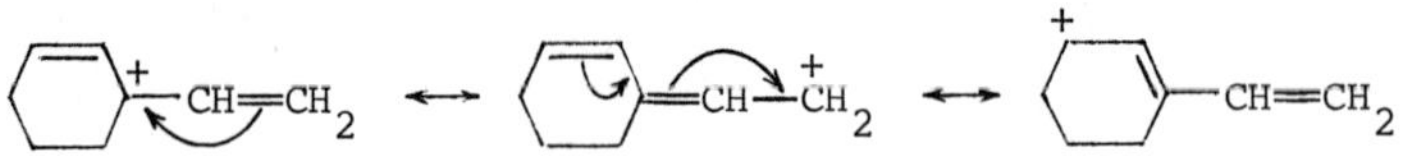 [resonance structures of the cyclohexenyl vinyl cation: $CH=CH_2$ ↔ $=CH-\overset{+}{C}H_2$ ↔ $CH=CH_2$]

(d) 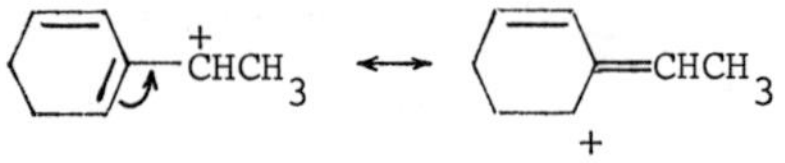 [cyclohexadienyl $\overset{+}{C}HCH_3$ ↔ $=CHCH_3$ with ring +]

In (d), note that the positive change is delocalized by only one of the two double bonds.

5.46 (a) 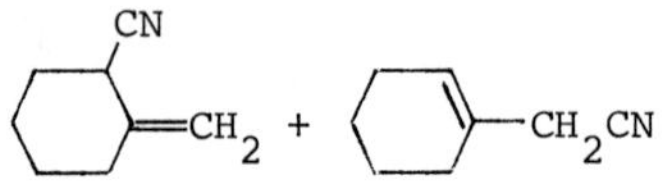 [2-cyanomethylenecyclohexane: CN, $=CH_2$] + [cyclohexenyl CH_2CN] (b) 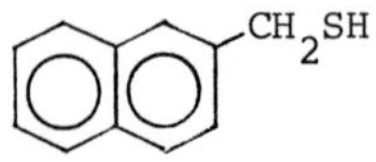 [naphthalene-CH_2SH] (c) $C_6H_5OCH_2C_6H_5$

(d) $CH_3CH=CHCH=CHCH_2OCH_3 + CH_3CH=CHCH(OCH_3)CH=CH_2 + CH_3CH(OCH_3)CH=CHCH=CH_2$

(e) 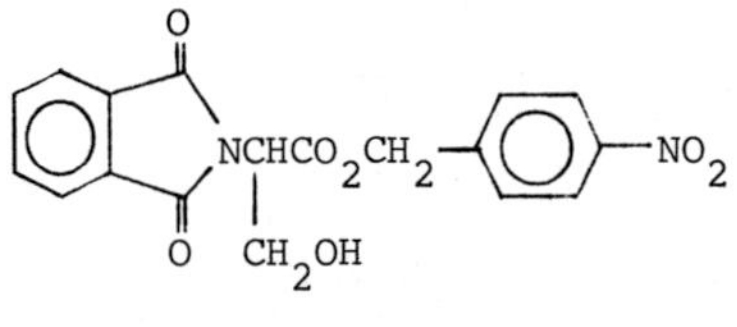 [phthalimide $NCHCO_2CH_2-C_6H_4-NO_2$ with CH_2OH substituent]

5.47 (a) $CH_3CHI(CH_2)_3CH_3 \xrightarrow[(1)]{-I^-} CH_3\overset{+}{C}H(CH_2)_3CH_3 \xrightarrow[(2)]{-H^+} CH_3CH=CH(CH_2)_2CH_3$ trans

(b) Step 1 (c) cis-$CH_3CH=CH(CH_2)_2CH_3$ and $CH_2=CH(CH_2)_3CH_3$

(d) Step 2

5.48 (a) 2-butene (b) 2,3-dimethyl-2-butene (c) 2-methyl-2-pentene

(d) 1-methyl-1-cyclohexene

In each case, the more substituted alkene is the more stable one.

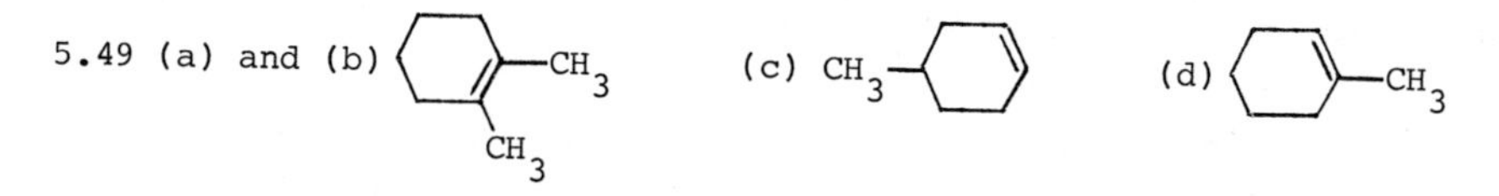

In each case, elimination gives the more substituted alkene if more than one alkene could be formed.

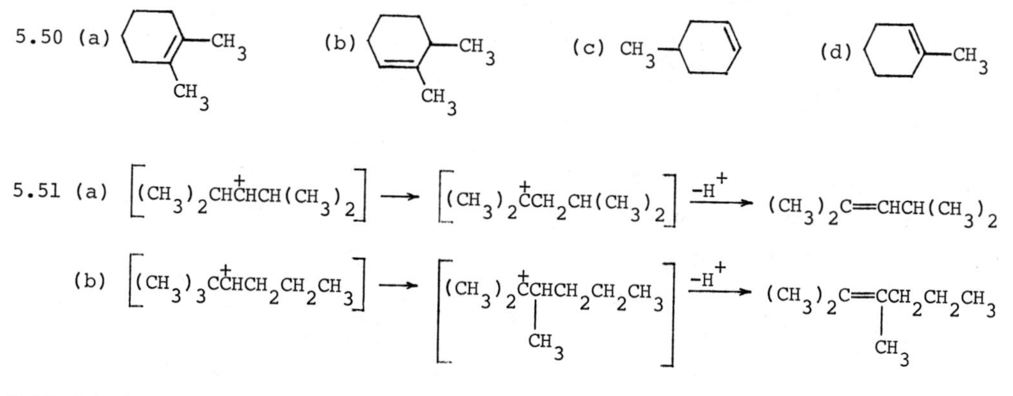

5.52 (a) $(CH_3)_2CBrCH_2CH_2CH_3$, the 3° halide

(b) $(CH_3)_2CHCHICH_3$, the 2° halide (c) CH_3CHICH_3, the iodide

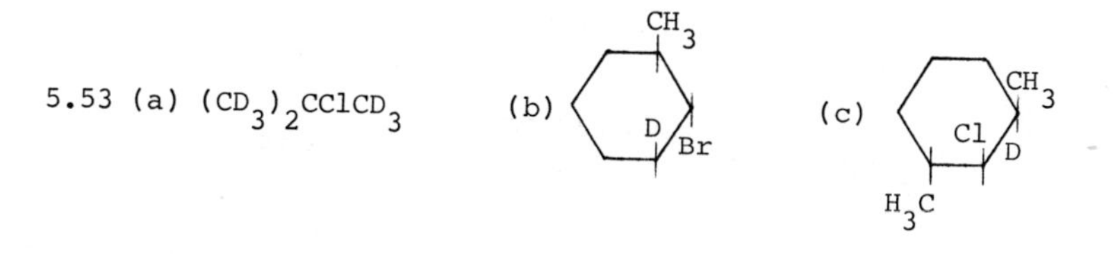

5.54 (a) Hofmann (b) and (c) Saytseff

5.55 (a) <u>trans</u>-$CH_3CH{=}CHCH_2CH_3$ (b) (2<u>E</u>,5<u>E</u>)-$CH_3CH{=}CHCH_2CH{=}CHCH_3$

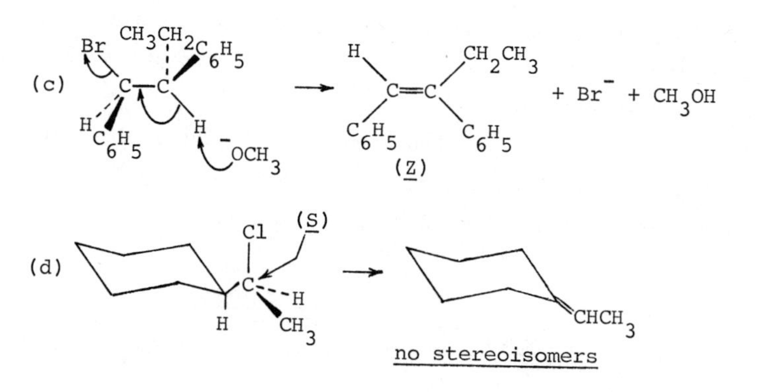

5.56 (a) $(CH_3)_2CHI$ (S_N2 because I^- is a weak base.)

(b) $CH_3CH{=}CH_2$ (E2 because OH^- is a strong base.)

(c) $(CH_3)_2CHOCH_2CH_3$ (S_N1 because CH_3CH_2OH is a weak nucleophile.)

(d) $(CH_3)_3COCH_3$ (S_N1)

(e) $C_6H_5OCH_3$ (S_N2 because CH_3I is a methyl halide.)

(f) $(CH_3)_2C{=}CHCH_3$ (E_2 because $^-OCH_3$ is a strong base.)

(g) trans-$CH_3CH{=}CHCH_2CH_3$ (E2)

(h) $CH_2{=}C(CH_3)CH_2CH_3$ + $(CH_3)_2C{=}CHCH_3$ (E2. With the bulky base, the Hofmann product predominates.)

5.57 (b), because the lower concentration of nucleophile causes a decrease in the rate of reaction by an S_N2 path. Therefore, reaction by an S_N1 path (with rearrangement) has a greater chance of occurring.

5.58 (a) $CH_2{=}CHCH_2Cl + NaOH \longrightarrow$ (b) $(CH_3)_2CHCH_2Cl + Na^+\ ^-SH \longrightarrow$

(c) $C_6H_5CH_2CHBrC_6H_5$ + KOH in $CH_3CH_2OH \longrightarrow$

(d) 1-chloro-2,2-dimethylcyclopentane (ring with CH_3, H_3C, Cl) $+ Na^+\ ^-OCH_3 \longrightarrow$

(e) Cl-substituted benzene ring bearing two CH_2Cl groups $+ 2\ Na^+\ CN^- \longrightarrow$ (The Cl on the ring does not undergo displacement.)

(f) $C_6H_5CH_2Cl + C_6H_5CO^-\ Na^+$ (with C=O) $\longrightarrow$

(g) $BrCH_2CH_2CH_2Br + 2\ (CH_3)_3N: \longrightarrow$

(h) (2R,3S)-$C_6H_5CHBrCHBrCH_3 + 2\ Na^+\ ^-CN \longrightarrow$ (Both chiral carbons undergo inversion.)

5.59 (a) The ring system prevents S_N2 backside attack on the 3° carbon bonded to the Br. The ring system does not allow a planar carbocation to form; thus the S_N1 path is also blocked.

(b) The transition state in the reaction with quinuclidine is less sterically hindered because the alkyl groups on the N are "tied back." Steric hindrance is less important in an acid-base reaction because of the small size of H^+.

(c)

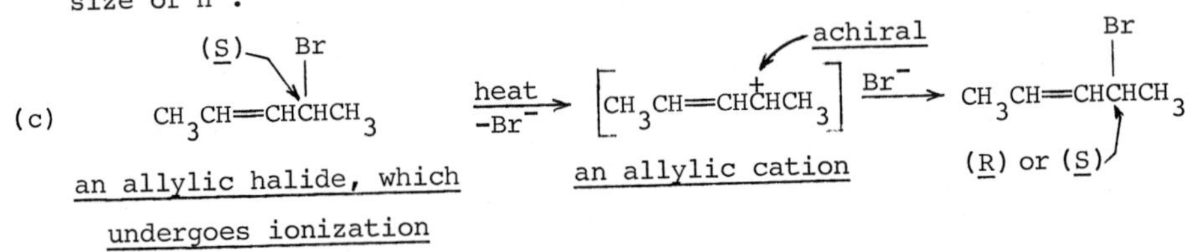

Note that the allylic cation can be attacked at two positions. Attack at either position yields the same racemic product.

$$CH_3\overset{+}{C}H-CH=CHCH_3 \longleftrightarrow CH_3CH=CH-\overset{+}{C}HCH_3$$

(d) Although $ClCH_2OCH_3$ is a primary alkyl halide, the carbocation is resonance-stabilized.

$$\overset{+}{C}H_2-\ddot{O}CH_3 \longleftrightarrow CH_2=\overset{+}{O}CH_3$$

(e) The reaction proceeds by an S_N2 mechanism and every product molecule is (S). For every one molecule that reacts, a pair of molecules becomes racemic.

$$2\ (R)\text{-RI} + {}^{*}I^{-} \xrightarrow{S_N2} (R)\text{-RI} + (S)\text{-R}^{*}I + I^{-}$$

When half the molecules of (R)-RI have undergone reaction, the mixture is racemic. If the reaction had proceeded by an S_N1 mechanism, the rate of racemization would have equalled the rate of $*I^-$ uptake because both (R) and (S) products are obtained.

$$2\ (R)\text{-RI} + 2\ {}^{*}I^{-} \xrightarrow{S_N1} (R)\text{-R}^{*}I + (S)\text{-R}^{*}I + 2\ I^{-}$$

5.60 $CH_3CH_2CHClCH_3 \xrightarrow{\text{base}} CH_3CH=CHCH_3 + CH_3CH_2CH=CH_2$

the alkyl halide — cis and trans

All other butyl chlorides would yield only one alkene upon elimination.

5.61 (a) Add aqueous $AgNO_3$.

$$Cl-C_6H_4-CH_3 \xrightarrow[H_2O]{Ag^+} \text{no reaction}$$

$$C_6H_5-CH_2Cl \xrightarrow[H_2O]{Ag^+} C_6H_5-CH_2OH + H^+ + AgCl\downarrow$$

Benzyl halides undergo solvolysis in H_2O to yield $C_6H_5CH_2OH$ and Cl^-. A precipitate of AgCl would be observed.

(b) Add aqueous $AgNO_3$.

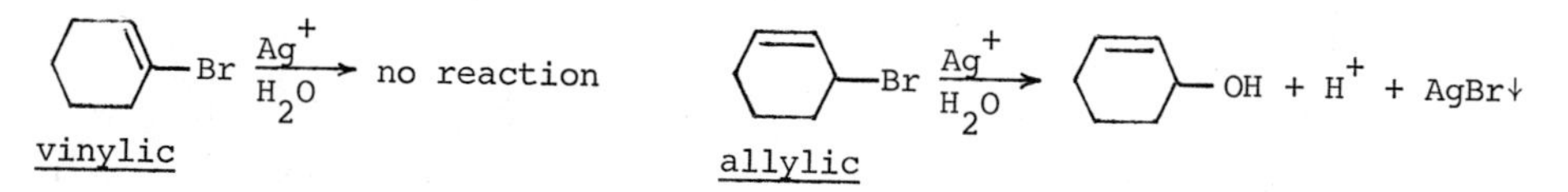

(c) Add KOH in CH_3CH_2OH and heat. Compare the physical constants of the product with those of known alkenes.

(2$\underline{S}$,3$\underline{R}$)-isomer ⟶ ($\underline{Z}$)-3-methyl-2-pentene

(2$\underline{S}$,3$\underline{S}$)-isomer ⟶ ($\underline{E}$)-3-methyl-2-pentene

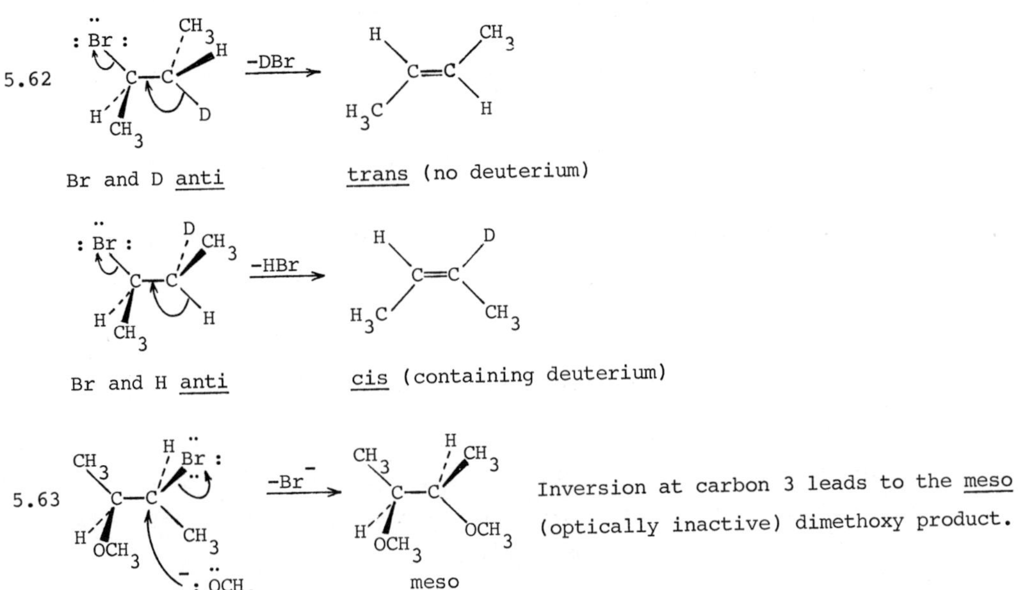

5.64 (a) The slowest step is the rate-determining step. Although the rate of the overall reaction depends on the concentration of AB, this concentration is proportional to the concentrations of both A and B.

$$\text{overall rate} = \underline{k}[AB] = \underline{k}'[A][B]$$

Therefore, the reaction would show <u>second-order kinetics</u>.

(b) Only one particle, AB, is involved in the transition state of the rate-determining step. Therefore, this reaction is <u>unimolecular</u> (even though it is second order in rate).

5.65 $BrCH_2CH_2Br \xrightarrow[-2\ Br^-]{2\ OH^-} HOCH_2CH_2OH \xrightarrow{2\ Na} {}^-OCH_2CH_2O^- \xrightarrow[-2\ Br^-]{BrCH_2CH_2Br}$ (1,4-dioxane)

5.66 (a) $(CH_3)_2CHCl \xrightarrow[-AgCl]{Ag^+} [(CH_3)_2\overset{+}{C}H] \xrightarrow{H_2\ddot{O}:} (CH_3)_2CH\overset{+}{\ddot{O}}H_2 \underset{}{\overset{-H^+}{\rightleftharpoons}} (CH_3)_2CH\ddot{\underset{..}{O}}H$

(b) (1-methylcyclobutyl)methyl chloride (CH_3, CH_2–$\ddot{Cl}:$) $\xrightarrow[-AgCl]{Ag^+}$ [1-methylcyclopentyl cation, $+$ CH_3] $\xrightarrow{H_2\ddot{O}:}$ (CH_3, $\overset{+}{\ddot{O}}H_2$) $\overset{-H^+}{\rightleftharpoons}$ (CH_3, $\ddot{\underset{..}{O}}H$)

(c) ▷–$CH_2Cl \xrightarrow[-Cl^-]{H_2O}$ ▷–$CH_2\overset{+}{\ddot{O}}H_2 \xrightarrow{-H^+}$ ▷–$CH_2\ddot{\underset{..}{O}}H$

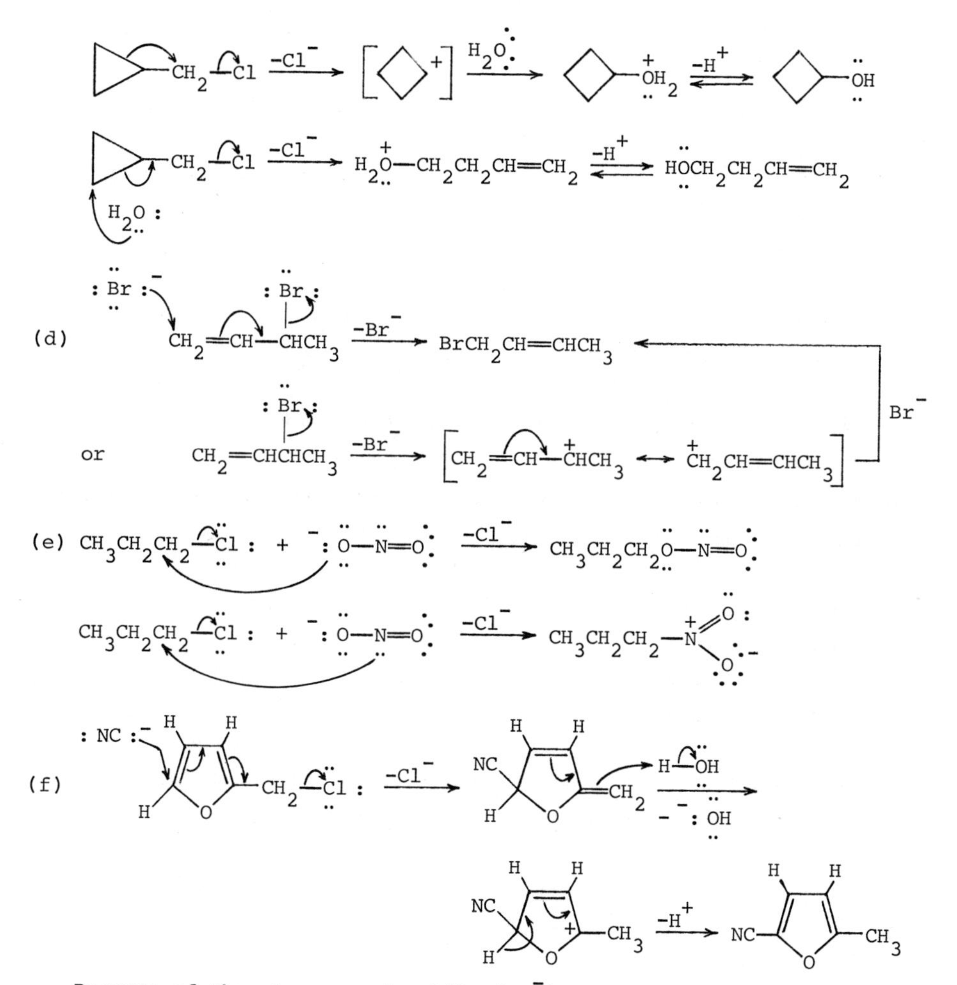

Because of the strong nucleophile (CN^-), the first part of the reaction probably does not proceed by an S_N1 type of mechanism.

5.67 (a) Consider the conformations in which Cl and a β H are trans and diaxial (the necessary conformations for elimination by an E2 path).

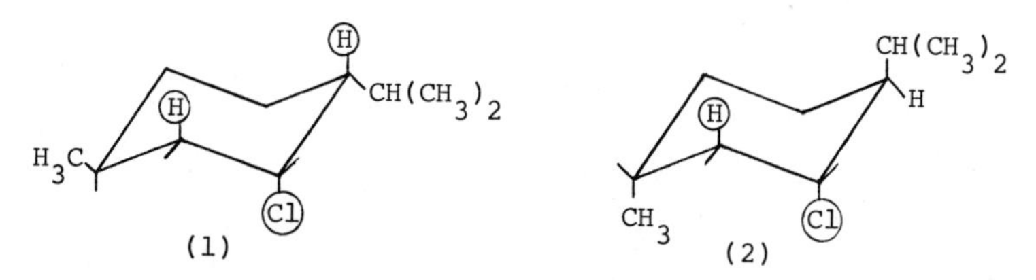

In the necessary conformation, structure (1) is also in a preferred conformation with CH_3 and $CH(CH_3)_2$ equatorial. Structure (2) is in a nonpreferred conformation with CH_3 and $CH(CH_3)_2$ axial. The concentration of the conformation that can undergo E2 reaction is greater for (1); therefore, (1) undergoes the faster reaction.

(b) When structure (1) undergoes reaction, either of two H's could be lost.

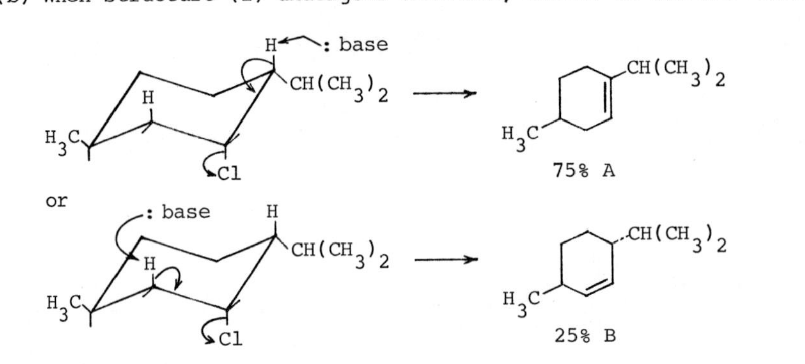

Alkene A predominates because it is the more highly substituted, more stable alkene.

(c) When structure (2) undergoes reaction, only one alkene can be formed because it contains only one trans, axial β H.

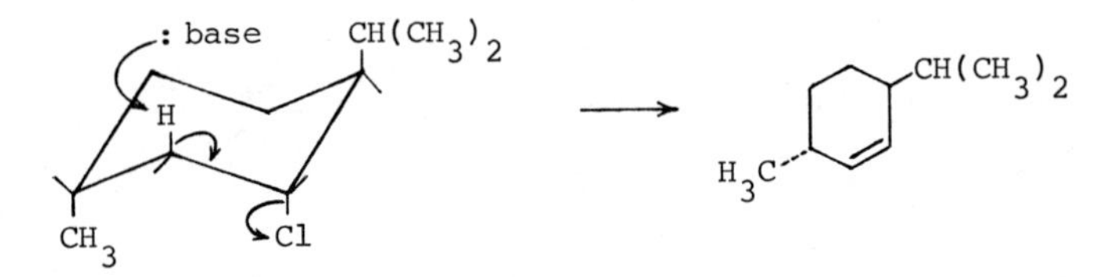

The reaction is slow both for the reason discussed in (a) and because the less substituted (higher-energy) alkene is the only possible product.

CHAPTER 6

Free-Radical Reactions; Organometallic Compounds

Some Important Features

A free radical is an atom or group of atoms containing an unpaired electron; most free radicals are highly reactive intermediates. An alkyl free radical is achiral around the reacting carbon.

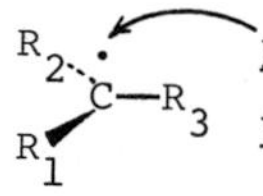

R_2, R_1, C—R_3 : planar around C·, therefore achiral

Free-radical reactions occur stepwise: (1) initiation (initial generation of free radicals); (2) propagation (reaction of free radicals yielding new reactive free radicals); (3) termination (destruction of free radicals, often by the joining together of two free radicals or by the formation of a relatively stable, nonreactive free radical).

Different hydrogen atoms in a structure may be replaced by halogen atoms at different rates. If the intermediate free radical is stabilized, then reaction rate at that position is enhanced.

CH_4 RCH_3 R_2CH_2 R_3CH allylic and benzylic

Increasing rate of free-radical reaction and increasing stability of free radical when H· is removed →

Because of the enhanced reactivity of some H atoms, we can often predict the principal free-radical product.

$$\overset{1^\circ\ \ 2^\circ}{CH_3CH_2CH_3} \xrightarrow[-H\cdot]{X_2,\underline{h}\nu} [CH_3\dot{C}HCH_3 \text{ favored over } CH_3CH_2\dot{C}H_2] \xrightarrow{-X\cdot} CH_3\underset{}{\overset{X}{\overset{|}{C}}}HCH_3$$

The reagent also affects the product ratio. For example, a less reactive free-radical reagent is more selective.

$$CH_3CH_2CH_3 \xrightarrow[\text{(more reactive)}]{Cl_2,\underline{h}\nu} CH_3CHClCH_3 + CH_3CH_2CH_2Cl$$

$$CH_3CH_2CH_3 \xrightarrow[\text{(less reactive)}]{Br_2,\underline{h}\nu} CH_3CHBrCH_3$$

The mechanisms of free-radical reactions are discussed in the text, including the reasons for the reactivities of different H atoms in a structure, the kinetic isotope effect, the different reactivities of halogenating agents, and halogenation with NBS. Other types of free-radical reactions (pyrolysis, autooxidation, and the quinone-hydroquinone reaction) are also covered in this chapter. Free-radical initiators and inhibitors are mentioned.

Organometallic compounds are compounds containing carbon-metal bonds. One of the most important organometallic compounds is the Grignard reagent, RMgX. The Grignard reagent contains a carbanion-like organic group that reacts with a compound containing a partially positive hydrogen (as in H_2O or ROH) or a partially positive carbon (as in a carbonyl compound).

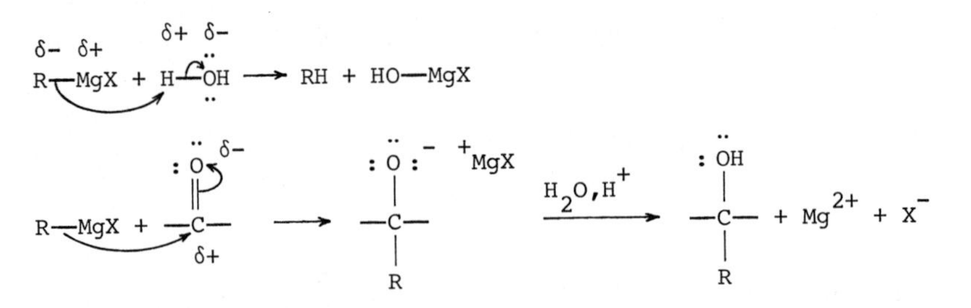

Lithium reagents (RLi) behave like Grignard reagents. Cuprates (R_2CuLi) are used to prepare alkanes.

$$2\ RLi \xrightarrow{CuI} R_2CuLi \xrightarrow{R'X} R{-}R'$$

Reminders

Be sure you know how to write resonance structures for free-radical intermediates. In these cases, we shift only one electron (not two, as in carbocation intermediates).

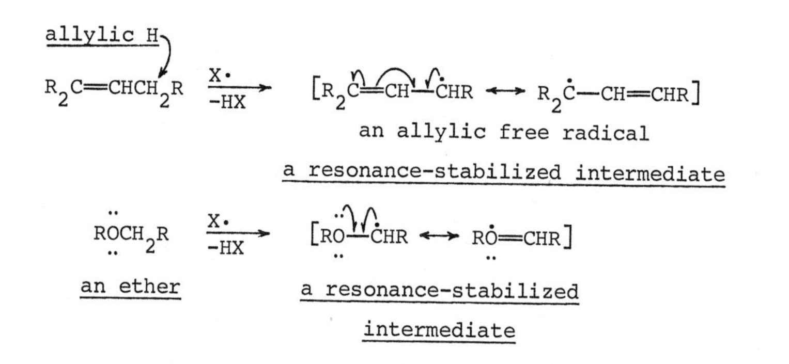

Remember that a stabilized organic intermediate (relative to reactants) means a faster reaction. (Allylic hydrogens are easily removed by free radicals because the allylic free radical is resonance-stabilized.) Conversely, a stabilized reactant (relative to intermediates) means a slower reaction. (I· is more stable and thus less reactive than Cl·.)

$$X\cdot + RH \longrightarrow R\cdot + HX$$

stabilization means a slower reaction (X·)   stabilization means a faster reaction (R·)

A Grignard reagent is a very strong base that reacts with acidic hydrogens. Acid-base reactions, such as $RMgX + H_2O \longrightarrow RH + HOMgX$, are fast compared to most organic reactions, such as that of RMgX with a carbonyl compound.

Answers to Problems

6.21 (a)
```
  H  H
H:C:C:O·
  H  H
```
(b)
```
 H
·C::C:C:H
 H   H H
```

6.22 (a) and (e) are propagation steps (b) is initiation

(c) and (d) are termination

These different steps can be distinguished as follows: If a free radical is formed from a nonradical reactant, the reaction is an initiation step. If both a reactant and a product are free radicals, the reaction is a propagation step. If free-radical reactants lead to nonradical products, the reaction is a termination step.

6.23 (1) $Cl_2 \xrightarrow{h\nu} 2\ Cl\cdot$ (2) Cl· + (cyclopentane) ⟶ (cyclopentyl radical) + HCl

(3) (cyclopentyl radical) + Cl_2 ⟶ (chlorocyclopentane, Cl) + Cl·   (4) (chlorocyclopentane, Cl) + Cl· ⟶ (chlorocyclopentyl radical, Cl) + HCl

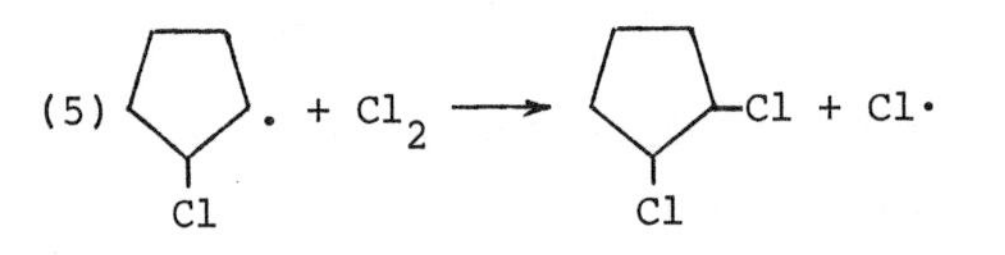

Termination could occur by the combination of any two radicals

6.24 The ratio would be the same as the ratio of the types of H being extracted: two propane CH_2 hydrogens, six propane CH_3 hydrogens, twelve cyclohexane CH_2 hydrogens, or 1 : 3 : 6. The product ratio would therefore be as follows:

1 part $CH_3CHClCH_3$ 3 parts $CH_3CH_2CH_2Cl$ 6 parts (cyclohexyl)—Cl

6.25 (a) $ClCH_2CHClCH_2Cl$, (R)-$Cl_2CH\overset{*}{C}HClCH_3$, $ClCH_2CCl_2CH_3$

(b) (S)-$ClCH_2\overset{*}{C}HClCH_2CH_3$, $CH_3CCl_2CH_2CH_3$,

(R)-$CH_3\overset{*}{C}HClCHClCH_3$, (R)-$CH_3\overset{*}{C}HClCH_2CH_2Cl$

In each case, chlorination at a chiral carbon yields a racemic or achiral product, while chlorination at an achiral carbon does not affect the stereochemistry of other, chiral carbons. In (a), note that chirality can be lost by the formation of identical groups on a chiral carbon. In (b), note that the order of priority of the groups around the chiral carbon in the first compound has changed.

6.26 $(CH_3)_4C$

6.27 (d), (b), (a), (c), (e)

6.28 (b), (a), (c), ranked in the order of stability of the most stable intermediate free radical that could be formed from the hydrocarbon.

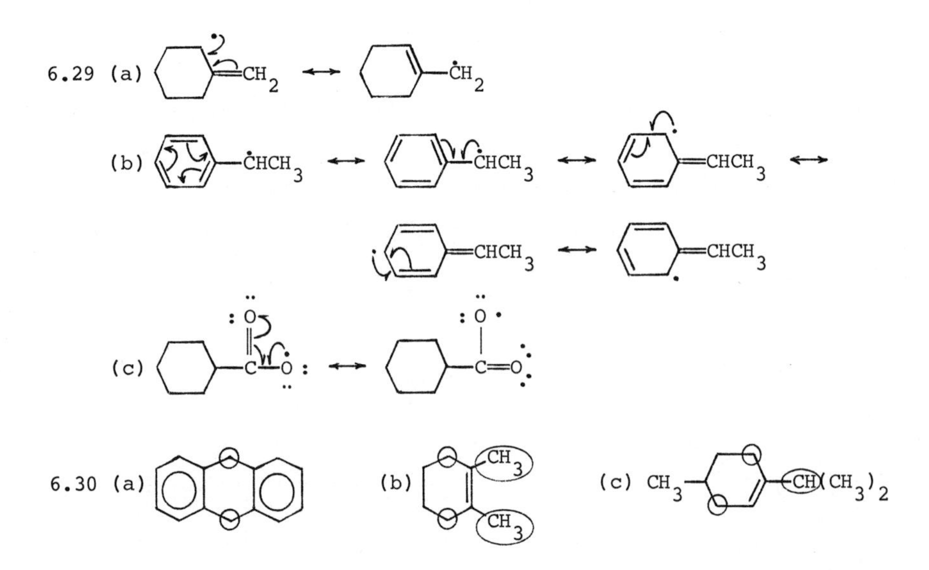

(d) 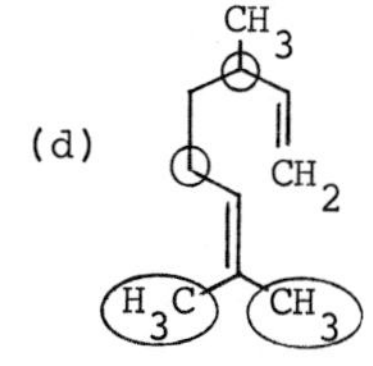

In each case, the benzylic or allylic positions would be attacked.

6.31 (a) [1-cyclohexenyl]$-CH_2Br$ + [3-bromo-1-methylcyclohexene]$-CH_3$ + [2-bromo-1-methylenecyclohexane]$=CH_2$ + [3-bromo-3-methylcyclohexene] (CH_3, Br) + [6-bromo-1-methylcyclohexene] ($-CH_3$, Br)

In (a), three different free radicals can be formed. Each is resonance-stabilized; therefore, bromination can occur at five positions.

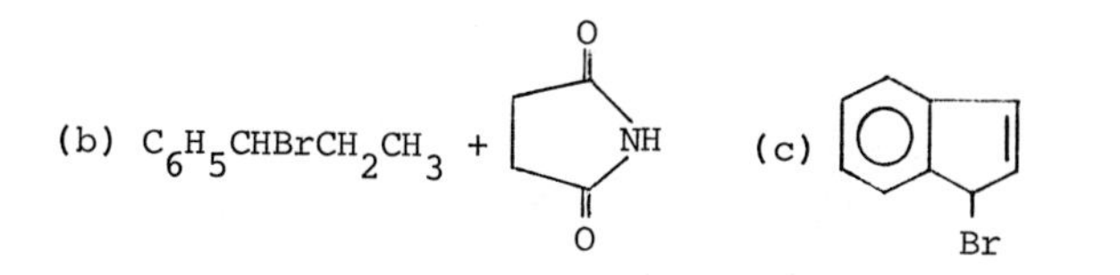

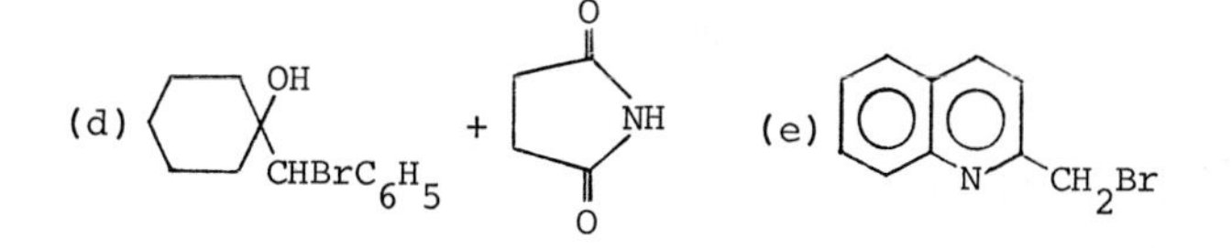

6.32 [cyclohexene*] $\xrightarrow{Br\cdot}$ [radical ↔ radical] $\xrightarrow[-Br\cdot]{Br_2}$ [3-bromocyclohexene*]

↳ $\xrightarrow{Br\cdot}$ [radical ↔ radical] $\xrightarrow[-Br\cdot]{Br_2}$ [bromocyclohexene*] + [3-bromocyclohexene*]

6.33 (a) $CH_3CH_2CH_2CH_3$ + $CH_3CH_2CH{=}CH_2$ (b) $CH_3CH_2CH_3$ + $CH_3CH{=}CH_2$

6.34 (a) $CH_3(CH_2)_6CH_3$ (b) $(CH_3)_2CHCH(CH_3)_2$

6.35 (a) $(CH_3)_2\dot{C}-CH_2-CH_2CH_3 \xrightarrow{\beta\ \text{cleavage}} (CH_3)_2C{=}CH_2 + \cdot CH_2CH_3$

$\xrightarrow{\text{disproportionation}} (CH_3)_2C{=}CHCH_2CH_3 + (CH_3)_2CHCH_2CH_2CH_3$

(b) [cyclohexyl radical] $\xrightarrow{\beta\ \text{cleavage}}$ $\dot{C}H_2CH_2CH_2CH_2CH{=}CH_2$

$\xrightarrow{\text{disproportionation}}$ [cyclohexene] + [cyclohexane]

6.36 (b) and (c). Compound (a) has neither hydrogens on an ether carbon, nor hydrogens in a benzylic position.

6.37 (a) and (c) both have carbon-metal bonds.

6.38 (a) phenylmagnesium iodide (b) n-propylmagnesium chloride
(c) n-heptyllithium

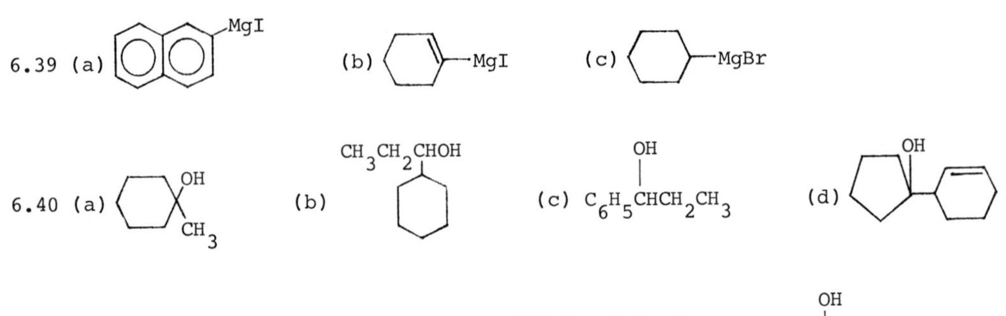

6.41 (a) $(CH_3)_2CHBr \xrightarrow[\text{ether}]{Mg} (CH_3)_2CHMgBr \xrightarrow[(2)\ H_2O, H^+]{(1)\ (CH_3)_2C{=}O} (CH_3)_2CHC(OH)(CH_3)_2$

(b) $(CH_3)_2CHMgBr \xrightarrow[(2)\ H_2O,\ H^+]{(1)\ CH_3CHO} (CH_3)_2CHCH(OH)CH_3$

6.42 (a) contains no acidic hydrogen

(b) $HOCH_2CH_2OH + 2\ CH_3MgI \longrightarrow {}^-OCH_2CH_2O^- + 2\ Mg^{2+} + 2\ I^- + 2\ CH_4$

(c) $(CH_3CH_2)_2NH + CH_3MgI \longrightarrow (CH_3CH_2)_2N^- + Mg^{2+} + I^- + CH_4$

(d) $CH(CO_2H)_3 + 3\ CH_3MgI \longrightarrow CH(CO_2^-)_3 + 3\ Mg^{2+} + 3\ I^- + 3\ CH_4$

6.43 (a) cyclohexyl–Li + LiBr (b) $C_6H_5CH_2CH_2CO_2H + Li^+ + H_2O$

(c) cyclopentane + piperidide $N^-\ Li^+$

6.44 (a) $CH_3CH_2Br \xrightarrow[(2)\ CuI]{(1)\ Li} (CH_3CH_2)_2CuLi \xrightarrow{CH_3CH_2CH_2Br}$

(b) $CH_3(CH_2)_3Br \xrightarrow[(2)\ CuI]{(1)\ Li} [CH_3(CH_2)_3]_2CuLi \xrightarrow{CH_3CH_2CH_2CH_2Br}$

(c) $(CH_3)_2CHBr \xrightarrow[(2)\ CuI]{(1)\ Li} [(CH_3)_2CH]_2CuLi \xrightarrow{CH_3CH_2CH_2CH_2Br}$

(d) $CH_2{=}CHCH_2Br \xrightarrow[\text{ether}]{Mg} CH_2{=}CHCH_2MgBr \xrightarrow{D_2O}$

(e) $(CH_3)_2CHBr \xrightarrow[\text{ether}]{Mg} (CH_3)_2CHMgBr \xrightarrow{D_2O}$

(f) $[(CH_3)_2CH]_2CuLi$ from (c) + $CH_2{=}CHCH_2Cl \longrightarrow$

(There may be other correct answers.)

6.45 If HBr is present in the reaction mixture, an exchange reaction leading to $C_6H_5CH_3$ occurs. The presence of $C_6H_5CH_3$ leads to a decreased isotope effect.

$$C_6H_5\dot{C}H_2 + HBr \longrightarrow \underset{\text{no isotope effect}}{C_6H_5CH_3} + Br\cdot$$

6.46 (a) has more steric hindrance than (b) and is therefore of higher energy than (b). Some of the steric hindrance is relieved in the free radical. (See Section 5.6E.)

6.47 The free-radical intermediate in the formation of the allylic Grignard reagent is resonance-stabilized; therefore, two Grignard reagents (and two Grignard products) can be formed.

$$CH_3CH{=}CHCH_2Cl \xrightarrow{Mg} [CH_3CH{=}CH\dot{C}H_2 \longleftrightarrow CH_3\dot{C}HCH{=}CH_2 + \dot{M}gCl]$$

$$\longrightarrow CH_3CH{=}CHCH_2MgCl + CH_3CH(MgCl)CH{=}CH_2$$

$$\xrightarrow[(2)\ H_2O,H^+]{(1)\ CH_3C(=O)CH_3} CH_3CH{=}CHCH_2C(CH_3)_2OH$$

and

$$CH_2{=}CHCH(CH_3){-}C(CH_3)_2OH$$

6.48 anode: $CH_3C(=\ddot{O}:)\ddot{O}:^- \xrightarrow{-e^-} CH_3{-}C(=O){-}\dot{O}: \longrightarrow \dot{C}H_3 + \dot{O}{=}C{=}\dot{O}$

$2\ \cdot CH_3 \longrightarrow CH_3CH_3$

cathode: $2\ H_2O \longrightarrow H_2 + 2\ OH^-$

6.49 (a) Yes, inversion has occurred.

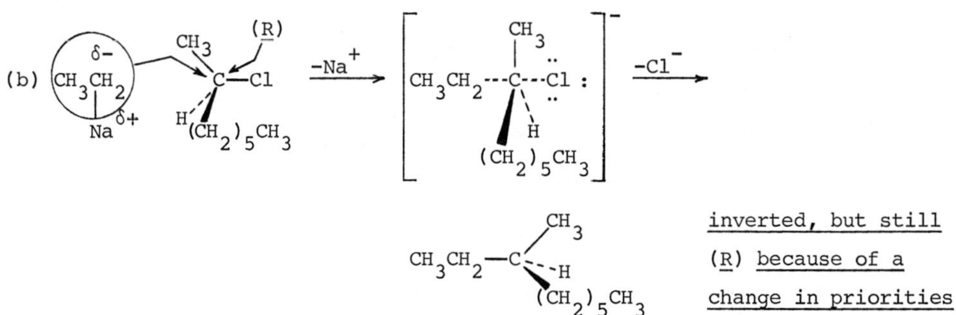

CH_3CH_2—C(CH_3)(H)($(CH_2)_5CH_3$) inverted, but still (R) because of a change in priorities

6.50 This type of problem is called a road map problem and is typical of what a chemist encounters in structure determination. To approach a road map problem, first write out a flow diagram.

$$CH_3CH_2CH_2CH_2CH_3 \xrightarrow[h\nu]{Br_2} A + B \xrightarrow[E2]{CH_3O^-} C$$

n-pentane — two bromopentanes — a pentene

We have added what is immediately apparent about the products of the two reactions: A and B are bromopentanes, while C must be an elimination product (and therefore a pentene). All we need do is fill in the positions of substitution. n-Pentane is most likely to yield 2- and 3-bromopentane (A and B). Will these two compounds yield the same alkene (C)? The answer is yes. Now, we fill in the flow diagram with structures.

$$CH_3CH_2CH_2CH_2CH_3 \xrightarrow[h\nu]{Br_2} \underset{A}{CH_3CHBrCH_2CH_2CH_3} + \underset{B}{CH_3CH_2CHBrCH_2CH_3}$$

$$\xrightarrow{E2} \underset{C}{CH_3CH{=}CHCH_2CH_3}$$

6.51 Typical synthetic paths follow. There may be other correct answers.

(a) $(CH_3)_3CBr \xrightarrow[ether]{Mg} (CH_3)_3CMgBr \xrightarrow[(2)\ H_2O,H^+]{(1)\ CH_3CHO}$

or $CH_3I \xrightarrow[ether]{Mg} CH_3MgI \xrightarrow[(2)\ H_2O,H^+]{(1)\ (CH_3)_3CCHO}$

Because of steric hindrance, a substitution reaction is not practical. Rearrangement and elimination would probably be observed.

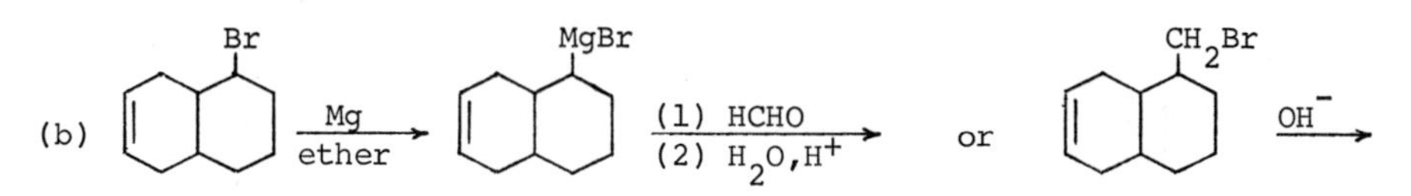

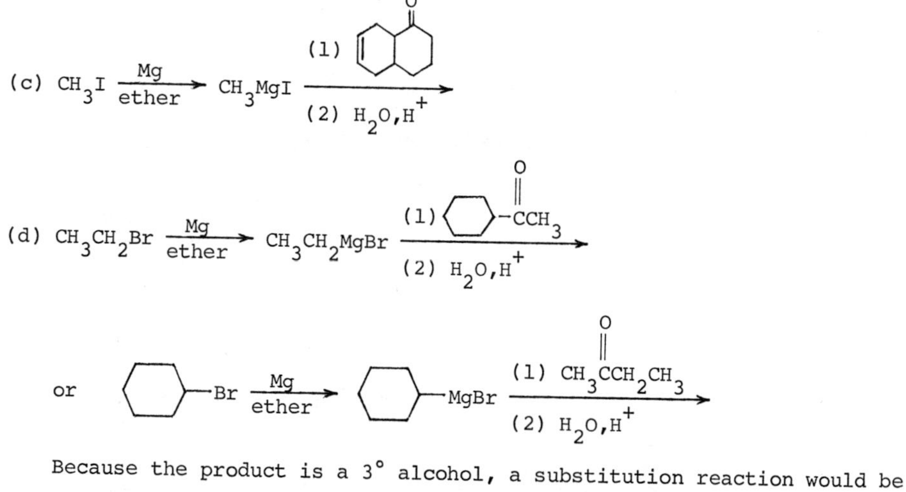

Because the product is a 3° alcohol, a substitution reaction would be impractical.

(e) $(CH_3)_2CHBr \xrightarrow[\text{ether}]{\text{Mg}} (CH_3)_2CHMgBr \xrightarrow[(2)\ H_2O, H^+]{(1)\ \text{cyclopentyl-CHO}}$

or cyclopentyl-Br $\xrightarrow[\text{ether}]{\text{Mg}}$ cyclopentyl-MgBr $\xrightarrow[(2)\ H_2O, H^+]{(1)\ (CH_3)_2CHCHO}$

(f) $C_6H_5Br \xrightarrow[\text{ether}]{\text{Mg}} C_6H_5MgBr \xrightarrow[(2)\ H_2O, H^+]{(1)\ (C_6H_5)_2C{=}O}$

6.52

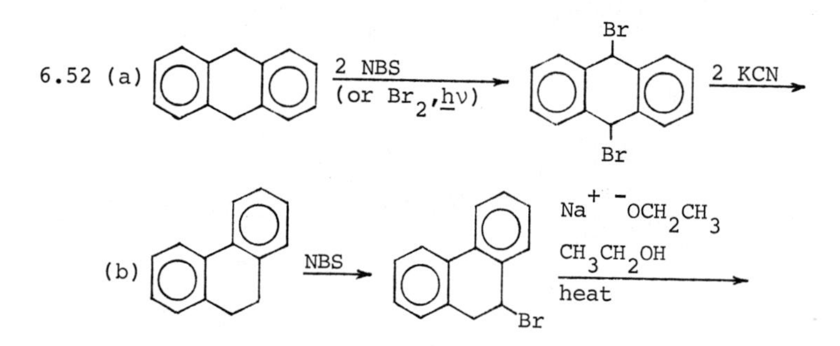

In the final step, almost any base could be used. Because of conjugation in the product, the alkene would be formed instead of the substitution product.

(c) $CH_3CH{=}CHCH_3 \xrightarrow{\text{2 NBS}} BrCH_2CH{=}CHCH_2Br \xrightarrow{2\ CH_3CO_2^-\ Na^+}$

(d) $$C_6H_5CH_2CH_2CH_3 \xrightarrow{NBS} C_6H_5CHBrCH_2CH_3 \xrightarrow[\text{heat}]{Na^+\ ^-OCH_2CH_3,\ CH_3CH_2OH}$$

$$C_6H_5CH{=}CHCH_3 \xrightarrow{NBS} C_6H_5CH{=}CHCH_2Br \xrightarrow[\text{ether}]{Mg}$$

$$C_6H_5CH{=}CHCH_2MgBr \left(+\ C_6H_5CH(MgBr)CH{=}CH_2\right) \xrightarrow[(2)\ H_2O,H^+]{(1)\ CH_3CHO}$$

(see Problem 6.47)

$$C_6H_5CH{=}CHCH_2CH(OH)CH_3 \left(+\ C_6H_5CH(CH_3CHOH)CH{=}CH_2\right)$$

(e) $$C_6H_5CH_3 \xrightarrow{NBS} C_6H_5CH_2Br \xrightarrow[(2)\ CuI]{(1)\ Li} (C_6H_5CH_2)_2CuLi \xrightarrow{C_6H_5CH_2Br}$$

(f) tetralin $\xrightarrow{NBS}$ 1-bromotetralin $\xrightarrow[\text{heat}]{Na^+\ ^-OCH_2CH_3,\ CH_3CH_2OH}$ 1,2-dihydronaphthalene $\xrightarrow{NBS}$

3-bromo-1,2-dihydronaphthalene (Br) $\xrightarrow[(2)\ CuI]{(1)\ Li}$ (dihydronaphthyl)$_2$CuLi $\xrightarrow{CH_3CH_2CH_2Br}$

dihydronaphthalene–$CH_2CH_2CH_3$ $\xrightarrow{NBS}$ bromo(Br)-dihydronaphthalene–$CH_2CH_2CH_3$

$\xrightarrow[\text{heat}]{Na^+\ ^-OCH_2CH_3,\ CH_3CH_2OH}$ naphthalene–$CH_2CH_2CH_3$ $\equiv$ naphthalene–$CH_2CH_2CH_3$

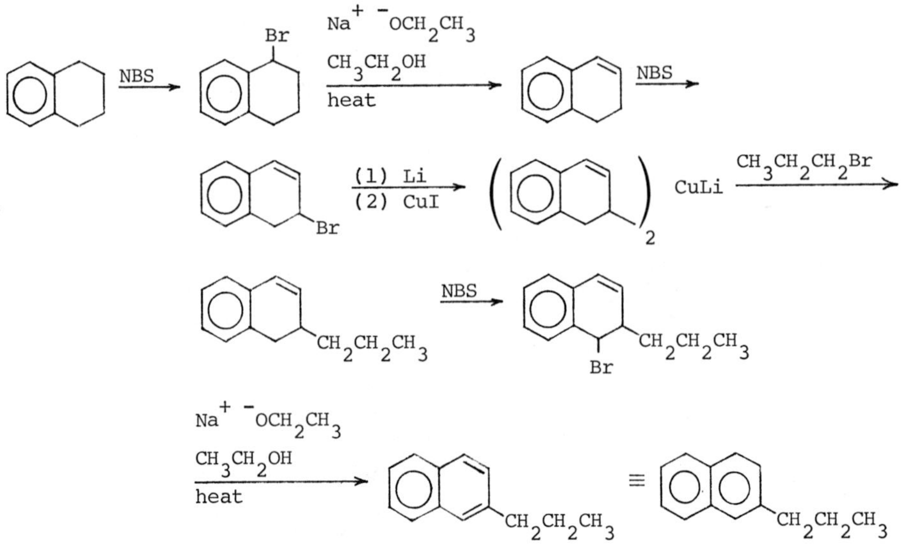

CHAPTER 7

Alcohols, Ethers, and Related Compounds

Some Important Features

Because the O in $R\ddot{\underset{..}{O}}H$ or $R\ddot{\underset{..}{O}}R$ is electronegative and has unshared electrons, alcohols, phenols, and ethers undergo hydrogen bonding with compounds that contain partially positive hydrogens $\left(R\ddot{\underset{..}{O}}H\text{---}:\underset{..}{\overset{H|}{O}}R \text{ or } R_2\ddot{O}:\text{---}H_2\ddot{O}:\right)$. Alcohols and ethers are protonated in acidic solution. Alcohols and, to a limited extent, ethers undergo substitution reactions with HX ($3°$ ROH > $2°$ ROH > $1°$ ROH). Because $3°$ and $2°$ alcohols react by an S_N1 path with HX, PCl_3 and $SOCl_2$ are often used as reagents for converting alcohols to alkyl halides. Both PCl_3 and $SOCl_2$ result in stereospecific reactions without rearrangement.

Dehydration of alcohols yields Saytseff products. (Again, the relative rates are $3°$ ROH > $2°$ ROH > $1°$ ROH.)

Alcohols act as acids when treated with extremely strong bases (such as RMgX) or with alkali metals (Na or K). Phenols are more acidic than alcohols because the phenoxide ion is resonance-stabilized; phenols react with NaOH.

Other important topics covered in this chapter are the reactions of epoxides in acidic and basic solutions; inorganic esters of alcohols; oxidation of $1°$ and $2°$ alcohols; and thiols and their derivatives.

Reminders

In acidic solution, an alcohol is protonated and can be attacked by a nucleophile such as Br^-.

In dilute base, alcohols are not protonated and undergo no appreciable reaction.

Reaction of a 3° or 2° alcohol in acidic solution yields racemic or achiral products.

Answers to Problems

7.31 (a) $(CH_3)_2NCH_2CH_2OCH(C_6H_5)_2$ ← ether

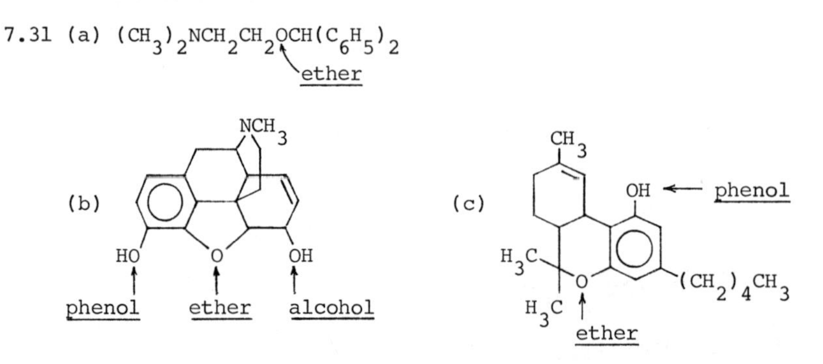

7.32 (a) t-butyl alcohol, because its alkyl group is the most branched and thus the least hydrophobic.

(b) tetrahydrofuran, because of its compactness and because its oxygen is more exposed.

(c) 1-octanol, because of the hydrophilic OH group.

(d) 1,5-pentanediol, because it has two OH groups.

7.33 (a) 3-methyl-1-butanol, 1° (b) trans-3-methyl-1-cyclohexanol, 2°

(c) 2,5-heptanediol, both 2° (d) cis-3-penten-1-ol, 1°

7.34 (a) 1,2-dimethoxyethane

(b) isopropyl n-propyl ether, 2-n-propoxypropane, or 1-isopropoxypropane

(c) cis-2,3-dimethyloxirane (d) trans-2,3-dimethyloxirane

7.35 (a) $(CH_3)_2CHCH_2CH_2OH$ (b) $CH_2{=}C(CH_3)_2$ (see Chapter 5)

(c)

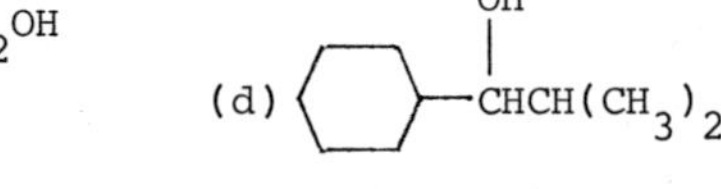

7.36 (a)

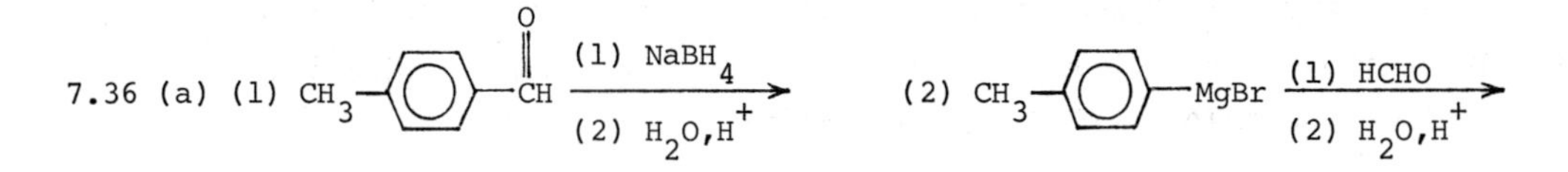

(b) (1) $(CH_3)_2CHCH_2\overset{O}{\overset{\|}{C}}CH_3 \xrightarrow[(2)\ H_2O,H^+]{(1)\ NaBH_4}$ (2) $(CH_3)_2CHCH_2MgBr \xrightarrow[(2)\ H_2O,H^+]{(1)\ CH_3CHO}$

(c) (1)

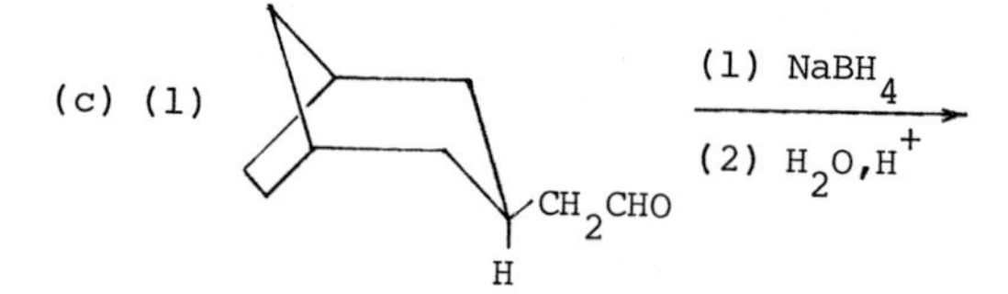

(2)

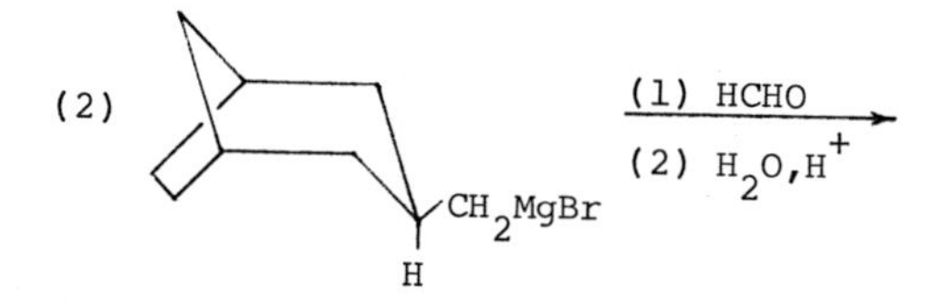

In (b), you may have used $CH_3MgI + (CH_3)_2CHCH_2CHO$.

In (c), you should not have used:

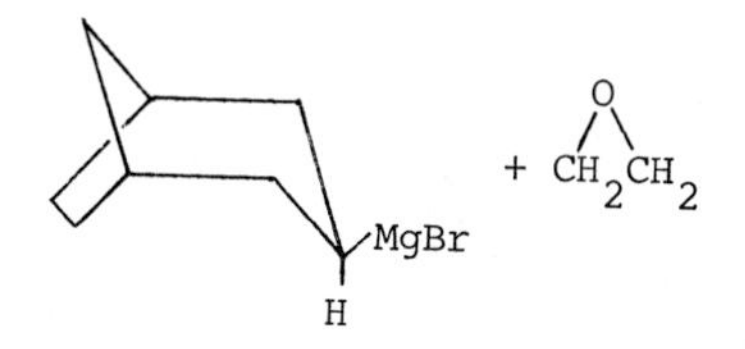

The reason is that, in the preparation of this Grignard reagent, both cis and trans RMgX would be formed.

7.37 (a) $(CH_3)_2CH\ddot{O}H \overset{H^+}{\rightleftharpoons} (CH_3)_2CH\overset{+}{\ddot{O}}H_2 \overset{-H_2\ddot{O}:}{\rightleftharpoons} [(CH_3)_2CH^+] \xrightarrow{I^-} (CH_3)_2CHI$

2° alcohol, S_N1

(b) $C_6H_{11}\text{—}\ddot{O}H \overset{H^+}{\rightleftharpoons} C_6H_{11}\text{—}\overset{+}{\ddot{O}}H_2 \overset{-H_2\ddot{O}:}{\rightleftharpoons} [C_6H_{11}^+] \xrightarrow{I^-} C_6H_{11}\text{—}I$

(c) $CH_3CH_2CH_2CH_2\ddot{O}H \overset{H^+}{\rightleftharpoons} CH_3CH_2CH_2CH_2\overset{+}{\ddot{O}}H_2 \xrightarrow{I^-}$

1° alcohol, S_N2

$$\left[\begin{array}{c} \delta+ \\ :OH_2 \\ \vdots \\ CH_3CH_2CH_2CH_2 \\ \vdots\ \ \delta- \\ I \end{array}\right] \xrightarrow{-H_2\ddot{O}:} CH_3CH_2CH_2CH_2I$$

7.38 (a) $(CH_3)_2CHCH_2CH_2Br + H_2O$ (b) $C_6H_{11}\text{—}Cl + H_2O$

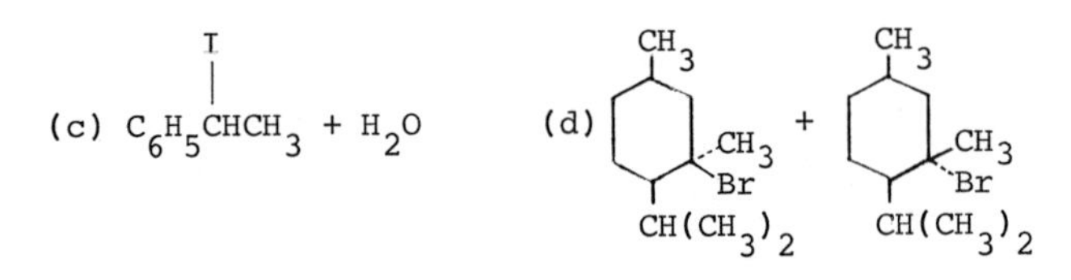

In (c), we would also expect $C_6H_5CH{=}CH_2$. In (d), two stereoisomers are obtained because the reaction proceeds by way of a carbocation.

7.39 (a) (naphthalene with CH_2Br substituent) + H_2O (fastest because it is a benzylic alcohol)

(b) $CH_3CH_2CH_2Br$ + H_2O (slowest because it is a 1° alcohol)

(c) (cyclohexyl)—Br + H_2O

7.40

7.41 (a) $(CH_3)_3CCHOH$ with CH_3 $\xrightarrow[\;-Cl^-\;]{ZnCl_2}$ $(CH_3)_3CCH\overset{+}{\frown}OZnCl$ with CH_3 $\xrightarrow{-HOZnCl}$

$\left[(CH_3)_2\overset{+}{C}CHCH_3 \text{ with } CH_3 \text{ shift} \right]$ a 2° carbocation $\xrightarrow{\text{methyl shift}}$ $\left[(CH_3)_2\overset{+}{C}CHCH_3 \text{ with } CH_3 \right]$ a 3° carbocation $\xrightarrow{Cl^-}$ $(CH_3)_2CClCH(CH_3)_2$

(b) $(C_6H_5)_2CHCH_2\overset{..}{O}H$ $\underset{\longrightarrow}{\overset{H^+}{\rightleftharpoons}}$ $(C_6H_5)_2C-CH_2\overset{+}{\underset{..}{O}}H_2$ $\xrightarrow[-H_2O]{\text{hydride shift}}$

$\left[(C_6H_5)_2\overset{+}{C}CH_3 \right]$ $\xrightarrow{I^-}$ $(C_6H_5)_2CICH_3$

resonance-stabilized
by two rings

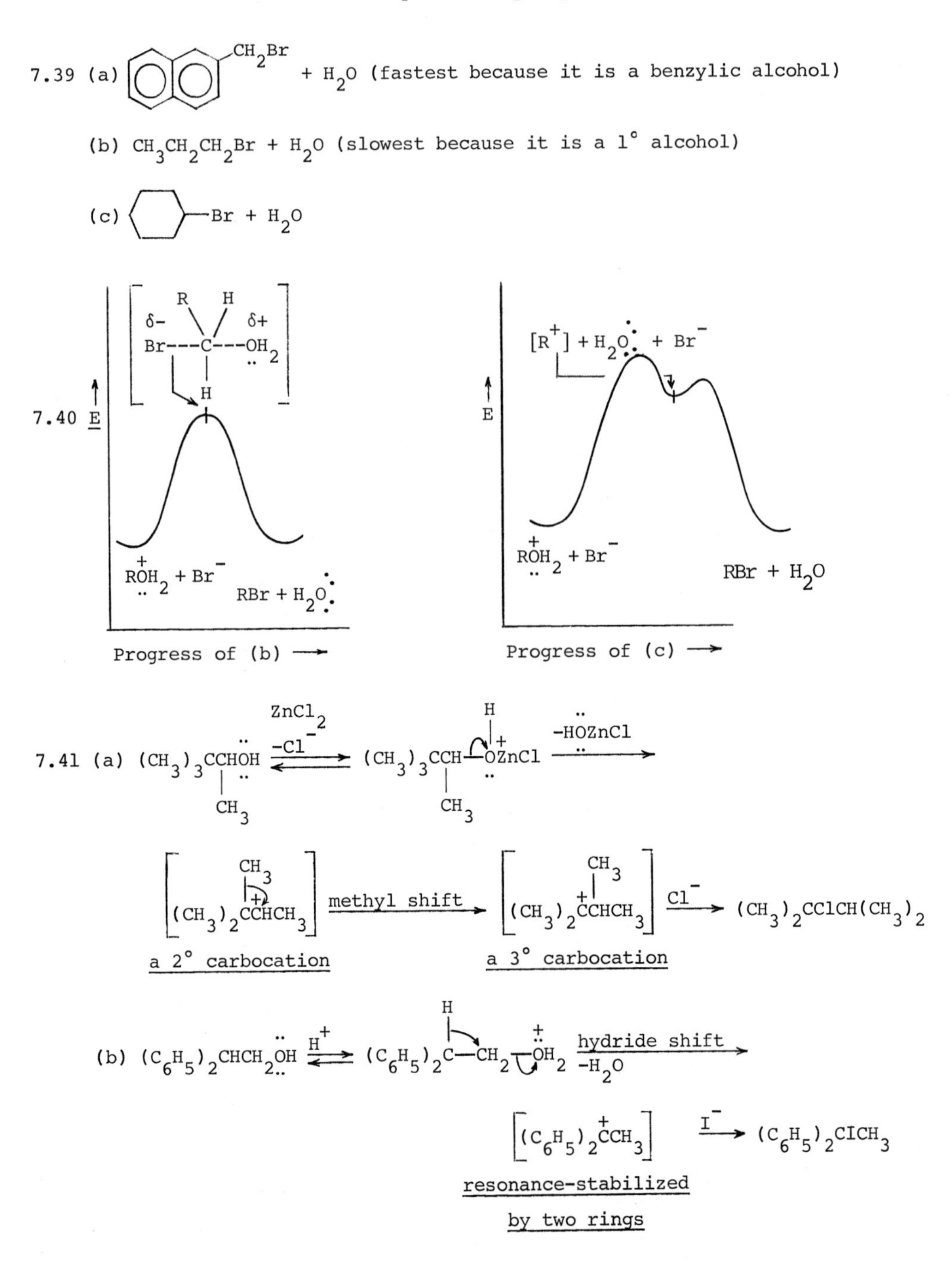

7.42 (a) $(CH_3)_2C{=}C(CH_3)_2$, Saytseff product from a 3° carbocation.

(b) $(C_6H_5)_2C{=}CH_2$, only possible alkene from the rearranged carbocation.

7.43 (a) (<u>R</u>)-$CH_3CHCl(CH_2)_3CH_3$ (inverted) (b) (<u>S</u>)-$CH_3CHCl(CH_2)_3CH_3$ (retention)

(c) $CH_3CHCl(CH_2)_3CH_3$ (racemic from S_N1 reaction)

(d) (<u>R</u>)-$CH_3CHCl(CH_2)_3CH_3$ (inverted)

7.44 (a) <u>trans</u>-$CH_3CH{=}CHCH_2CH_2CH_3$ (b) <u>trans</u>-$C_6H_5CH{=}CHCH_3$

(c) $CH_2{=}CHCH_2CH_3$ (d) <u>trans</u>-$CH_3CH{=}CHCH_3$ (e) $HOCH_2CH_2CH{=}C(CH_3)_2$

In each case, the more substituted, <u>trans</u> alkene is formed, if possible. In (e), the 4-OH is eliminated preferentially because 3° alcohols undergo elimination more readily than 1° alcohols.

7.45

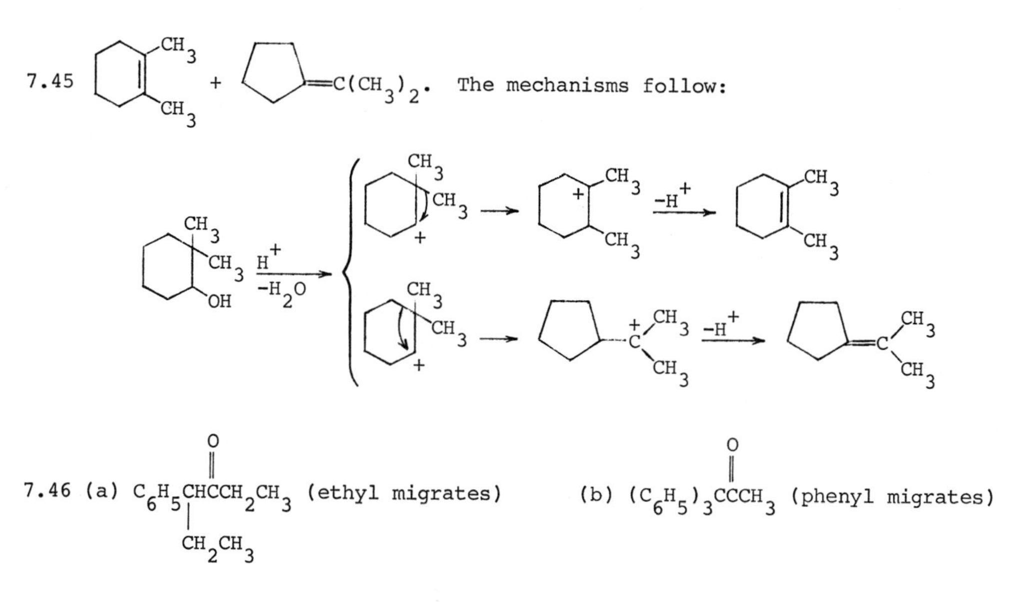

7.47 (a) $CH_3CH_2\overset{+}{O}H_2$ (b) cyclohexyl–$\ddot{O}$:$^-$ + $H_2\ddot{O}$: (c) 1-naphthoxide ($:\ddot{O}:^-$) + $H_2\ddot{O}$:

(d) $(CH_3)_3C\ddot{O}H$ + $:\ddot{O}H^-$

In (b), "no appreciable reaction" is also an acceptable answer because alcohols are weaker acids than water and very little alkoxide would be formed.

7.48 (a) $(CH_3)_2CHOCH_2CH_3$ + $CH_2{=}CHCH_3$, primarily by S_N2 and E2 reactions of an alkyl halide with a strong nucleophile and strong base.

(b) Na^+ ^-OH + CH_3CH_2OH, acid-base reaction

(c) $CH_3CO_2^- \; Na^+ + CH_3CH_2OH$, acid-base reaction.

(d) $C_6H_5O^- \; Na^+ + CH_3CH_2OH$, acid-base reaction.

7.49 (a) $CH_3CH_2CH_2CH_2O^- \; \overset{+}{Mg}I + CH_4$ (b) $CH_3CH_2CH_2CH_2O^- \; Li^+ + C_6H_6$

(c) no appreciable reaction (d) no reaction (e) $CH_3(CH_2)_3Br + H_2O$

(f) $CH_3CH_2CH_2CH_2O^- \; K^+ + H_2$

The reagents in (a) and (b) are both strong bases and remove a proton from 1-butanol. The reagents in (c) and (d) are not strong enough bases to react.

7.50 (a)

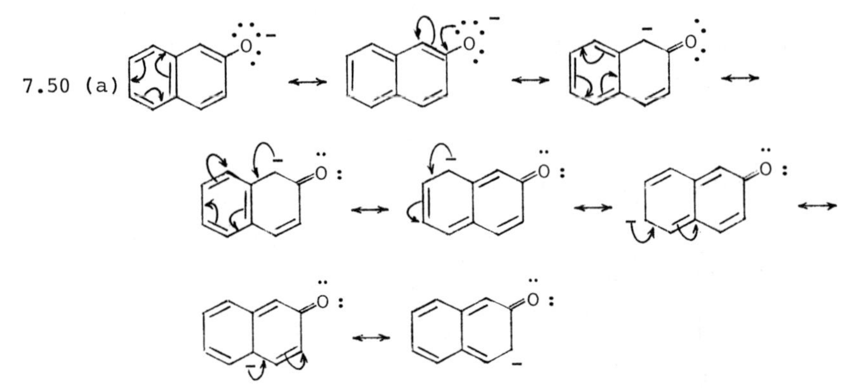

(b) The alkoxide ion does not have the negative charge in an allylic position; there is no resonance-stabilization.

$$CH_2{=}CH{-}CH_2{-}\ddot{\underset{..}{O}}:^-$$

(c) no resonance-stabilization.

7.51 (a) achiral $CH_3(CH_2)_4\overset{O}{\overset{\|}{C}}CH_3$

(b) racemic $CH_3(CH_2)_4CHICH_3$ (by an S_N1 reaction)

(c) ($\underline{R}$)-$CH_3(CH_2)_4\overset{O^-}{\overset{|}{C}}HCH_3 \; Li^+$ (chiral carbon unaffected)

(d) achiral $\underline{trans}$-$CH_3(CH_2)_3CH{=}CHCH_3$

(e) ($\underline{R}$)-$CH_3(CH_2)_4\overset{O^-}{\overset{|}{C}}HCH_3 \; \overset{+}{Mg}I + CH_4$ (chiral carbon unaffected)

(f) no reaction (g) no appreciable reaction (h) ($\underline{R}$)-$CH_3(CH_2)_4CHClCH_3$

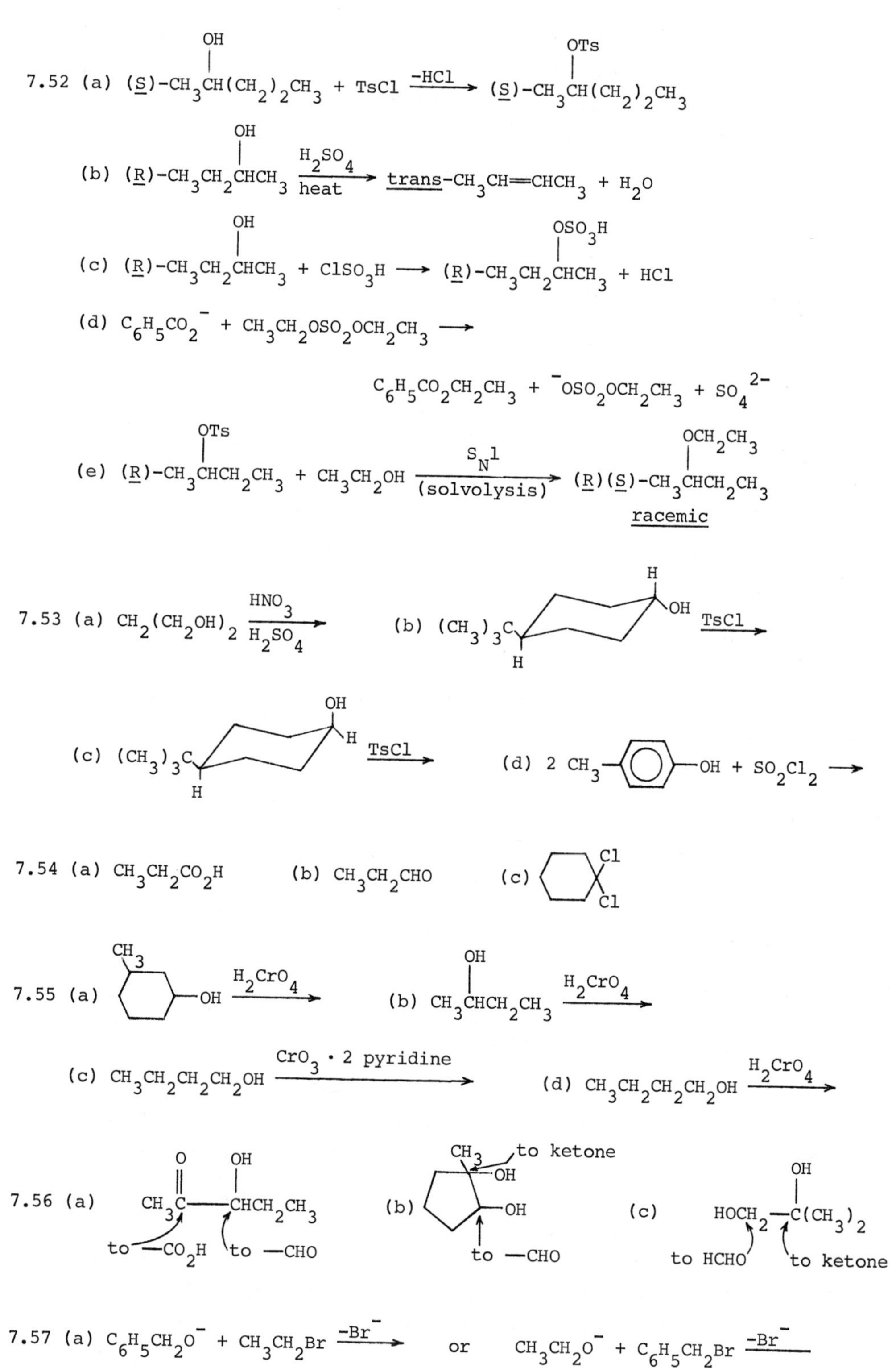
7.52 (a) (S)-CH3CH(CH2)2CH3 (OH) + TsCl —HCl→ (S)-CH3CH(CH2)2CH3 (OTs)
(b) (R)-CH3CH2CHCH3 (OH) —H2SO4, heat→ trans-CH3CH=CHCH3 + H2O
(c) (R)-CH3CH2CHCH3 (OH) + ClSO3H → (R)-CH3CH2CHCH3 (OSO3H) + HCl
(d) C6H5CO2⁻ + CH3CH2OSO2OCH2CH3 →
C6H5CO2CH2CH3 + ⁻OSO2OCH2CH3 + SO4 2-
(e) (R)-CH3CHCH2CH3 (OTs) + CH3CH2OH —SN1 (solvolysis)→ (R)(S)-CH3CHCH2CH3 (OCH2CH3)
racemic
7.53 (a) CH2(CH2OH)2 —HNO3, H2SO4→
(b) (CH3)3C … OH, H, H —TsCl→
(c) (CH3)3C … OH, H, H —TsCl→
(d) 2 CH3—C6H4—OH + SO2Cl2 →
7.54 (a) CH3CH2CO2H
(b) CH3CH2CHO
(c) Cl, Cl
7.55 (a) CH3 … OH —H2CrO4→
(b) CH3CHCH2CH3 (OH) —H2CrO4→
(c) CH3CH2CH2CH2OH —CrO3 · 2 pyridine→
(d) CH3CH2CH2CH2OH —H2CrO4→
7.56 (a) CH3C(=O)—CHCH2CH3 (OH)
to —CO2H
to —CHO
(b) CH3, OH, OH
to ketone
to —CHO
(c) HOCH2—C(CH3)2 (OH)
to HCHO
to ketone
7.57 (a) C6H5CH2O⁻ + CH3CH2Br —-Br⁻→ or CH3CH2O⁻ + C6H5CH2Br —-Br⁻—

(b) $C_6H_5O^- + CH_3CH_2CH_2Br \xrightarrow{-Br^-}$ but not $CH_3CH_2CH_2O^- + C_6H_5X$ because halobenzenes do not undergo S_N2 reactions.

7.58 (a) $CH_3CHI(CH_2)_4OH$ (b) $CH_3CHI(CH_2)_4I$

For (a) and (b), [2-methyltetrahydropyran] $\xrightleftharpoons{H^+}$ [protonated ether, $\overset{+}{C}$, :OH] $\rightleftharpoons$ [$\overset{+}{C}H-CH_3$, :ÖH] $\xrightarrow{I^-}$

2° carbocation

[iodo alcohol, :ÖH] $\xrightleftharpoons{H^+}$ [iodo, $:\overset{+}{O}H_2$] $\xrightarrow[S_N2]{I^-}$ [diiodide] $+ H_2O$

7.59 (a) $CH_3CH(OH)CH_2NH_2$ (b) $CH_3(CH_2)_4OCH(CH_3)CH_2OH + CH_3CH(Cl)CH_2OH$

(c) $CH_3CH(OH)CH_2CH_2CH{=}CH_2$ (d) $CH_3CH(OH)CH_2C_6H_5$ (e) $CH_3CH(OH)CH_2OC_6H_5$

In (e), it is the phenoxide ion ($C_6H_5O^-$) that attacks. This ion is formed by the reaction of phenol and NaOH.

7.60 [cyclohexane ring: Br and CH_3 on one carbon, OH on adjacent carbon] The attack on the unsymmetrical epoxide is by an acidic reagent; consequently, Br^- attacks at the 3° carbon. Since the bromide attacks from the opposite side of the bridging three-membered ring, a trans-bromohydrin is the product.

7.61 (a) $CH_3CH_2S(=O)_2CH_2CH_3$ (b) [cyclic sulfoxide, S=O in six-membered ring] (c) [cyclohexyl]$-S^-\ Na^+$

7.62 (d). Only (a) and (d) undergo a rapid acid-base reaction with a Grignard reagent:

$$ROH + CH_3MgI \longrightarrow ROMgI + CH_4$$

Of these two alcohols, (a) undergoes rapid reaction with Lucas reagent because it is a 3° alcohol. (d) undergoes slow reaction with Lucas reagent because it is a less-reactive 1° alcohol.

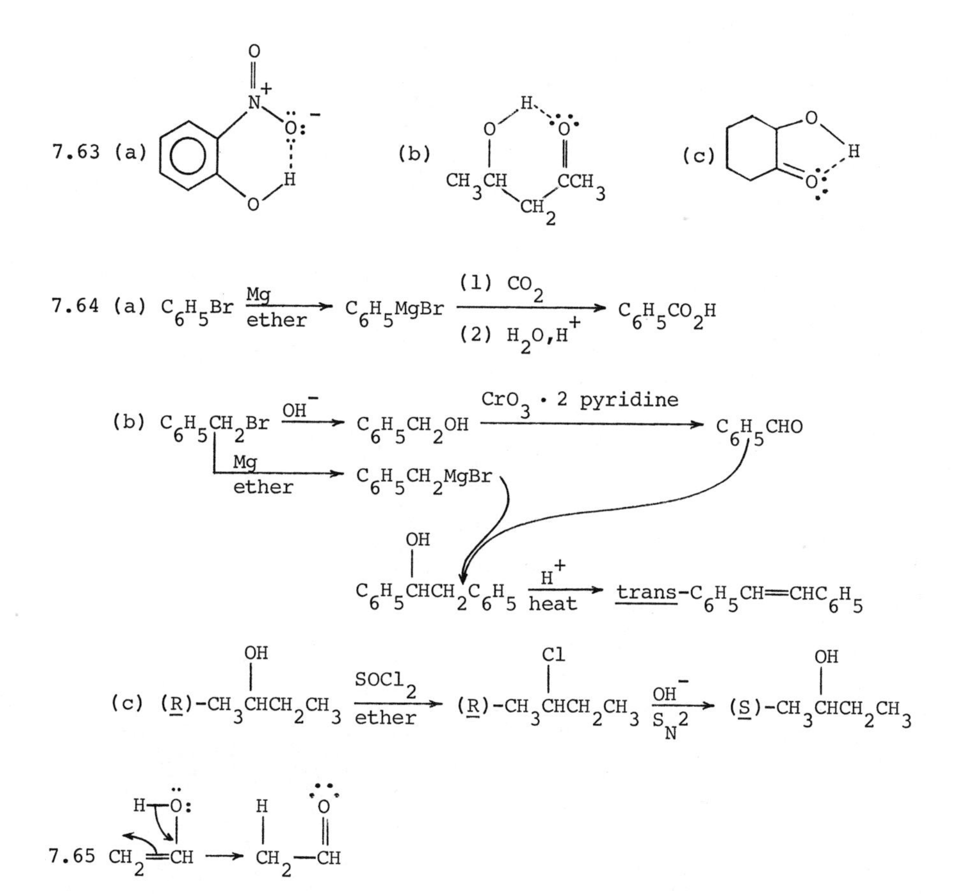

7.63 (a) (b) (c)

7.64 (a) $C_6H_5Br \xrightarrow{Mg}{ether} C_6H_5MgBr \xrightarrow[\text{(2) } H_2O,H^+]{\text{(1) } CO_2} C_6H_5CO_2H$

(b) $C_6H_5CH_2Br \xrightarrow{OH^-} C_6H_5CH_2OH \xrightarrow{CrO_3 \cdot 2 \text{ pyridine}} C_6H_5CHO$

$\xrightarrow[\text{ether}]{Mg} C_6H_5CH_2MgBr$

$\underset{OH}{C_6H_5CHCH_2C_6H_5} \xrightarrow[\text{heat}]{H^+} trans\text{-}C_6H_5CH=CHC_6H_5$

(c) $(\underline{R})\text{-}CH_3\overset{OH}{CHCH_2CH_3} \xrightarrow[\text{ether}]{SOCl_2} (\underline{R})\text{-}CH_3\overset{Cl}{CHCH_2CH_3} \xrightarrow[S_N2]{OH^-} (\underline{S})\text{-}CH_3\overset{OH}{CHCH_2CH_3}$

7.65

Although this is an acceptable answer, this reaction is probably not intramolecular as shown here. More likely, the proton is lost to another molecule of $CH_2{=}CHOH$, which then transfers its OH proton to another molecule. One possible mechanism for this transfer follows:

$$2 \ CH_2=CHOH \longrightarrow CH_2=C\overset{+}{O}H_2 + [CH_2=CHO^- \longleftrightarrow {}^-CH_2CH=O]$$

$$CH_2=CHOH + CH_3CH=O$$

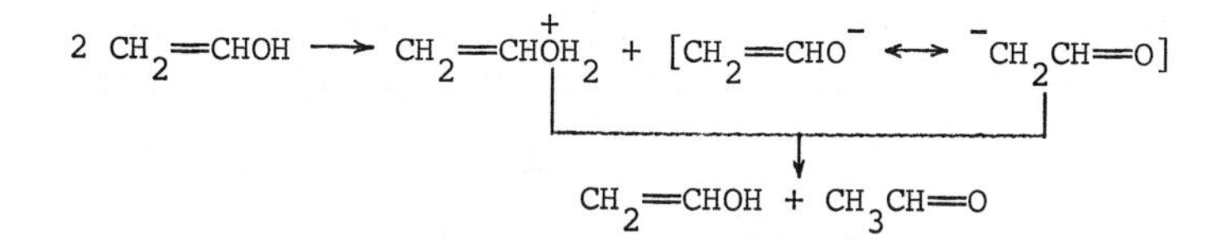

7.66 (a) The cis-compound cannot undergo intramolecular S_N2 displacement of the chlorine to yield an epoxide.

(b)

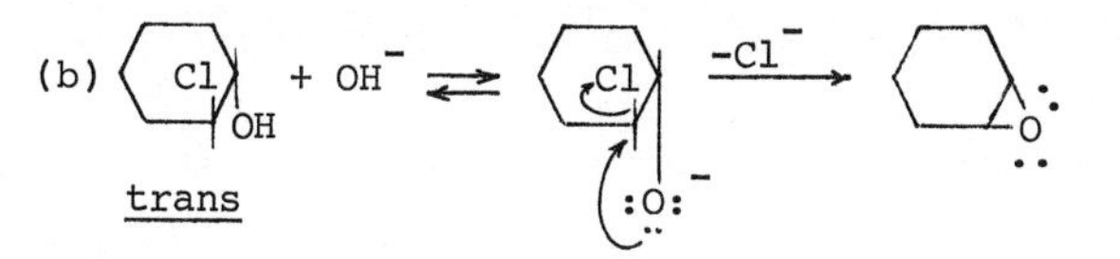

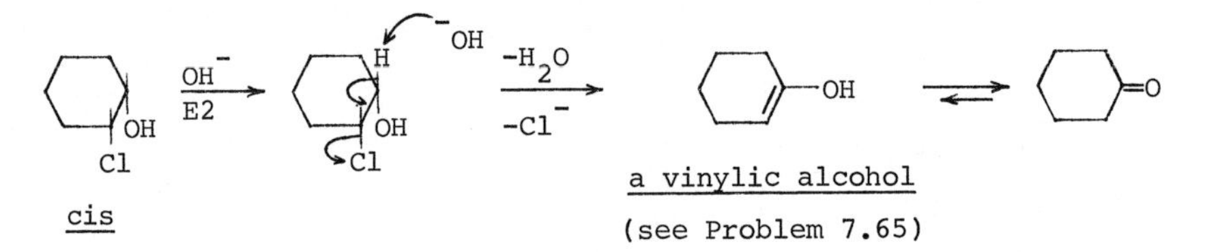

7.67 A 1,2-diol undergoes a pinacol rearrangement in H_2SO_4 solution. Either the cis or trans diol could yield cyclohexanone or cyclopentanecarboxaldehyde:

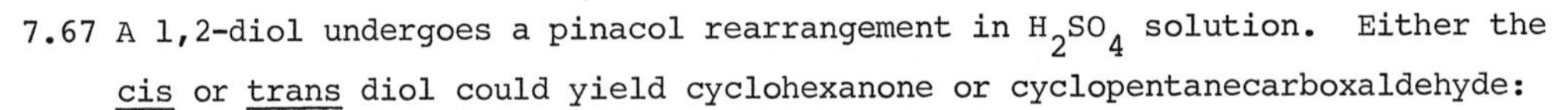

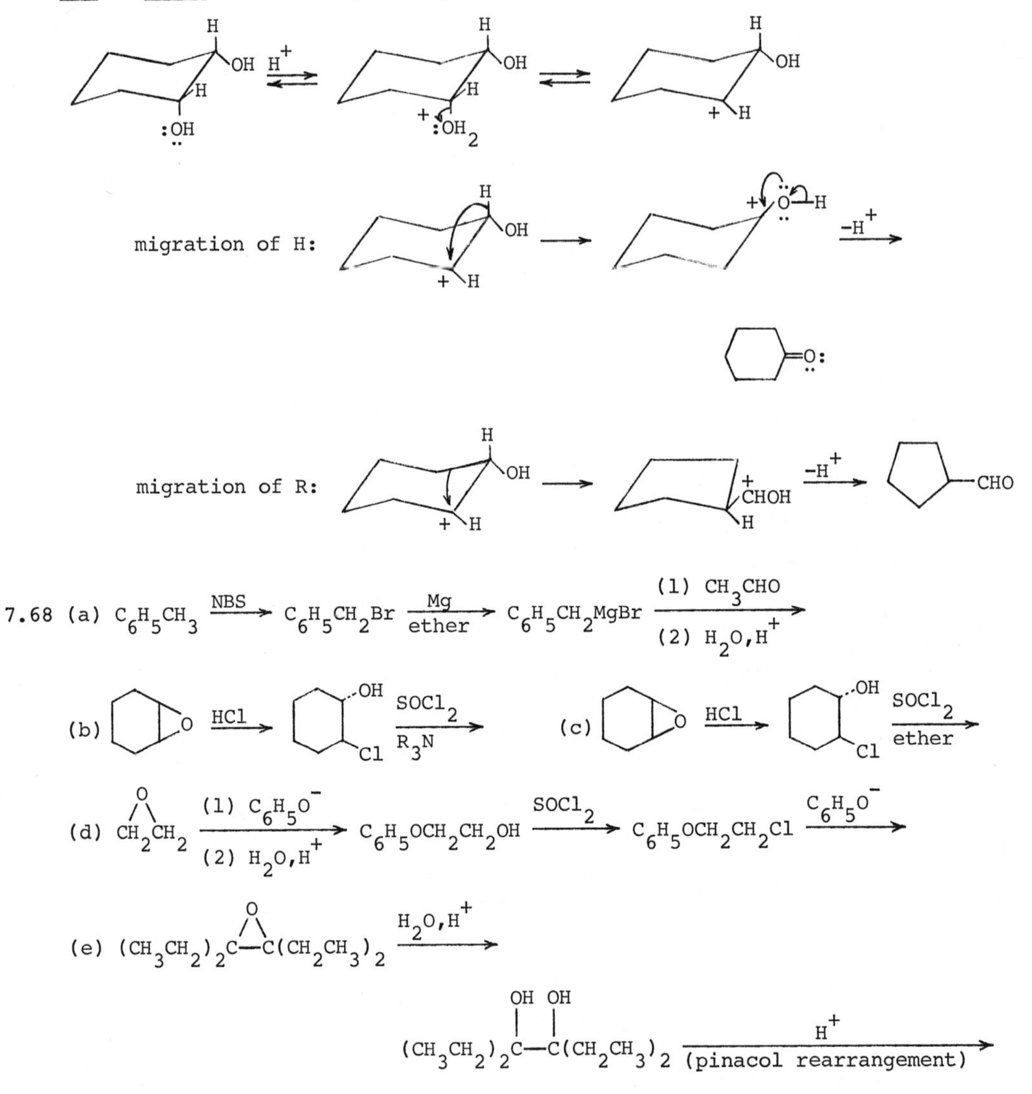

7.69 In a road-map problem, first outline the reactions, then find your clue (in this case, the reaction of the Grignard reagent of B with CH_3CHO to yield 5-methyl-2-heptanol). Work the problem backwards.

$$\text{A} \xrightarrow{SOCl_2} \text{B} \xrightarrow[\substack{(2)\ CH_3CHO \\ (3)\ H_2O,H^+}]{(1)\ Mg} CH_3CH_2\overset{\overset{CH_3}{|}}{C}HCH_2CH_2\overset{\overset{OH}{|}}{C}HCH_3$$

$CH_3CH_2\overset{CH_3}{CH}CH_2CH_2$ portion: from $CH_3CH_2\underset{\underset{CH_3}{|}}{C}HCH_2CH_2MgCl$

$CHCH_3$ portion: from CH_3CHO

B is $CH_3CH_2\overset{\overset{CH_3}{|}}{C}HCH_2CH_2Cl$ and therefore A is $CH_3CH_2\overset{\overset{CH_3}{|}}{C}HCH_2CH_2OH$.

7.70 (a) $CH_2{=}CHCH_2OH \xrightarrow{PCl_3} CH_2{=}CHCH_2Cl \xrightarrow{\text{excess } Na^+\ {}^-SH}$

$$CH_2{=}CHCH_2SH \xrightarrow{I_2} \text{product}$$

7.71 There may be correct answers other than those that follow.

(a) $CH_3CH_2\overset{\overset{CH_3}{|}}{C}HOH \xrightarrow{HBr} CH_3CH_2\overset{\overset{CH_3}{|}}{C}HBr \xrightarrow[\text{ether}]{Mg} CH_3CH_2\overset{\overset{CH_3}{|}}{C}HMgBr \xrightarrow[(2)\ H_2O,H^+]{(1)\ HCHO}$

$$CH_3CH_2\overset{\overset{CH_3}{|}}{C}HCH_2OH \xrightarrow{K} CH_3CH_2\overset{\overset{CH_3}{|}}{C}HCH_2O^- \xrightarrow{CH_3I}$$

(b) $(CH_3)_3COH \xrightarrow{K} (CH_3)_3CO^-\ K^+ \xrightarrow{CH_3CH_2Br}$

(c) $(CH_3)_2CHOH \xrightarrow{HBr} (CH_3)_2CHBr \xrightarrow[\text{ether}]{Mg} (CH_3)_2CHMgBr \xrightarrow[(2)\ H_2O,H^+]{(1)\ CH_3CHO}$

$$(CH_3)_2CH\overset{\overset{OH}{|}}{C}HCH_3 \xrightarrow{H_2CrO_4}$$

(d) $\overset{CH_2}{/\backslash}$ $CH_2CHCH_2OH \xrightarrow{PBr_3} \overset{CH_2}{/\backslash}$ $CH_2CHCH_2Br \xrightarrow[\text{ether}]{Mg} \overset{CH_2}{/\backslash}$ $CH_2CHCH_2MgBr \xrightarrow[(2)\ H_2O,H^+]{(1)\ CH_3CH_2CHO}$

$$\overset{CH_2}{/\backslash}\ CH_2CHCH_2\overset{\overset{OH}{|}}{C}HCH_2CH_3 \xrightarrow{SOCl_2}$$

(e) $CH_3CH_2OH \xrightarrow{HBr} CH_3CH_2Br \xrightarrow[ether]{Mg} CH_3CH_2MgBr \xrightarrow[(2)\ H_2O,H^+]{(1)\ CH_3\overset{O}{\overset{\|}{C}}CH_3}$

$(CH_3)_2\overset{OH}{\overset{|}{C}}CH_2CH_3 \xrightarrow[heat]{H_2SO_4}$

(f) $(CH_3)_2CHCH_2OH \xrightarrow{PBr_3} (CH_3)_2CHCH_2Br \xrightarrow[ether]{Mg} (CH_3)_2CHCH_2MgBr \xrightarrow[(2)\ H_2O,H^+]{(1)\ \overset{O}{CH_2CH_2}}$

(g) $(CH_3)_2CHCH_2MgBr$ from (f) $\xrightarrow[(2)\ H_2O,H^+]{(1)\ CH_3CHO} (CH_3)_2CHCH_2\overset{OH}{\overset{|}{C}H}CH_3 \xrightarrow{HBr}$

$(CH_3)_2CHCH_2CHBrCH_3 \xrightarrow{CN^-}$

(h) $(CH_3)_2CHOH \xrightarrow{Na} (CH_3)_2CHO^-\ Na^+ \xrightarrow{\overset{O}{CH_2CH_2}} (CH_3)_2CHOCH_2CH_2OH \xrightarrow{SOCl_2}$

(i) $CH_3CH_2CH_2OH \xrightarrow{HBr} CH_3CH_2CH_2Br \xrightarrow[ether]{Mg} CH_3CH_2CH_2MgBr$

$\xrightarrow[(2)\ H_2O,H^+]{(1)\ (CH_3)_2CHCHO} CH_3CH_2CH_2\overset{OH}{\overset{|}{C}H}CH(CH_3)_2 \xrightarrow[H_2SO_4]{HNO_3}$

7.72 There are other possible synthetic schemes in each case.

(a) $CH_3CH_2Br \xrightarrow[ether]{Mg} CH_3CH_2MgBr \xrightarrow[(2)\ H_2O,H^+]{(1)\ CH_3CHO} CH_3\overset{OH}{\overset{|}{C}H}CH_2CH_3 \xrightarrow[heat]{H_2SO_4}$

$CH_3CH{=}CHCH_3 \xrightarrow{NBS}$

or $CH_3CH{=}CHI \xrightarrow[ether]{Mg} CH_3CH{=}CHMgI \xrightarrow[(2)\ H_2O,H^+]{(1)\ HCHO} CH_3CH{=}CHCH_2OH \xrightarrow{PBr_3}$

(b) $CH_3CH_2CO_2H \xrightarrow[heat]{CH_3CH_2OH,H^+} CH_3CH_2CO_2CH_2CH_3 \xrightarrow[(2)\ H_2O,H^+]{(1)\ 2\ CH_3CH_2MgBr \text{ from (a)}}$

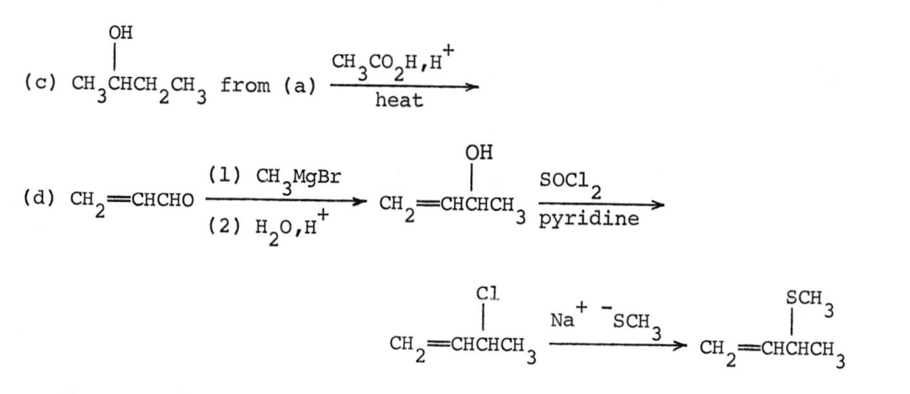

(c) CH$_3$CHCH$_2$CH$_3$ (OH) from (a) $\xrightarrow[\text{heat}]{\text{CH}_3\text{CO}_2\text{H,H}^+}$

(d) CH$_2$=CHCHO $\xrightarrow[\text{(2) H}_2\text{O,H}^+]{\text{(1) CH}_3\text{MgBr}}$ CH$_2$=CHCHCH$_3$ (OH) $\xrightarrow[\text{pyridine}]{\text{SOCl}_2}$

CH$_2$=CHCHCH$_3$ (Cl) $\xrightarrow{\text{Na}^+ \ ^-\text{SCH}_3}$ CH$_2$=CHCHCH$_3$ (SCH$_3$)

Because the chloride is allylic, a by-product might be $CH_3SCH_2CH{=}CHCH_3$.

(e) $CH_3CH_2Br \xrightarrow[\text{ether}]{\text{Mg}} CH_3CH_2MgBr \xrightarrow[\text{(2) } H_2O,H^+]{\text{(1) } \overset{O}{CH_2CH_2}}$

$CH_3CH_2CH_2CH_2OH \xrightarrow{\text{HBr}} CH_3CH_2CH_2CH_2Br \xrightarrow[\text{(2) CuI}]{\text{(1) Li}}$

$(CH_3CH_2CH_2CH_2)_2CuLi \xrightarrow{CH_3(CH_2)_3Br}$

7.73 A $\xrightarrow[\text{-HCHO}]{\text{HIO}_4}$ B $\xrightarrow{\text{KMnO}_4}$ adipic acid

to HCHO

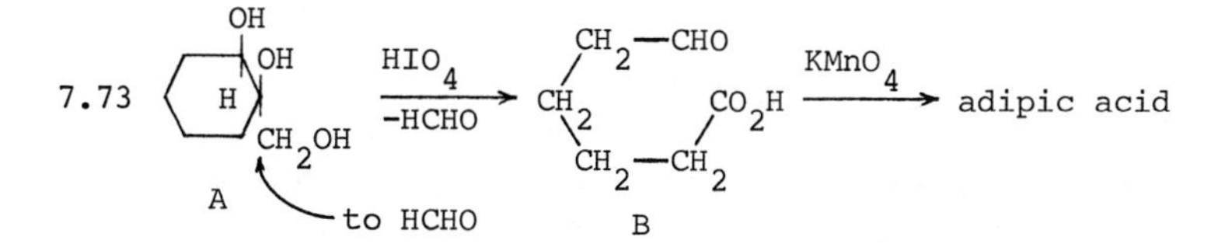

CHAPTER 8

Spectroscopy I: Infrared and Nuclear Magnetic Resonance

Some Important Features

Electromagnetic radiation can be absorbed by organic compounds, and this absorption of energy results in increased energy of the molecules. Different compounds absorb electromagnetic radiation of different energy, and thus of different wavelength (λ) or frequency (ν). Radiation of shorter wavelength is of higher frequency and higher energy than radiation of longer wavelength.

Absorption of infrared radiation results in increased amplitudes of bond vibrations. Polar groups usually exhibit stronger peaks in the infrared spectrum than nonpolar groups. The usual positions of absorption are shown inside the front cover of the text. The absorption positions (and peak appearances) of —OH, —NH and —NH_2 (a double peak), —CO_2H, and C=O are particularly distinctive.

An nmr spectrum results from the change in spin states of hydrogen nuclei (protons). The position of absorption by a proton (the chemical shift) is affected by neighboring atoms and by other groups in the molecule. A nearby electronegative atom results in deshielding, and absorption is observed farther downfield.

← downfield (deshielded)	upfield → (shielded)

Other groups in the molecule affect the position of absorption by anisotropic effects. Absorption by protons attached to aromatic rings, to aldehyde carbonyl groups, and to *sp*^{2} carbons of alkenes is observed downfield. Figure 8.27 in the text shows some typical positions of proton absorption in nmr spectra.

Recognizing the equivalence or nonequivalence of protons in nmr spectroscopy is important in structure determination. We suggest that you review Section 8.8A.

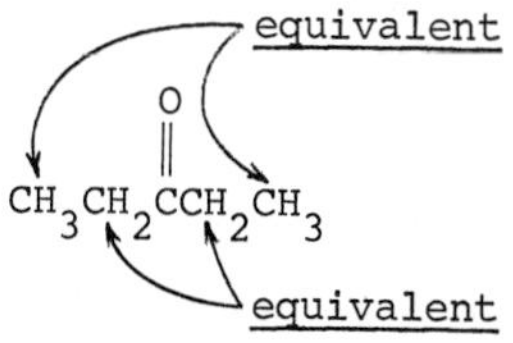

$$CH_3CH_2\overset{O}{\overset{\|}{C}}CH_2CH_3$$

(the two CH_3 groups: equivalent; the two CH_2 groups: equivalent)

The area under the signal for a proton is proportional to the number of protons giving rise to that signal. In the above example, we observe two principal signals: the CH_3 signal and the CH_2 signal. The area ratio is 6 : 4, or 3 : 2.

Absorption peaks in the nmr spectrum may be split into multiple peaks. A group of magnetically equivalent protons absorb radio waves at the same position of the nmr spectrum and do not split the signals of each other. Neighboring protons that are not magnetically equivalent to the proton in question do split the signal of the proton.

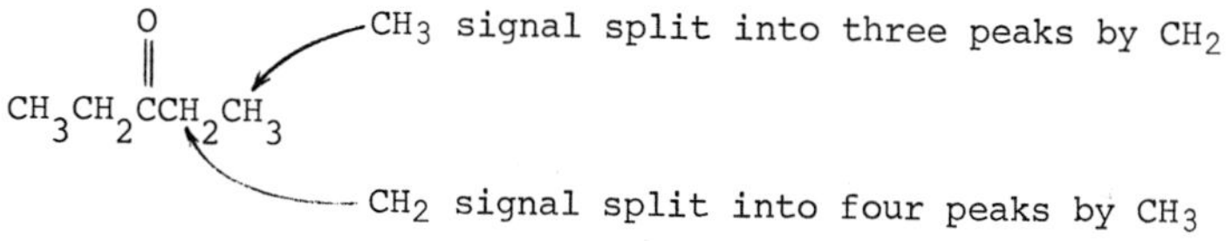

$$CH_3CH_2\overset{O}{\overset{\|}{C}}CH_2CH_3$$

CH_3 signal split into three peaks by CH_2

CH_2 signal split into four peaks by CH_3

If a signal is split by a group of neighboring protons equivalent to each other, then the *n* + 1 rule is followed. In the above example, CH_3 is split by CH_2 (two protons equivalent to each other but not equivalent to the CH_3 protons). The splitting pattern for CH_3 is 2 + 1, or a triplet.

Another topic covered in the nmr discussion is chemical exchange, which may result in no splitting of (or by) an OH or NH proton.

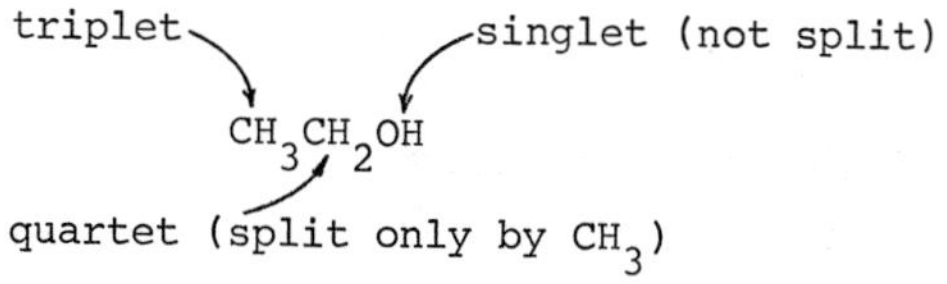

$$CH_3CH_2OH$$

triplet (CH_3)

singlet (not split) (OH)

quartet (split only by CH_3) (CH_2)

Section 8.13 in the text is an introduction to determination of structure from the molecular formula and spectral data. This section should be studied carefully, as well as the discussions in the answers to Problem 8.44 (after you have solved them or tried to solve them).

Before studying spectra in detail for clues about a structure, first examine the molecular formula and the spectra for obvious clues.

1. Does the structure contain an aromatic ring? Check the nmr spectrum for aryl protons.
2. Does the structure contain C═O, NH or OH, NH_2, or CO_2H? Check the infrared spectrum for these distinctive peaks.
3. How many rings or double bonds does the structure contain? Check the molecular formula. Remember that the C═O group is a site of unsaturation. Butanone has the general formula C_4H_8O, or $C_nH_{2n}O$. A phenyl group contains three sites of unsaturation and a ring. $C_6H_5CH_2CH_3$ has the general formula C_nH_{2n-6}.
4. Does the structure contain an ethyl group bonded to an electronegative atom? Check the nmr spectrum for an upfield triplet and a downfield quartet.

Answers to Problems

8.18 (a) 3.33 μm (b) 5.68 μm (c) 1785 cm^{-1} (d) 1220 cm^{-1}
(e) 3×10^{-5} MHz
(a) through (d) are solved by the following equations:

$$\text{wavenumber in cm}^{-1} = \frac{1}{\lambda \text{ in cm}}$$

$$1\ \mu m = 10^{-4}\ cm$$

8.19 (a) C═O (b) C═C—Cl (c) O—H
In each case, the vibration that results in the greater change in dipole moment gives stronger absorption.

8.20 (a) $CH_3CH_2CH_2NH_2$ shows NH absorption, while $CH_3CH_2CH_2N(CH_3)_2$ does not.
(b) $CH_3CH_2CH_2CO_2H$ shows OH absorption, while $CH_3CH_2CH_2CO_2CH_3$ does not.
(c) $CH_3CH_2CO_2CH_3$ shows C—O absorption, while $CH_3CH_2C(=O)CH_3$ does not.

8.21 cyclohexanol (—OH) $\xrightarrow{[O]}$ cyclohexanone (═O)

The organic content of the reaction mixture can be isolated and its infrared spectrum determined. As the reaction proceeds, the OH absorption of the organic extract gradually disappears. The carbonyl absorption of cyclohexanone is not useful in this case because it appears early in the course of the reaction.

8.22 Any cyclic ether containing a total of five carbon atoms is consistent with the data. For example:

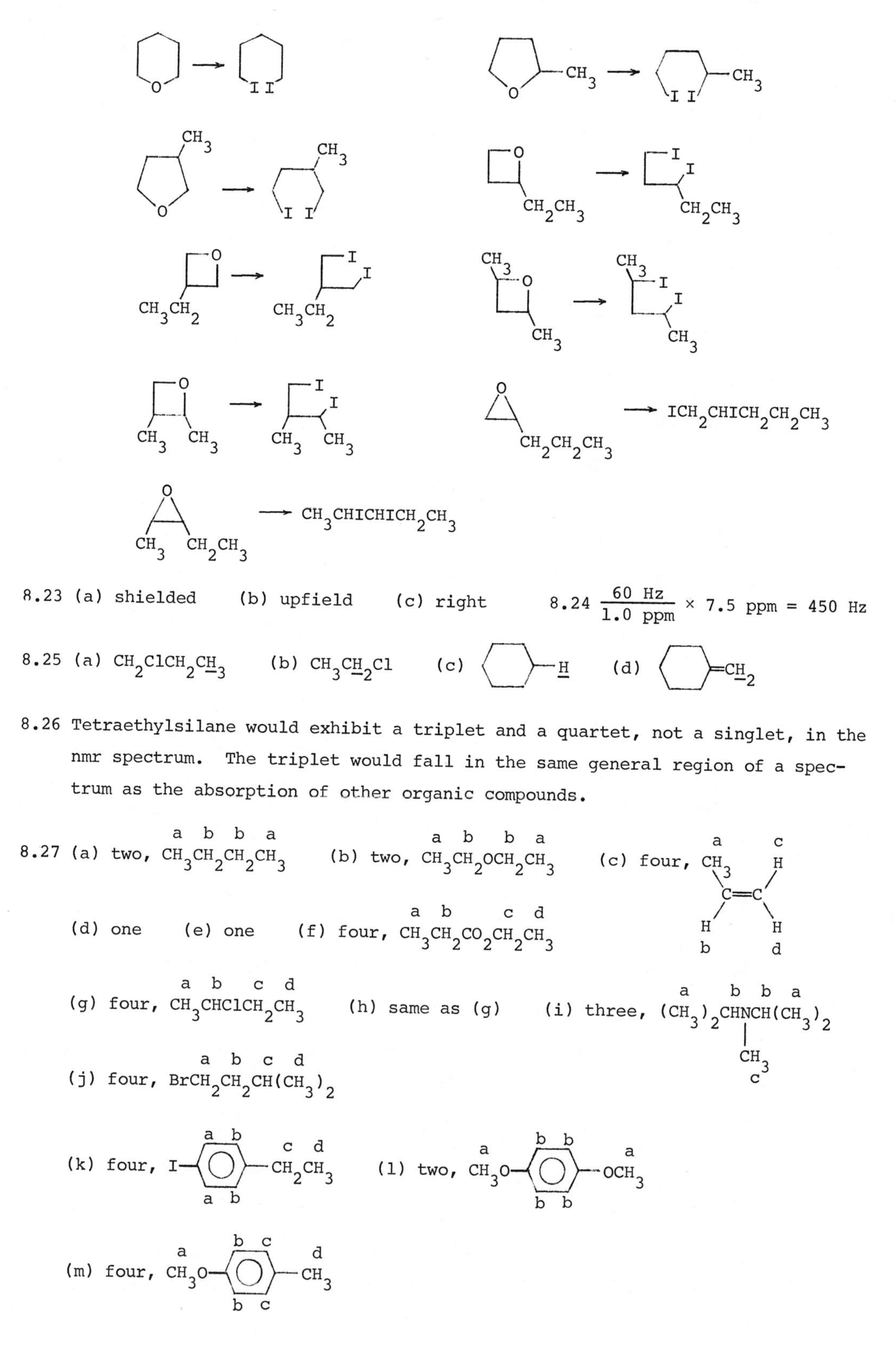

8.23 (a) shielded (b) upfield (c) right

8.24 $\frac{60 \text{ Hz}}{1.0 \text{ ppm}} \times 7.5 \text{ ppm} = 450 \text{ Hz}$

8.25 (a) $CH_2ClCH_2\underline{CH_3}$ (b) $CH_3\underline{CH_2}Cl$ (c) (d)

8.26 Tetraethylsilane would exhibit a triplet and a quartet, not a singlet, in the nmr spectrum. The triplet would fall in the same general region of a spectrum as the absorption of other organic compounds.

8.27 (a) two, $\overset{a}{CH_3}\overset{b}{CH_2}\overset{b}{CH_2}\overset{a}{CH_3}$ (b) two, $\overset{a}{CH_3}\overset{b}{CH_2}O\overset{b}{CH_2}\overset{a}{CH_3}$ (c) four

(d) one (e) one (f) four, $\overset{a}{CH_3}\overset{b}{CH_2}CO_2\overset{c}{CH_2}\overset{d}{CH_3}$

(g) four, $\overset{a}{CH_3}\overset{b}{CHCl}\overset{c}{CH_2}\overset{d}{CH_3}$ (h) same as (g) (i) three, $(\overset{a}{CH_3})_2\overset{b}{CH}N\overset{b}{CH}(\overset{a}{CH_3})_2$ with N–$\overset{c}{CH_3}$

(j) four, $Br\overset{a}{CH_2}\overset{b}{CH_2}\overset{c}{CH}(\overset{d}{CH_3})_2$

(k) four, (l) two, (m) four,

8.28 The number of principal signals is equal to the number of equivalent protons. The relative areas are equal to the ratios of protons giving rise to the principal signals.

(a) two principal signals (area ratio, 3 : 2) (b) two (3 : 2)

(c) four (3 : 1 : 1 : 1) (d) and (e) one (f) four (3 : 3 : 2 : 2)

(g) and (h) four (3 : 3 : 2 : 1) (i) three (12 : 3 : 2)

(j) four (6 : 2 : 2 : 1) (k) four (3 : 2 : 2 : 2) (l) two (3 : 2)

(m) four (3 : 3 : 2 : 2)

8.29 (d) and (e). None of the other possibilities have chemically nonequivalent protons in the ratio of 3 : 1. Instead, the following ratios are observed:

(a) 3 : 1 : 1 : 1 (b) 3 : 1 : 1 : 1 (c) 3 : 2 : 2 and (f) 1 : 1

8.30 Divide all areas by the smallest one:

$$\frac{81.5}{28} = 2.9 \qquad \frac{28}{28} = 1.0 \qquad \frac{55}{28} = 2.0 \qquad \frac{80}{28} = 2.9$$

Rounding, the ratios are 3 : 1 : 2 : 3

8.31 (a) $\overset{3}{CH_3}\overset{4}{CH_2}CO_2\overset{1}{CH_3}$ (<u>areas</u>, 3 : 2 : 3) (b) $CH_3OCH_2CH_2OCH_3$ (<u>areas</u>, 3 : 2); CH_2CH_2: 1, CH_3…CH_3: 1

(c) $\overset{1}{CH_3}\overset{\overset{O}{\|}}{C}\overset{4}{CH_2}\overset{3}{CH_3}$ (<u>areas</u>, 3 : 2 : 3) (d) $\overset{1}{CH_3}O-C_6H_4-Cl$ (ring protons a, b) (<u>areas</u>, 3 : 2 : 2) a = 2; b = 2

(e) $Cl_2\overset{3}{CH}\overset{2}{CH_2}Br$ (<u>areas</u>, 1 : 2)

8.32 A, $(CH_3)_2CHCl$ — CH_3: doublet; CH: septet B, $CH_3CH_2CH_2Cl$ — CH_3 and CH_2Cl: triplets; middle CH_2: sextet

8.33 (a) $CH_3CH_2CH_2CH_2CH{=}CH_2$ shows a large number of peaks, while $(CH_3)_2C{=}C(CH_3)_2$ shows a singlet

(b) CH_3CH_2CHO shows three principal signals (including offset absorption for an aldehyde proton), while CH_3COCH_3 shows a singlet

(c) CH_3COCH_3 shows one singlet, while $CH_3CO_2CH_3$ shows two singlets

(d) $CH_3CH_2{-}C_6H_5$ shows a quartet and a triplet between the TMS peak and aryl absorption, while p-$CH_3{-}C_6H_4{-}CH_3$ shows a singlet

8.34 (a) In infrared, 1-propanol shows OH; propylene oxide does not.

(b) In nmr, diisopropyl ether shows a doublet and a septet; di-n-propyl ether shows two triplets and a sextet.

(c) In nmr, ethanol shows one singlet, one triplet, and one quartet; 1,2-ethanediol shows two singlets.

(d) In infrared, [cyclohexane ring]NH shows NH; [cyclohexane ring]NCH_3 does not.

(e) In infrared, ethanol shows OH; chloroethane does not.

(f) In infrared, acetic acid shows broad OH; acetone does not show OH.

8.35 (a) (b) (c)

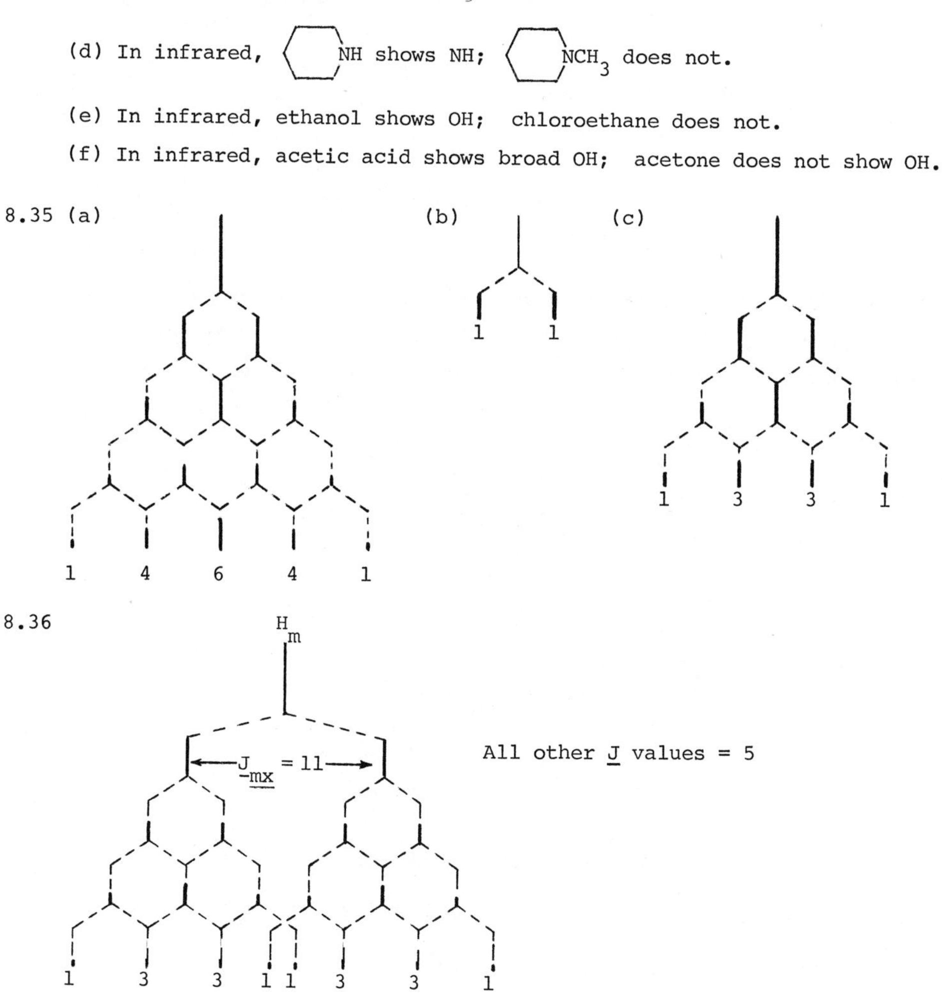

8.36

All other $\underline{J}$ values = 5

8.37 (a) seven

(b) aryl, doublet (area 2) plus triplet (area 1), but all the aryl protons may appear as a singlet (area 3); CH_3, singlet (area 6); NH, singlet (area 1); CH_2, singlet (area 2); CH_3CH_2, triplet (area 4) plus quartet (area 6)

(c) NH absorption (∿ 3500 cm^{-1}), C—H absorption (∿ 3000 cm^{-1}), and C═O absorption (∿ 1700 cm^{-1}) are the most characteristic peaks we have discussed in this chapter.

8.38

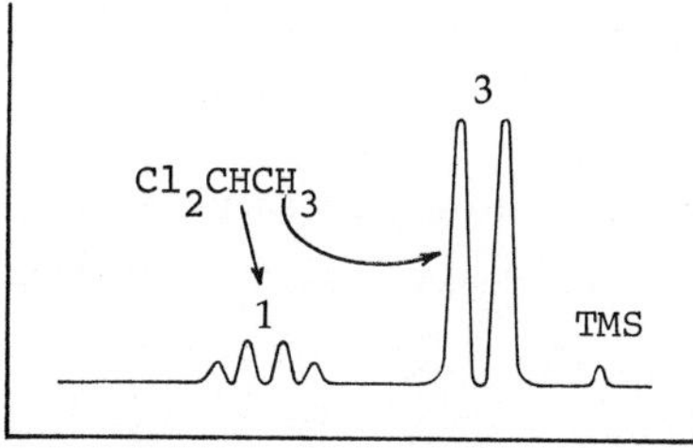

8.39

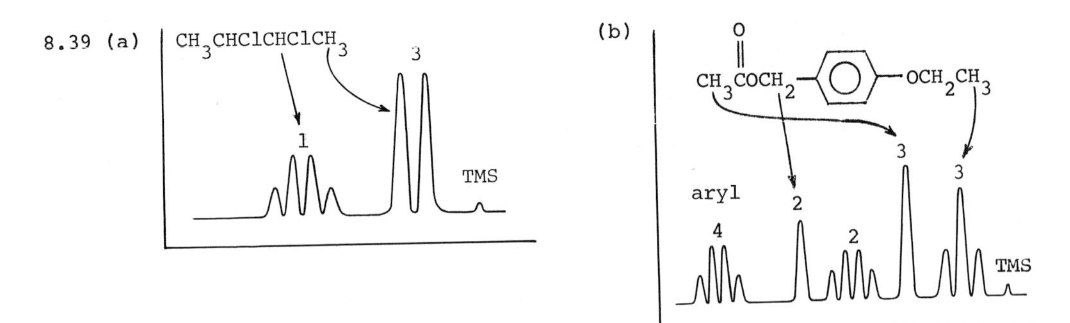

8.40 The outside protons, like those of benzene, are deshielded and absorb downfield in the aromatic region. The inner protons are highly shielded and absorb upfield. (This absorption is observed upfield even of TMS.)

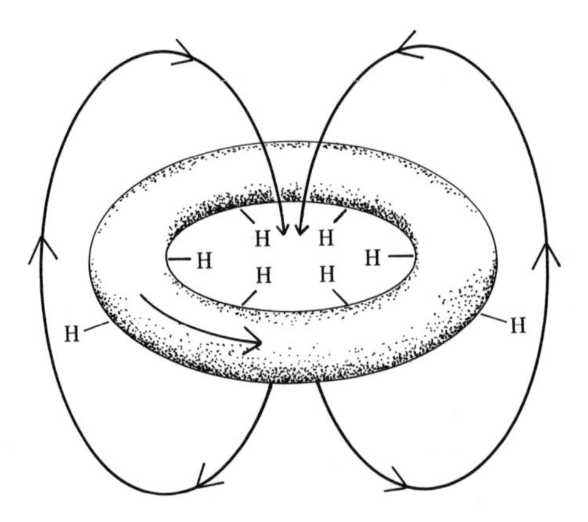

All outer H's not shown.

8.41 As in [18]annulene, the CH_3 protons are shielded by the field induced by the ring current.

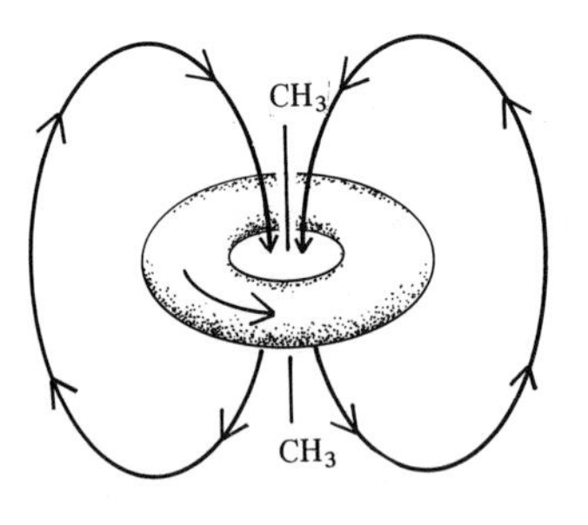

8.42

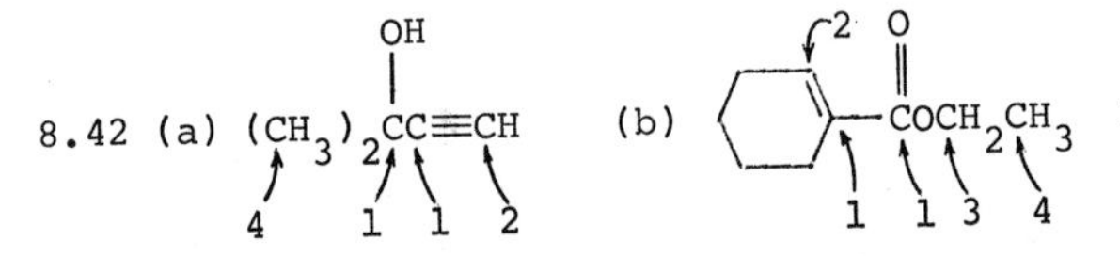

Each other ring carbon: 3

8.43 First let us consider the alkyl bromide. The molecular formula $C_5H_{11}Br$ indicates an open-chain saturated skeleton. A singlet (area 6) arises from

$(CH_3)_2C\!<$. A quartet (area 2) and a triplet (area 3) arise from CH_3CH_2—. Therefore, the bromide is $CH_3CH_2CBr(CH_3)_2$.

The spectrum of the alcohol is not consistent with the carbon skeleton of this bromide; therefore, a carbocation rearrangement must have occurred. The singlet (area 1) must come from OH. The doublet (area 3) is from $CH_3CH\!<$. The doublet (area 6) is from $(CH_3)_2CH$—. The multiplets (areas 1) are from —CH— and —CH—. The only feasible structure for the alcohol, plus the mechanism for its rearrangement, follow:

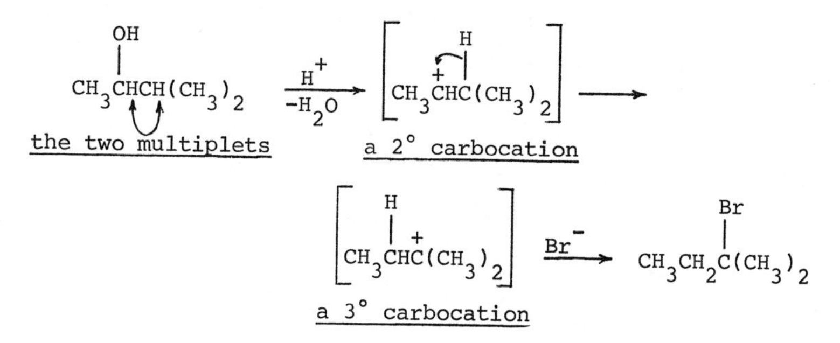

8.44 (a) The formula $C_8H_{10}O_2$ indicates extensive unsaturation or ring(s). In such a case, check the aryl region of the nmr spectrum. This compound does contain an aromatic ring. The infrared spectrum shows OH absorption at about 3370 cm^{-1} (3.0 μm), but no C═O absorption. At least one of the oxygens in the formula is accounted for. The second oxygen could be in a second OH group or in an OR group (an ether).

The nmr spectrum indicates a 1,4-disubstituted benzene ring (the two doublets in the $\delta = 7.0$ ppm region). The singlet at $\delta = 3.7$ ppm must arise from a CH_3 group (area 3) bonded to O (because the signal is shifted so far downfield). The pieces determined so far are:

—OH —C₆H₄— CH_3O—

Because of the CH_3O— group, there is only one OH in the molecule. In the nmr spectrum, this is the peak at $\delta = 3.1$ ppm. The remaining CH_2 in the formula gives rise to the singlet at 4.4 ppm. It must be bonded to O (because of its chemical shift).

The only way the pieces can be fit together is as follows:

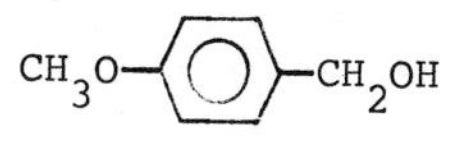

(b) This is another example of a compound containing an aromatic ring. The infrared spectrum shows C═O absorption (1680 cm^{-1}, 5.95 μm). Also note the CH peak at about 2700 cm^{-1} (3.7 μm). This absorption suggests an aldehyde. Indeed, the nmr spectrum shows offset absorption characteristic of RCHO or RCO_2H. (This compound, however, could not be RCO_2H. Why not?) The nmr spectrum also indicates a 1,4-disubstituted benzene ring (the pair of doublets in the aryl region.) The fragments determined so far are:

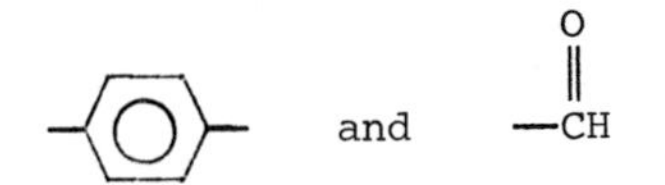

This leaves only C_3H_7O to account for. The triplet at $\delta = 1.0$ ppm must arise from a CH_3 group (area 3) bonded to a CH_2 group (because its multiplicity is 3). One CH_2 group in the molecule exhibits a sextet in the nmr spectrum: $CH_3\underline{CH_2}CH_2$—. The CH_2 group that shows a triplet is shifted downfield; therefore, it must be attached to the other O: $CH_3CH_2CH_2O$—. Now a complete formula may be written:

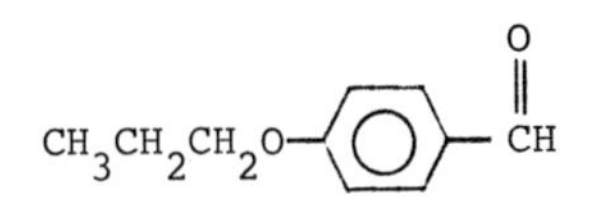

(c) The infrared spectrum shows strong carbonyl absorption at about 1730 cm^{-1} (5.8 μm), which accounts for one of the two oxygens. The second oxygen could be involved in a second C═O or in OR (as an ether or an ester), but not in an OH group. (Why not?)

Let us consider the nmr spectrum. The downfield septet arises from an isopropyl group, $(CH_3)_2\underline{CH}$—, which must be bonded to O (not to C═O because it is so far downfield). The upfield doublet is therefore from $(\underline{CH_3})_2CHO$—. The two triplets must come from —CH_2CH_2—. Because one of them is shifted downfield to $\delta = 3.75$ ppm, the group must be —CH_2CH_2Cl. (The downfield CH_2 group is not bonded to O, because the two O's have been accounted for.) The fragments of the molecule are:

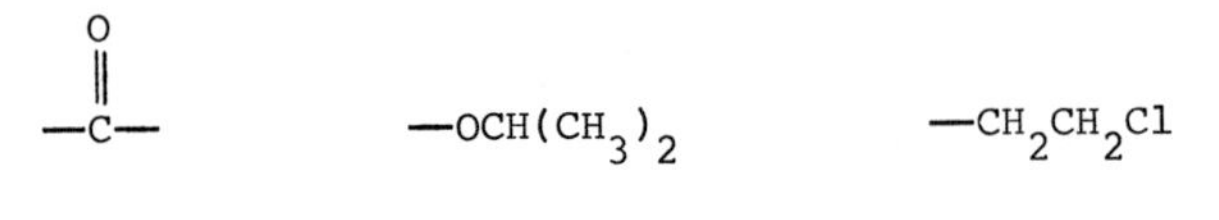

The formula for the compound is $ClCH_2CH_2\overset{O}{\overset{\|}{C}}OCH(CH_3)_2$.

(d) The infrared spectrum shows C═O, but no OH. The nmr spectrum shows CH_3 and CH_2, accounting for <u>half</u> of the ten protons in the molecule. Thus, the fragments are

—CH_3 —CH_3 (equivalent and not split by CH_2)   —CH_2— —CH_2— (equivalent and not split by CH_3) $—\overset{O}{\overset{\|}{C}}—$

The remaining C and O must be in a second C═O group, and the compound must be:

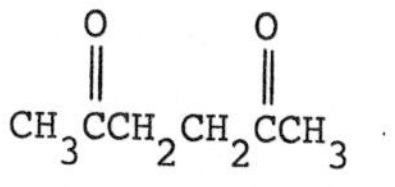

$$CH_3\overset{O}{\overset{\|}{C}}CH_2CH_2\overset{O}{\overset{\|}{C}}CH_3$$

Any other arrangement of the fragments would result in splitting between the CH_2 and CH_3 groups.

(e) The infrared spectrum shows strong OH absorption and strong C═O absorption. The compound is not a carboxylic acid (or an aldehyde) because the nmr spectrum shows no offset absorption. The nmr spectrum shows a singlet fairly far downfield at $\delta = 3.75$ ppm. This singlet (area 3) must be due to CH_3O— because of its large chemical shift. Now the three oxygens in the molecule have been identified: —OH, >C═O, and —OCH_3. The singlet at 3.3 ppm represents the OH proton. The upfield doublet (area 3) arises from ($\underline{CH_3}CH$<). The downfield quartet is thus from $CH_3\underline{CH}$<. This CH must also be bonded to O. (Why?) The fragments are:

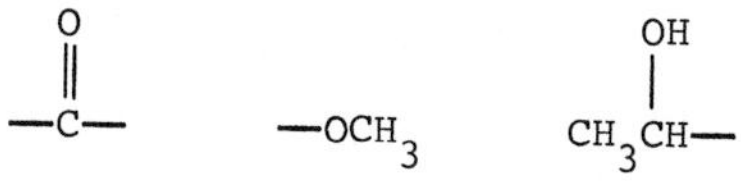

$$-\overset{O}{\overset{\|}{C}}- \qquad -OCH_3 \qquad CH_3\overset{OH}{\overset{|}{C}H}-$$

The compound is thus $CH_3\overset{OH}{\overset{|}{C}H}-\overset{O}{\overset{\|}{C}}OCH_3$.

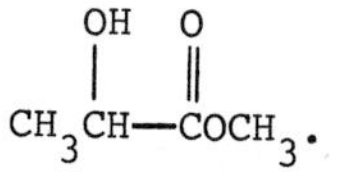

(f) The infrared spectrum shows C═O absorption as the only truly significant peak. The CH region of the spectrum indicates that the compound is not an aldehyde. The nmr spectrum confirms this. Thus, the compound must be a ketone.

The nmr spectrum indicates a phenyl group, C_6H_5— (the singlet at $\delta = 7.4$ ppm, area 5). The nmr spectrum also shows a CH_3 group not split by neighboring protons. The chemical shift of the CH_3 signal implies that the CH_3 group is bonded to the C═O group. The parts of the molecule determined so far are:

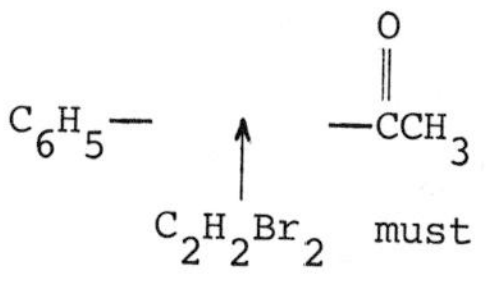

$$C_6H_5- \quad \uparrow \quad -\overset{O}{\overset{\|}{C}}CH_3$$

$C_2H_2Br_2$ must fit here

Because only two protons are unaccounted for, the absorption in the 5.0-ppm region must be two doublets and not a quartet. The middle of the molecule is thus —CHBrCHBr—, and the entire structure is:

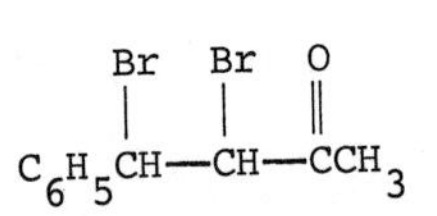

(g) The infrared spectrum shows NH_2 absorption, leaving us with a partial structural formula of $C_4H_9NH_2$. Because of the number of hydrogens ($2\underline{n} + 1$), we know that the alkyl portion of the molecule contains no double bonds or rings. The nmr spectrum is sufficiently complex that it cannot be completely analyzed. The downfield sextet arises from $CH_3\overset{|}{C}\underline{H}CH_2$—. Because of its shift, we can assume that the CH carbon is bonded to N. Therefore, the remaining CH_3 group is bonded to the CH_2:

$$\begin{array}{c} NH_2 \\ | \\ CH_3CHCH_2CH_3 \end{array}$$

CHAPTER 9

Alkenes and Alkynes

Some Important Features

The principal reactions discussed in Chapter 9 are additions of reagents to carbon-carbon double bonds. These addition reactions can be grouped together according to the types of intermediates formed. Addition of HX or H_2O usually proceeds by way of the more stable carbocation to yield Markovnikov products.

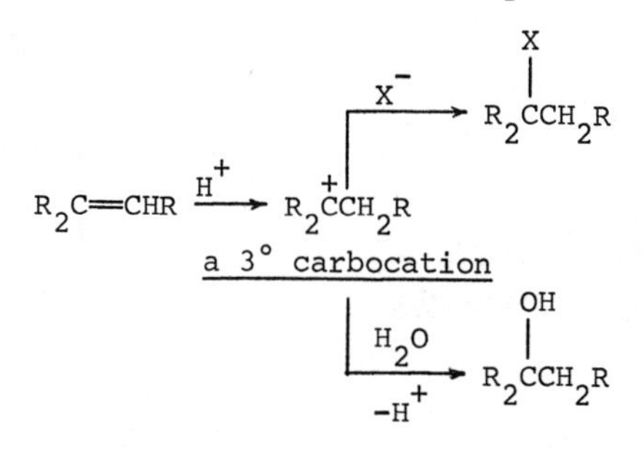

A carbocation can undergo rearrangement if a more stable carbocation will result. In predicting products from carbocation reactions, always inspect the structure of the intermediate carbocation to see if it can rearrange to a more stable intermediate.

1,4-Addition reactions yield allylic cations and often result in mixtures of products. To predict the principal products, look at the more stable carbocations and at their resonance structures. The principal products arise from the major contributors.

Addition reactions of X_2 or $Hg(O_2CCH_3)_2$ proceed through bridged intermediates, which result in anti-addition. No rearrangements occur, but Markovnikov's rule is still followed.

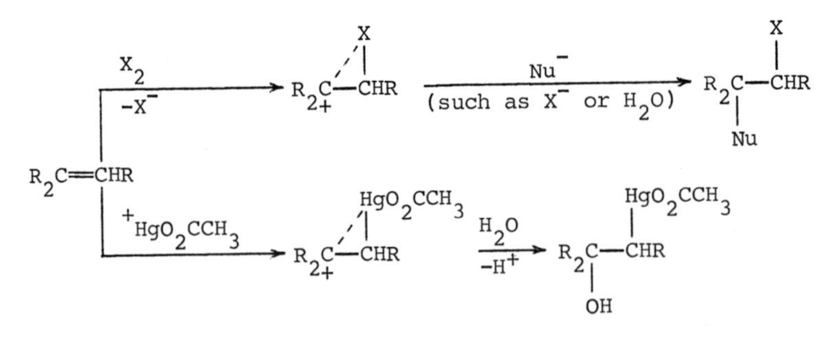

Other types of addition reactions are those of HBr with peroxides (a free-radical, anti-Markovnikov reaction); BH_3 (which yields what appear to be anti-Markovnikov products); singlet methylene (syn-addition); and H_2 (syn-addition).

Oxidation reactions of alkenes can lead to a variety of products, depending on the reagent and on the degree of alkene substitution.

$$R_2C{=}CHR \xrightarrow[\text{syn-addition}]{\text{cold } KMnO_4 \text{ solution or (1) } OsO_4\text{, (2) } Na_2SO_3} R_2C(OH){-}CHR(OH)$$

$$R_2C{=}CHR \xrightarrow{\text{hot } KMnO_4 \text{ solution}} R_2C{=}O + HOCR \;(\text{with } C{=}O)$$

A Diels-Alder reaction (Section 9.16) is a type of 1,4-addition reaction of conjugated polyenes. In a Diels-Alder reaction, a six-membered ring is formed and the position of the one double bond in this ring is different from either of the two starting positions.

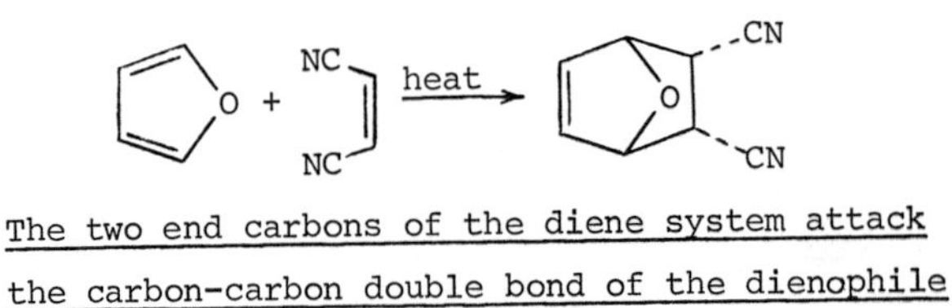

The two end carbons of the diene system attack the carbon-carbon double bond of the dienophile

Reminders

In an addition reaction that proceeds by way of a carbocation (or a bridged intermediate with carbocation character), check the relative stabilities of all possible intermediates. The nucleophile will attack predominantly the most positive carbon.

2° 3° allylic and benzylic →
increasing carbocation stability

Weak nucleophiles (like water, an alcohol, or an anion) can attack a carbocation.

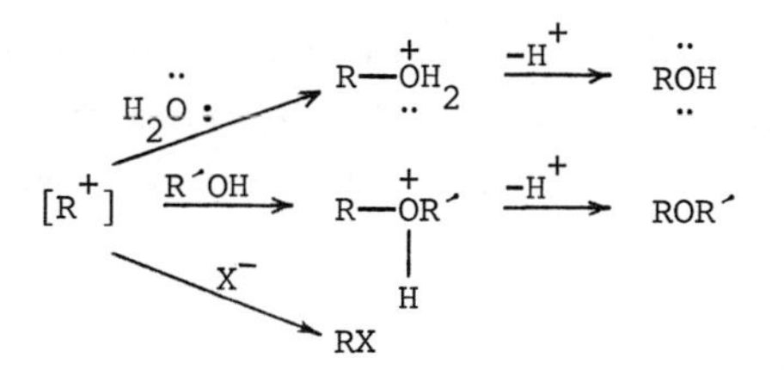

Reactions that proceed through true carbocations are not stereospecific. However, anti-addition reactions, which proceed by way of bridged ions, and syn-addition reactions are stereospecific.

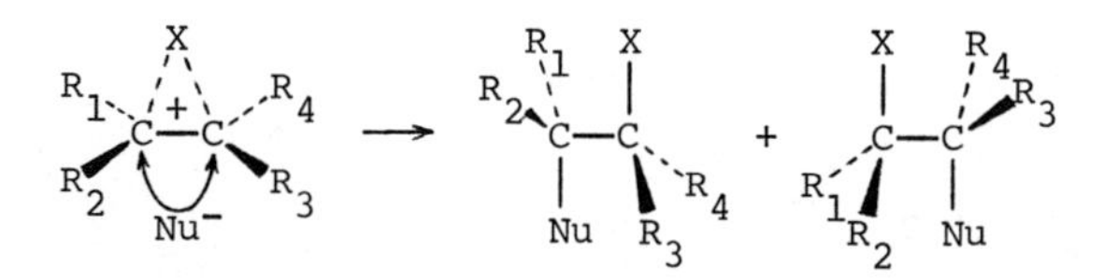

Answers to Problems

9.42 (a) 3-bromo-1-propene (b) 3-methyl-1-butyne (c) 2-methyl-3-penten-1-ol
(d) 1,3-butadiene (e) 5,5-dichloro-1,3-cyclohexadiene
(f) propenoic acid

9.43

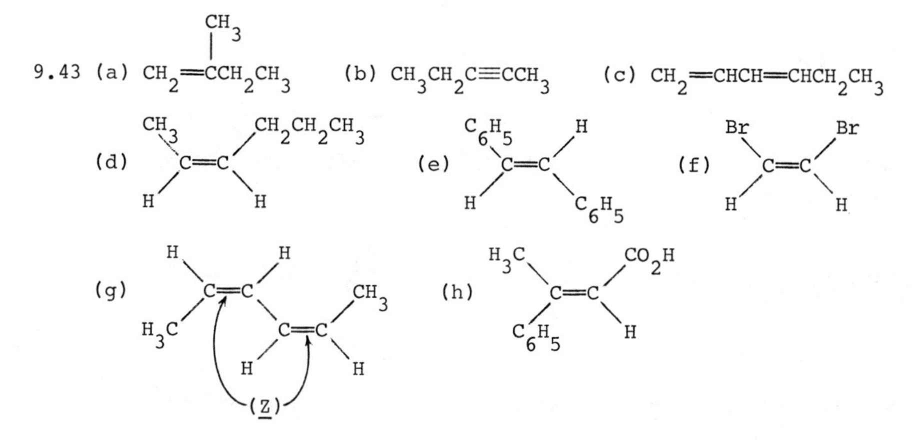

9.44 (a) neither (b) (Z) (c) (Z) (d) (E)

Compound (a) has no geometric isomers; therefore, the (E) and (Z) system is not applicable. For (c), the higher priority groups on each side of the double bond are circled:

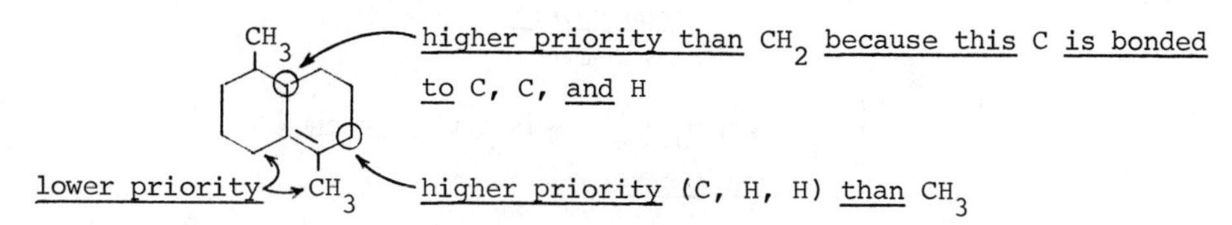

9.45 Elimination by way of (a) would yield the desired product in the higher yield (Saytseff rule). Elimination by way of (b) would also yield a large proportion of 3-methyl-1-cyclohexene, but only when the bromo and methyl groups are <u>cis</u> on the ring.

<u>trans</u>: (cyclohexane with CH_3, H, H, Br) $\xrightarrow{-HBr}$ (cyclohexene)—CH_3
<u>only product</u>

<u>cis</u>: (cyclohexane with CH_3, Br, H, H) $\xrightarrow{-HBr}$ (cyclohexene)—CH_3 + (cyclohexene)—CH_3
<u>major product</u>

9.46 (a) $CH_3C{\equiv}CH \xrightarrow{NaNH_2} CH_3C{\equiv}C^- \; Na^+ \xrightarrow{CH_3I} CH_3C{\equiv}CCH_3$

(b) $CH_3C{\equiv}CH \xrightarrow{CH_3MgI} CH_3C{\equiv}CMgI \xrightarrow[(2)\ H_2O,H^+]{(1)\ \text{cyclohexanone}}$ product

(c) $CH_3C{\equiv}CH \xrightarrow{NaNH_2} CH_3C{\equiv}C^- \; Na^+ \xrightarrow{D_2O}$ product or

$CH_3C{\equiv}CH \xrightarrow{CH_3MgI} CH_3C{\equiv}CMgI \xrightarrow{D_2O}$ product

9.47 (a) $CHCl{=}CClCH_2CH_2CH_3$ (b) $CHCl_2CCl_2CH_2CH_2CH_3$

(c) $CH{\equiv}CCH_2CH_2CH_3 \xrightarrow{HCl} CH_2{=}CClCH_2CH_2CH_3 \xrightarrow{HCl} CH_3CCl_2CH_2CH_2CH_3$
<u>Markovnikov product</u> <u>Markovnikov product</u>

(d) $BrMgC{\equiv}CCH_2CH_2CH_3 + C_6H_6$

(e) $CH{\equiv}CCH_2CH_2CH_3 \xrightarrow[-NH_3]{NaNH_2} Na^+ \; ^-C{\equiv}CCH_2CH_2CH_3 \xrightarrow[-NaI]{CH_3I} CH_3C{\equiv}CCH_2CH_2CH_3$

9.48 (a) $CH_3CHICH_2CH_2CH_3$

(b) The most stable carbocation is formed initially, followed by reaction with the nucleophile:

$CH_2{=}CHCH{=}CHCH_3 \xrightarrow[\text{Markovnikov}]{H^+}$

$\left[CH_3\overset{+}{C}HCH{=}CHCH_3 \longleftrightarrow CH_3CH{=}CH\overset{+}{C}HCH_3\right]$
<u>equivalent</u>

+

$\left[CH_2{=}CH\overset{+}{C}HCH_2CH_3 \longleftrightarrow \overset{+}{C}H_2CH{=}CHCH_2CH_3\right]$
<u>major contributor</u>

$\xrightarrow{I^-} CH_3CH{=}CHCHICH_3 + CH_2{=}CHCHICH_2CH_3$ + some $ICH_2CH{=}CHCH_2CH_3$

The other possible intermediate carbocation is not resonance-stabilized and thus is not formed in any appreciable amounts.

$$\left[CH_2{=}CHCH_2\overset{+}{C}HCH_3\right]$$

not resonance-stabilized

(c) As in (b), first determine the more stable carbocation intermediates and their resonance structures; then, determine the products:

$$(CH_3)_2CICH{=}CH_2 + CH_2{=}\underset{\displaystyle CH_3}{\underset{|}{C}}CHICH_3 + ICH_2\underset{\displaystyle CH_3}{\underset{|}{C}}{=}CHCH_3$$

and some $(CH_3)_2C{=}CHCH_2I$

(d) $(CH_3)_2\underset{\displaystyle CH_3}{\underset{|}{C}}ICH(CH_2)_3CH_3$ + some $(CH_3)_3CCH_2CHICH_2CH_2CH_3$

from rearrangement

9.49 (a) $CH_2{=}CHCl$, because it leads to the more stabilized carbocation (relatively speaking) $CH_3\overset{+}{C}H \rightarrow Cl$ and not $Cl \leftarrow \overset{+}{C}H \rightarrow CH_2Cl$.

(b) $CH_3CH{=}C(CH_3)_2$, because it yields a more stabilized tertiary carbocation.

9.50 (1) (a), (c), (b). The order is based upon the stabilities of the carbocations formed by protonation.

(2) (a) $\left[(CH_3)_2CH^+\right] \longrightarrow (CH_3)_2CHOSO_3H$

(b) $\left[(CH_3)_3C^+\right] \longrightarrow (CH_3)_3COSO_3H$

(c) $\left[CH_3\overset{+}{C}HCH_2CH_3\right] \longrightarrow CH_3\underset{}{\overset{\displaystyle OSO_3H}{\overset{|}{C}}}HCH_2CH_3$

9.51 (a) $(CH_3CH_2)_2\overset{\displaystyle OH}{\overset{|}{C}}CH_3$ (b) $(CH_3)_3COCH_2CH_3$ (c) $(CH_3)_2\overset{\displaystyle OH}{\overset{|}{C}}\underset{\displaystyle CH_3}{\underset{|}{C}}HCH_2CH_2CH_3$

(d) $CH_3\overset{\displaystyle OH}{\overset{|}{C}}HCH_2CH_3$ (e) 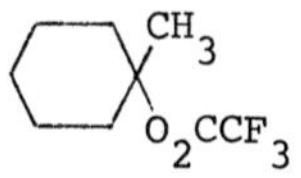 1-methylcyclohexyl with CH_3 and O_2CCF_3 on the same ring carbon

In (c), rearrangement occurs. In (d), $CH_3CHICH_2CH_3$ will be formed in small amounts. In (e), the alkene is protonated and then attacked by the nucleophilic $CF_3CO_2^-$.

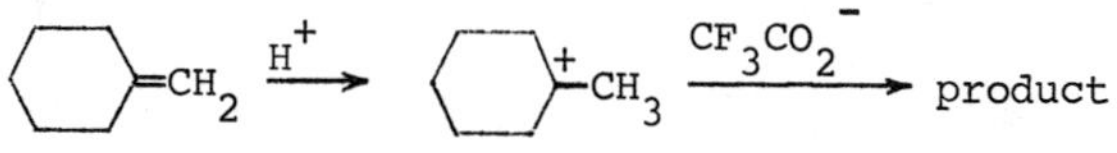

$$\text{cyclohexane}{=}CH_2 \xrightarrow{H^+} \text{cyclohexyl}^+{-}CH_3 \xrightarrow{CF_3CO_2^-} \text{product}$$

9.52

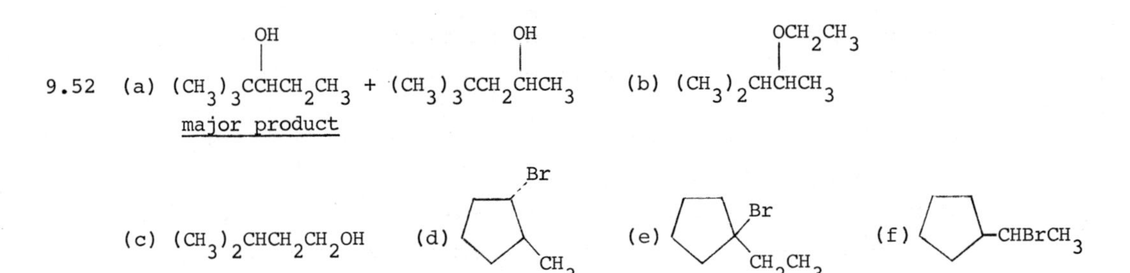

Reaction (f) is a free-radical reaction that proceeds through an intermediate tertiary free radical:

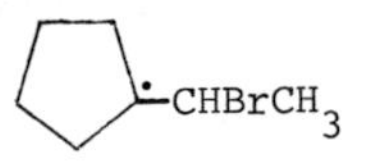

9.53 (a) (2R,3R)- and (2S,3S)-$CH_3CHClCHClCH_2CH_3$

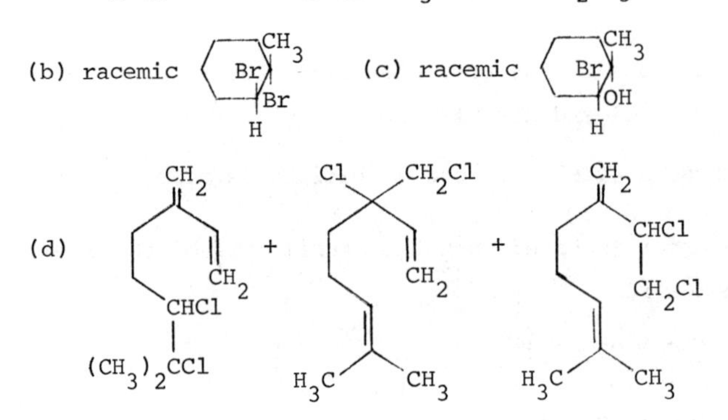

In (a), (b), and (c), the stereochemistry is a result of anti-addition (see Section 9.11). In (d), the predominant products are those from intermediates with tertiary or allylic carbocation character. For example:

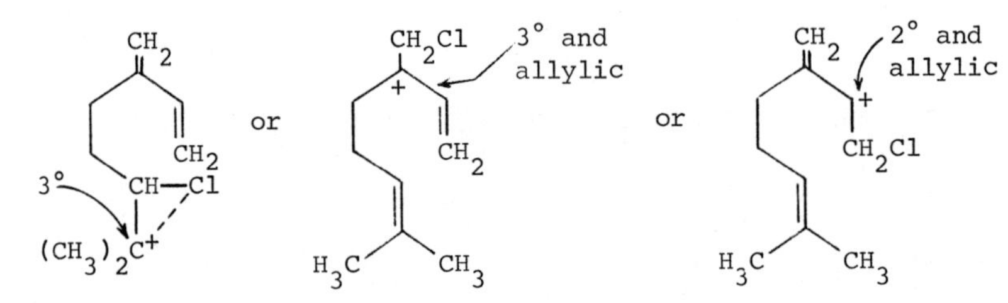

Some 1,4-addition products might also be observed in (d).

9.54

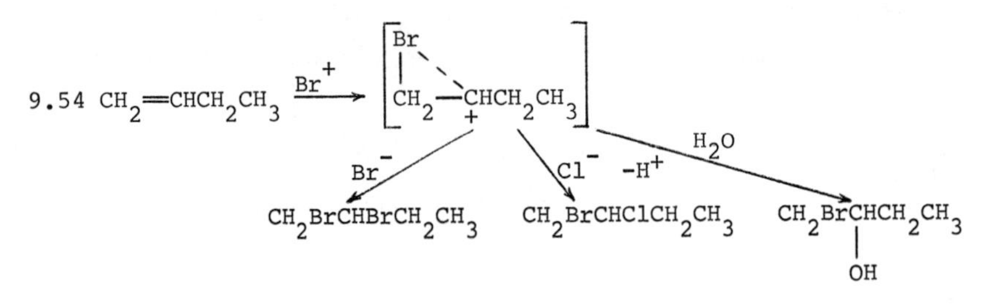

9.55 (a) H_2, Pt (b) $Br_2 + H_2O$

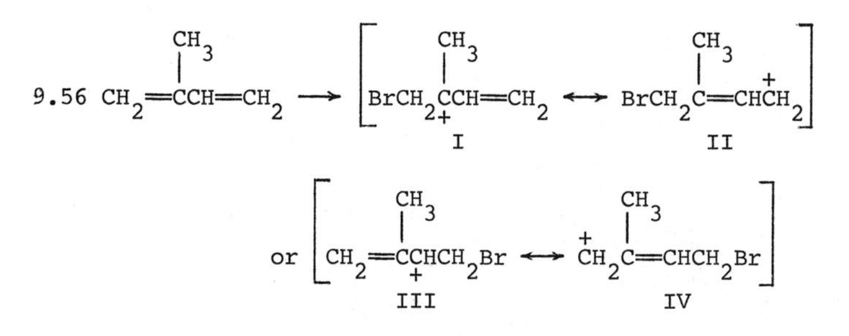

II and IV (primary carbocations) are not major contributors; therefore, the observed products are:

$$BrCH_2\overset{\displaystyle CH_3}{\overset{|}{C}}BrCH{=}CH_2 \; + \; CH_2{=}\overset{\displaystyle CH_3}{\overset{|}{C}}CHBrCH_2Br$$

(Note that II and IV would lead to the same product, and that cis and trans isomers are possible.)

9.57 $(CH_3)_3\overset{+}{N}{-}\overset{-}{\ddot{C}}H_2 \xrightarrow{\text{heat}} (CH_3)_3N: + :CH_2 \xrightarrow{\text{cyclohexene}}$ bicyclo[4.1.0]heptane

9.58 (a) methylenecyclohexane ($C_6H_{10}{=}CH_2$) $\xrightarrow[(2)\ NaBH_4]{(1)\ Hg(O_2CCH_3)_2,\ H_2O}$ (b) methylenecyclohexane ($C_6H_{10}{=}CH_2$) $\xrightarrow[(2)\ H_2O_2,\ OH^-]{(1)\ BH_3}$

(c) methylenecyclohexane ($C_6H_{10}{=}CH_2$) $\xrightarrow[\text{or (1) } BH_3 \text{ (2) } Br^-,\ OH^-]{HBr,\ \text{peroxide}}$ $C_6H_{11}{-}CH_2Br \xrightarrow[CH_3OH]{Na^+\ {}^-OCH_3}$

9.59 (a) $CH_3C{\equiv}CH \xrightarrow{H_2O,\ Hg^{2+}} \left[CH_3\overset{\displaystyle OH}{\overset{|}{C}}{=}CH_2\right] \longrightarrow CH_3\overset{\displaystyle O}{\overset{\|}{C}}CH_3$

(b) $CH_3C{\equiv}CH \xrightarrow[(2)\ CH_3I]{(1)\ NaNH_2} CH_3C{\equiv}CCH_3 \xrightarrow{H_2O,\ Hg^{2+}}$

(c) $CH_3C{\equiv}CH \xrightarrow{CH_3MgI} CH_3C{\equiv}CMgI \xrightarrow[(2)\ H_2O,\ H^+]{(1)\ CH_3CHO}$

$$CH_3C{\equiv}C\overset{\displaystyle OH}{\overset{|}{C}}HCH_3 \xrightarrow{Na} CH_3C{\equiv}C\overset{\displaystyle O^-\ Na^+}{\overset{|}{C}}HCH_3 \xrightarrow{CH_3I}$$

9.60 (a)

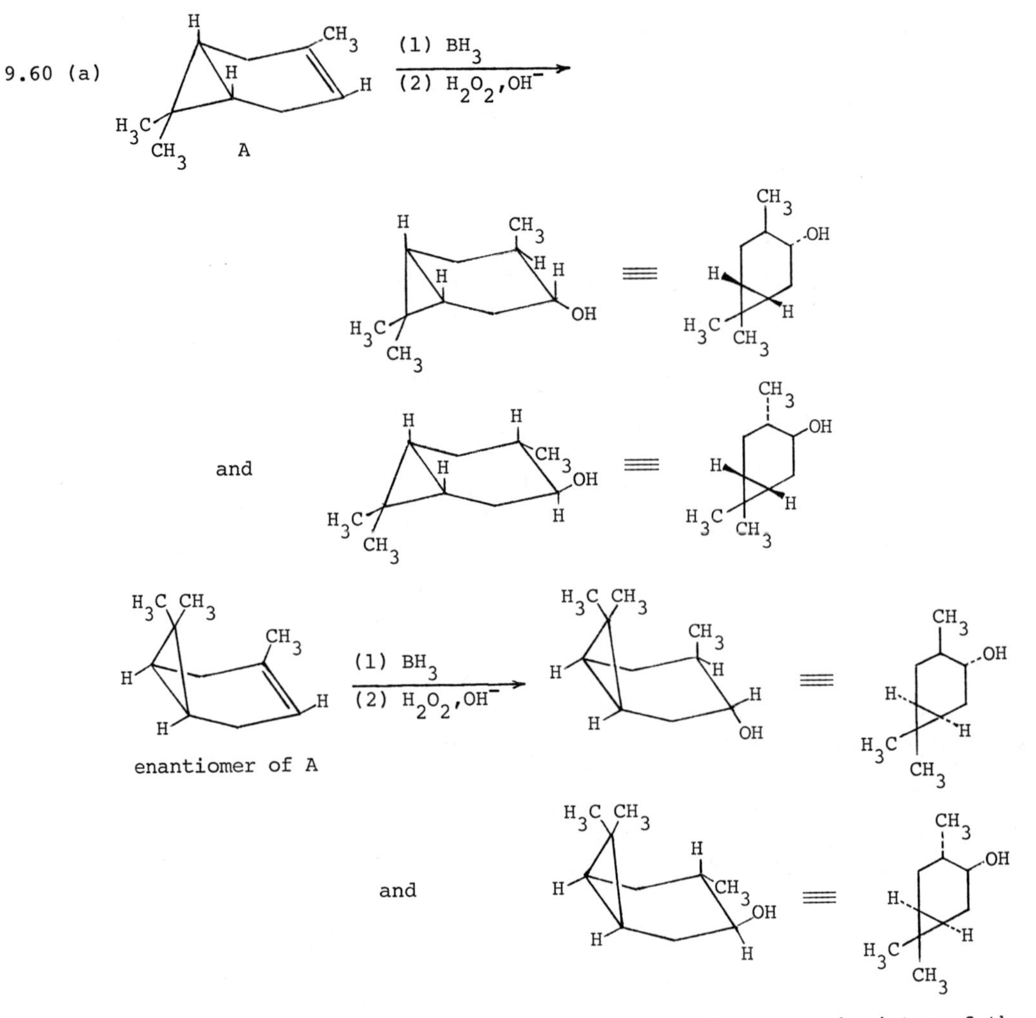

The stereochemistry of the products depends on the stereochemistry of the starting chiral cycloalkene.

(b)

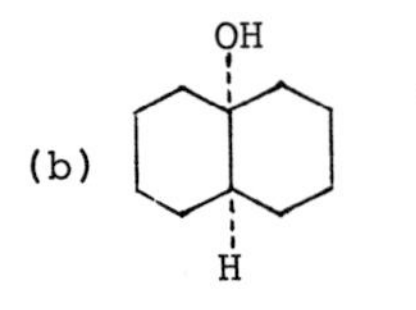

9.61 (a)

attack trans to bridge

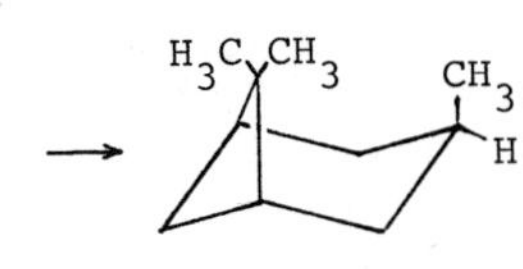

cis product

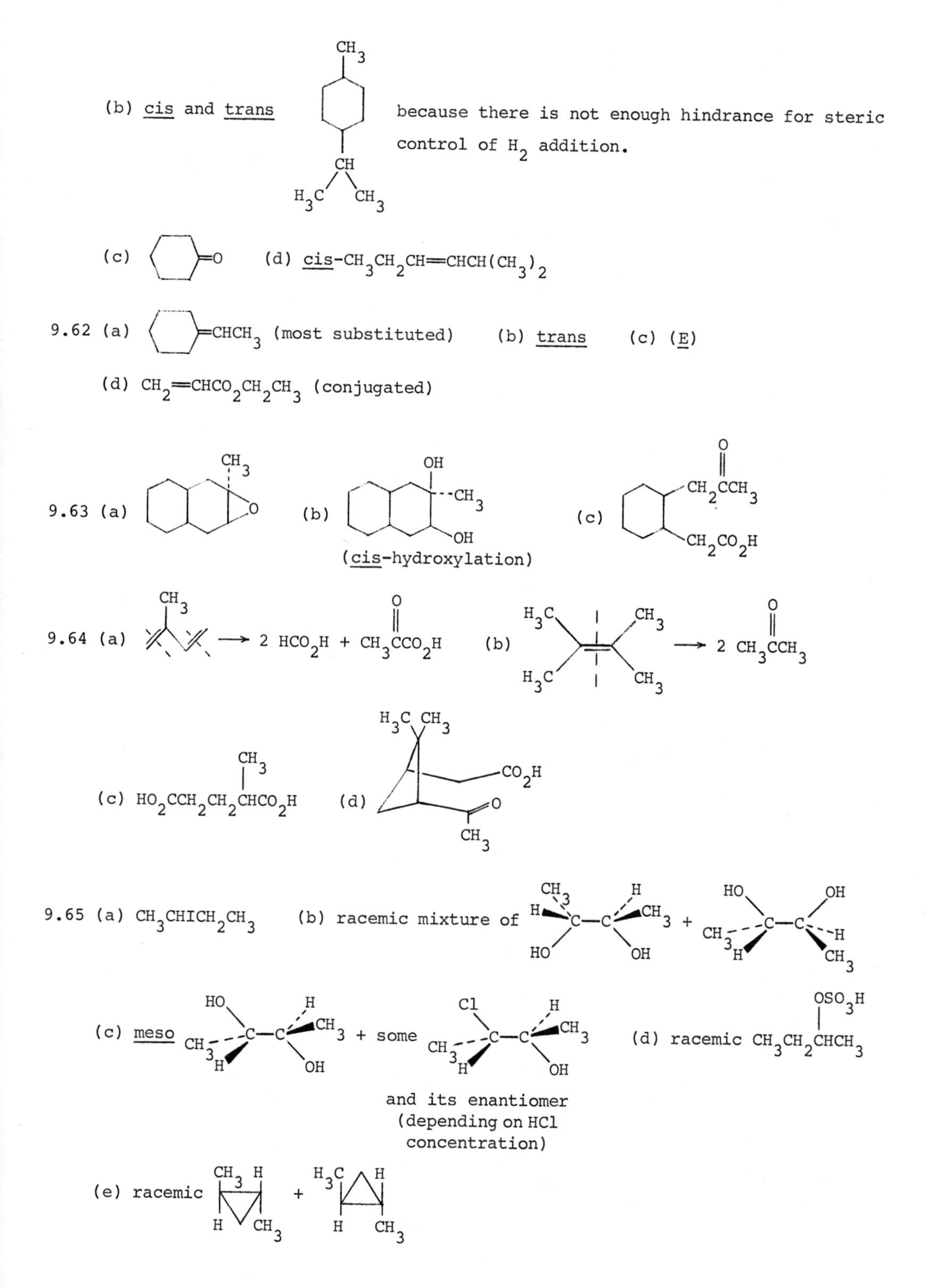
(b) cis and trans
because there is not enough hindrance for steric control of H2 addition.
(c)
(d) cis-CH3CH2CH=CHCH(CH3)2
9.62 (a) =CHCH3 (most substituted) (b) trans (c) (E)
(d) CH2=CHCO2CH2CH3 (conjugated)
9.63 (a) (b) (c)
(cis-hydroxylation)
9.64 (a) → 2 HCO2H + CH3CCO2H (b) → 2 CH3CCH3
(c) HO2CCH2CH2CHCO2H (d)
9.65 (a) CH3CHICH2CH3 (b) racemic mixture of
(c) meso + some
(d) racemic CH3CH2CHCH3
and its enantiomer
(depending on HCl
concentration)
(e) racemic

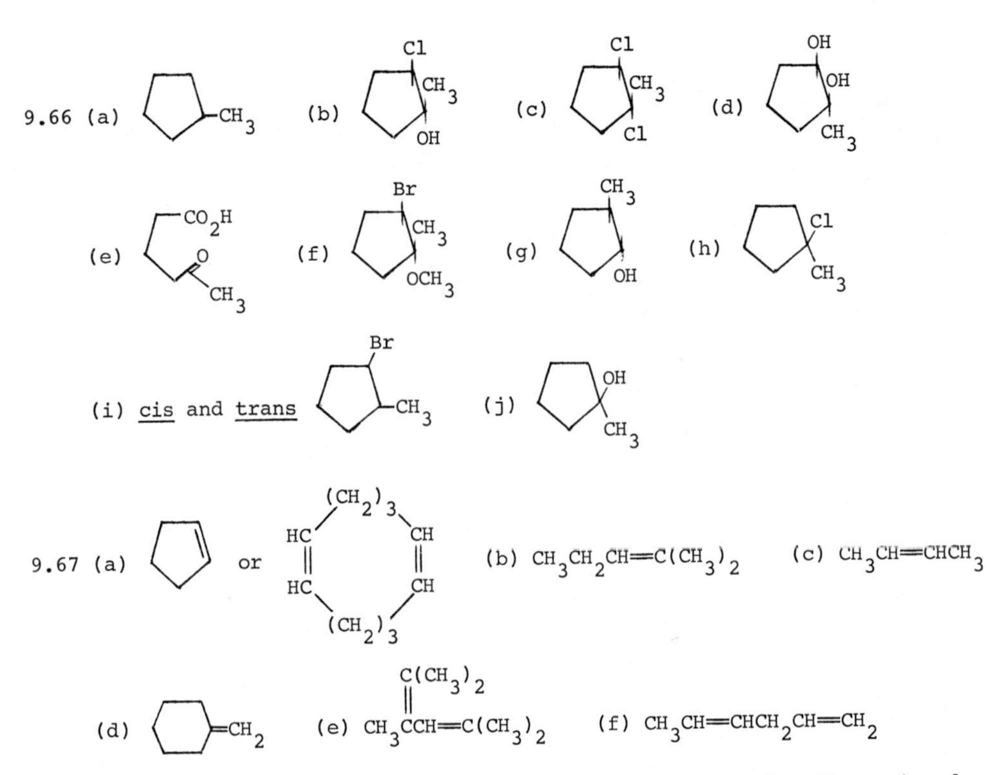

9.66 (a) ... —CH₃ (b) ... (c) ... (d) ...

(e) ... (f) ... (g) ... (h) ...

(i) <u>cis</u> and <u>trans</u> ... (j) ...

9.67 (a) ... or ... (b) $CH_3CH_2CH\!=\!C(CH_3)_2$ (c) $CH_3CH\!=\!CHCH_3$

(d) ...$=CH_2$ (e) $CH_3CCH\!=\!C(CH_3)_2$ with $C(CH_3)_2$ (f) $CH_3CH\!=\!CHCH_2CH\!=\!CH_2$

To solve this type of problem, redraw the products so that the carbonyl groups are close together (because two carbonyl groups result from one double bond). For example,

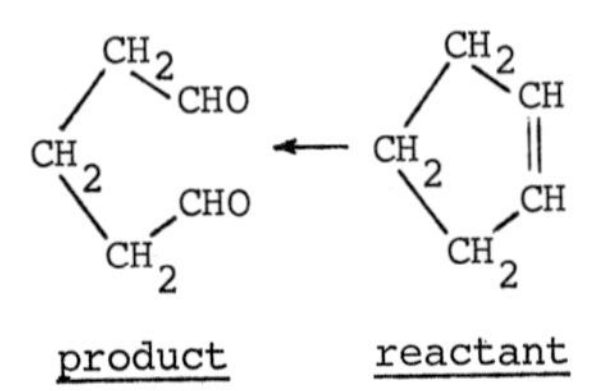

<u>product</u> <u>reactant</u>

9.68 ...—$CH\!=\!CH_2$

9.69 CH_3— ... —CH_3

9.70 (a) ... (b) no (c) ...

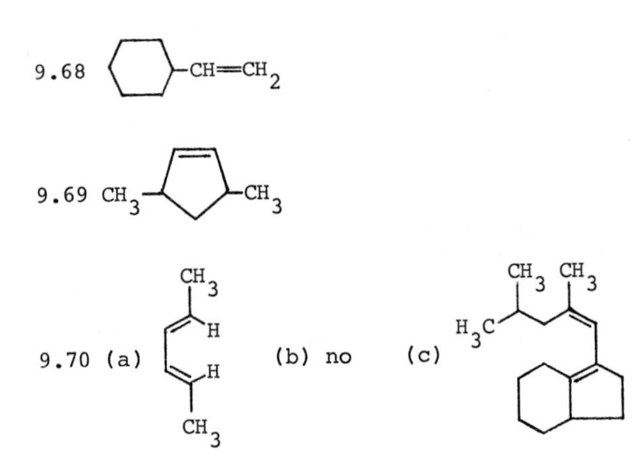

9.71

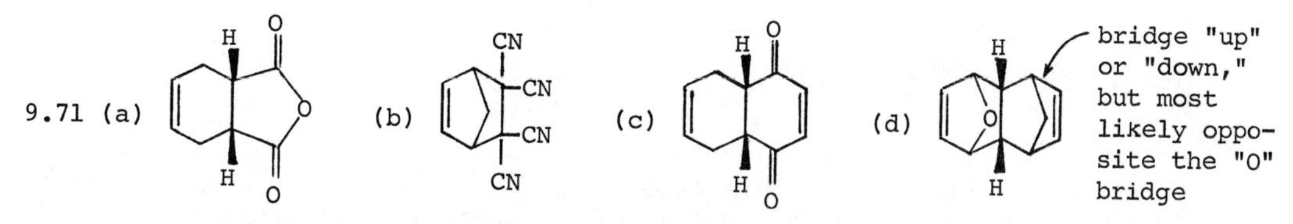

(e) A triple bond, rather than a double bond, is the reactive site of the dienophile. The product still contains a double bond at this position.

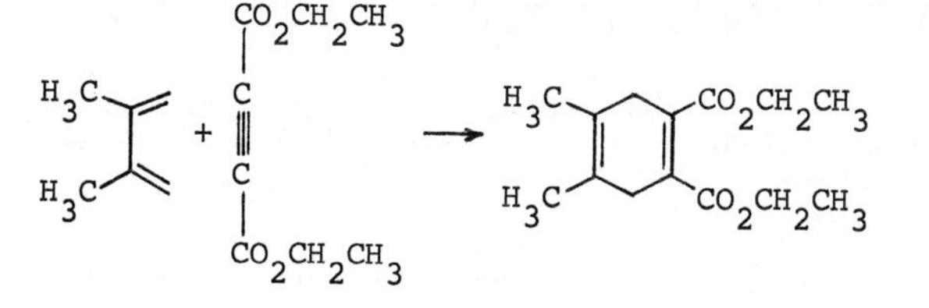

9.72

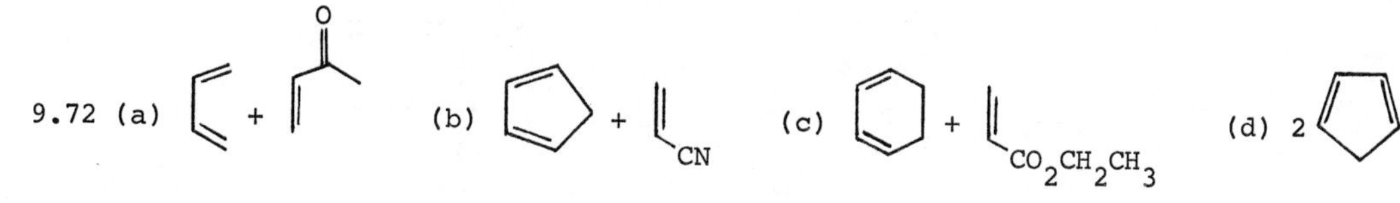

9.73

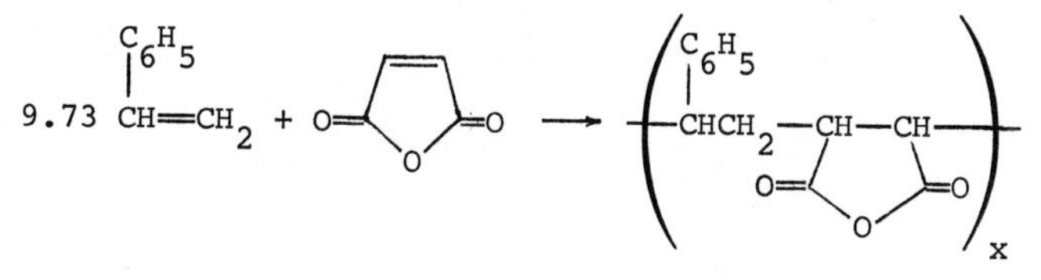

9.74 (a) $CH_3CH{=}CH_2 \xrightarrow{H^+} [(CH_3)_2\overset{+}{C}H] \xrightarrow{CH_2{=}CHCH_3}$

$[(CH_3)_2CHCH_2\overset{+}{C}HCH_3] \xrightarrow{-H^+} (CH_3)_2CHCH{=}CHCH_3$

(b) $CH_3CH_2C(CH_3){=}CH_2 \xrightarrow{H^+} [CH_3CH_2\overset{+}{C}(CH_3)_2] \xrightarrow{CH_2{=}C(CH_3)CH_2CH_3} [CH_3CH_2C(CH_3)_2{-}CH_2\overset{+}{C}(CH_3)CH_2CH_3] \xrightarrow{-H^+}$

$CH_3CH_2C(CH_3)_2CH{=}C(CH_3)CH_2CH_3 + CH_3CH_2C(CH_3)_2CH_2C(CH_3){=}CHCH_3$

9.75 (a) $CH_3CHBrCH_3 \xrightarrow[\text{heat}]{\text{KOH},\ CH_3CH_2OH} CH_3CH{=}CH_2 \xrightarrow{Br_2} CH_3CHBrCH_2Br$

(b) $CH_3CH_2CH_2OH \xrightarrow[\text{heat}]{H_2SO_4} CH_3CH{=}CH_2 \xrightarrow{HBr} CH_3CHBrCH_3$

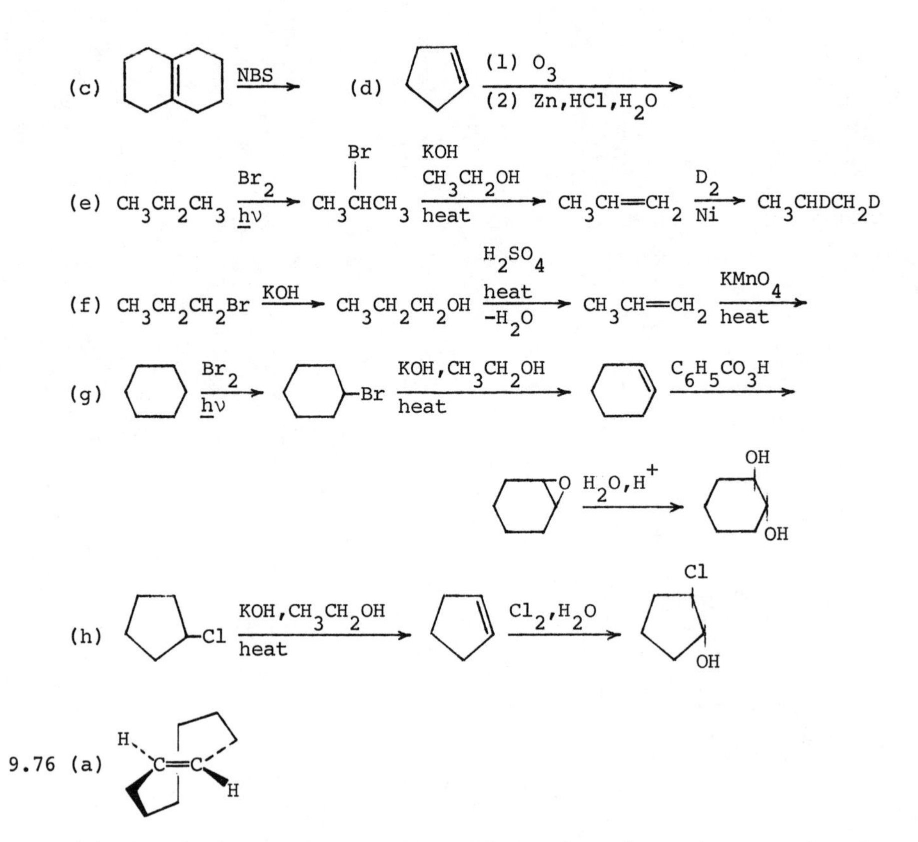

9.76 (a)

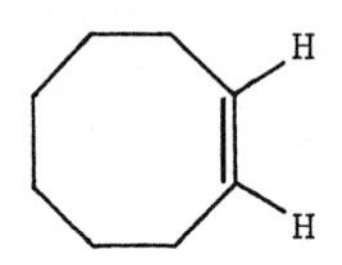

(b) The cis-isomer is superimposable on its mirror image; therefore, it does not exist as an enantiomer. (Try it with models.)

9.77 The products from cis-3-hexene are diastereomers of those from trans-3-hexene. A pair of enantiomers [(3R,4R) and (3S,4S)] are obtained from cis-3-hexene, while a different pair of enantiomers [(3S,4R) and (3R,4S)] are obtained from the trans isomer.

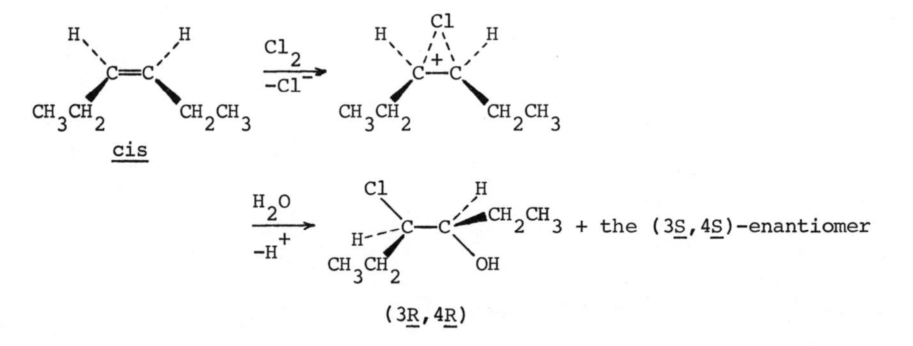

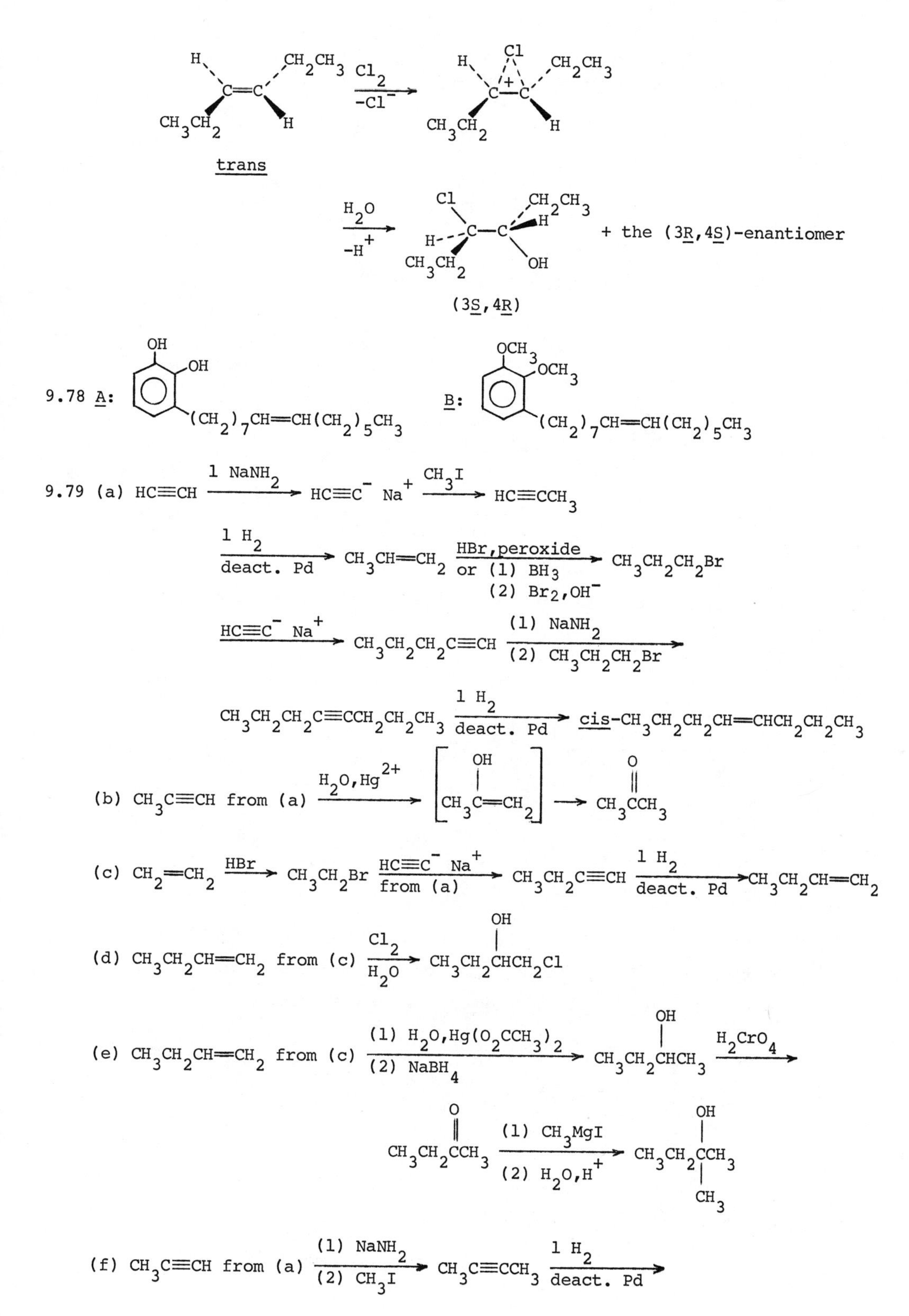
trans
Cl_2
$-Cl^-$
H_2O
$-H^+$
+ the (3R,4S)-enantiomer
(3S,4R)
9.78 A:
$(CH_2)_7CH=CH(CH_2)_5CH_3$
B:
$(CH_2)_7CH=CH(CH_2)_5CH_3$
9.79 (a) $HC\equiv CH$
1 $NaNH_2$
$HC\equiv C^-\ Na^+$
CH_3I
$HC\equiv CCH_3$
1 H_2
deact. Pd
$CH_3CH=CH_2$
HBr, peroxide
or (1) BH_3
(2) Br_2,OH^-
$CH_3CH_2CH_2Br$
$HC\equiv C^-\ Na^+$
$CH_3CH_2CH_2C\equiv CH$
(1) $NaNH_2$
(2) $CH_3CH_2CH_2Br$
$CH_3CH_2CH_2C\equiv CCH_2CH_2CH_3$
1 H_2
deact. Pd
cis-$CH_3CH_2CH_2CH=CHCH_2CH_2CH_3$
(b) $CH_3C\equiv CH$ from (a)
H_2O,Hg^{2+}
CH_3CCH_3
(c) $CH_2=CH_2$
HBr
CH_3CH_2Br
$HC\equiv C^-\ Na^+$
from (a)
$CH_3CH_2C\equiv CH$
1 H_2
deact. Pd
$CH_3CH_2CH=CH_2$
(d) $CH_3CH_2CH=CH_2$ from (c)
Cl_2
H_2O
$CH_3CH_2CHCH_2Cl$
(e) $CH_3CH_2CH=CH_2$ from (c)
(1) $H_2O,Hg(O_2CCH_3)_2$
(2) $NaBH_4$
$CH_3CH_2CHCH_3$
H_2CrO_4
$CH_3CH_2CCH_3$
(1) CH_3MgI
(2) H_2O,H^+
$CH_3CH_2CCH_3$
(f) $CH_3C\equiv CH$ from (a)
(1) $NaNH_2$
(2) CH_3I
$CH_3C\equiv CCH_3$
1 H_2
deact. Pd

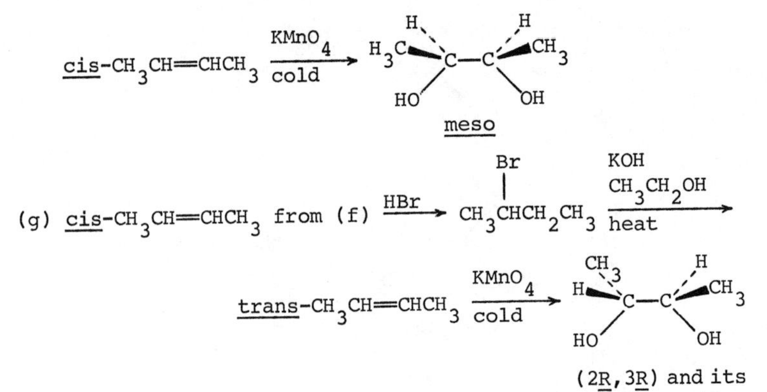

9.80 (a)

(b)

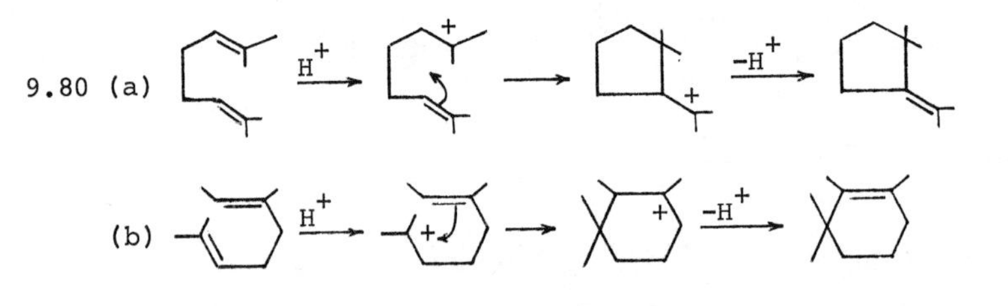

9.81 The diastereomers would be the (2R,3R) and the (2R,3S)-tetraols.

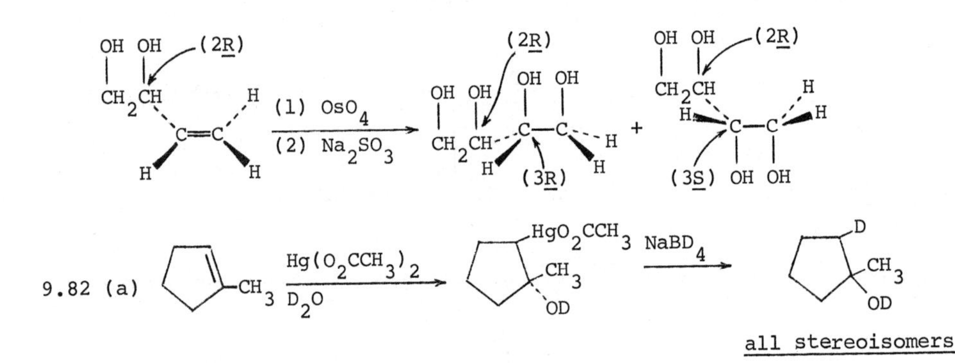

The reduction of the intermediate mercury compound is not stereospecific; therefore, D can be cis or trans to OD in the product.

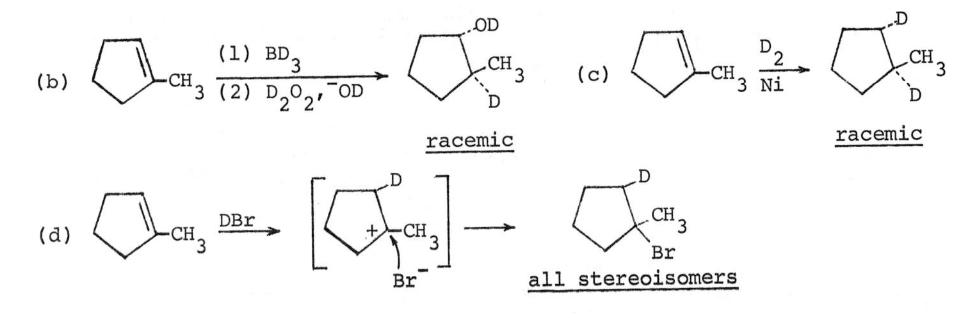

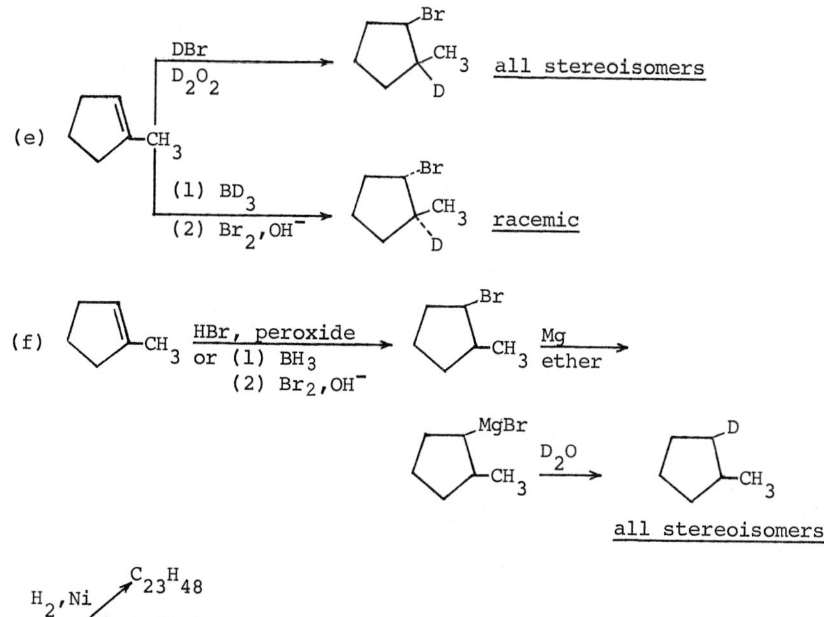

9.83 $C_{23}H_{46}$ $\xrightarrow{H_2,\,Ni}$ $C_{23}H_{48}$

$C_{23}H_{46}$ $\xrightarrow{\text{hot } KMnO_4}$ $CH_3(CH_2)_{12}CO_2H + HO_2C(CH_2)_7CH_3$

$C_{23}H_{46}$ $\xrightarrow{Br_2}$ $C_{23}H_{46}Br_2$ (a pair of enantiomers)

The formula $C_{23}H_{46}$ ($C_{\underline{n}}H_{2\underline{n}}$) tells us that the structure contains one double bond or one ring. The H_2 reaction and the $KMnO_4$ reaction tell us that the structure contains one double bond. The $KMnO_4$ reaction shows the location of the double bond:

$$CH_3(CH_2)_{12}CH{=}CH(CH_2)_7CH_3$$

The fact that the Br_2 reaction yields a pair of enantiomers tells us that the housefly sex attractant is either <u>cis</u>-$CH_3(CH_2)_{12}CH{=}CH(CH_2)_7CH_3$ or the <u>trans</u>-isomer, and not a mixture of both. (A mixture would lead to <u>two</u> pairs of enantiomers.) It has been determined by other means that this alkene is the <u>cis</u>-compound.

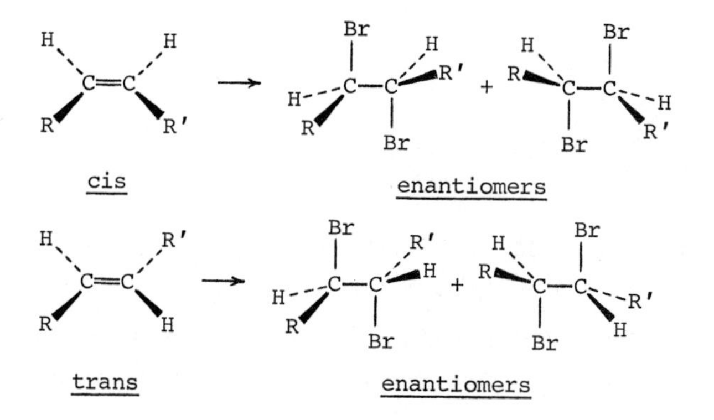

9.84 Each of the alkenes in (c) has the same number of alkyl substituents on the sp^2 carbon atoms. Also, in (c), one alkene contains an exocyclic double bond, and one contains an endocyclic double bond. The pair of alkenes in (c) is thus the best choice.

9.85 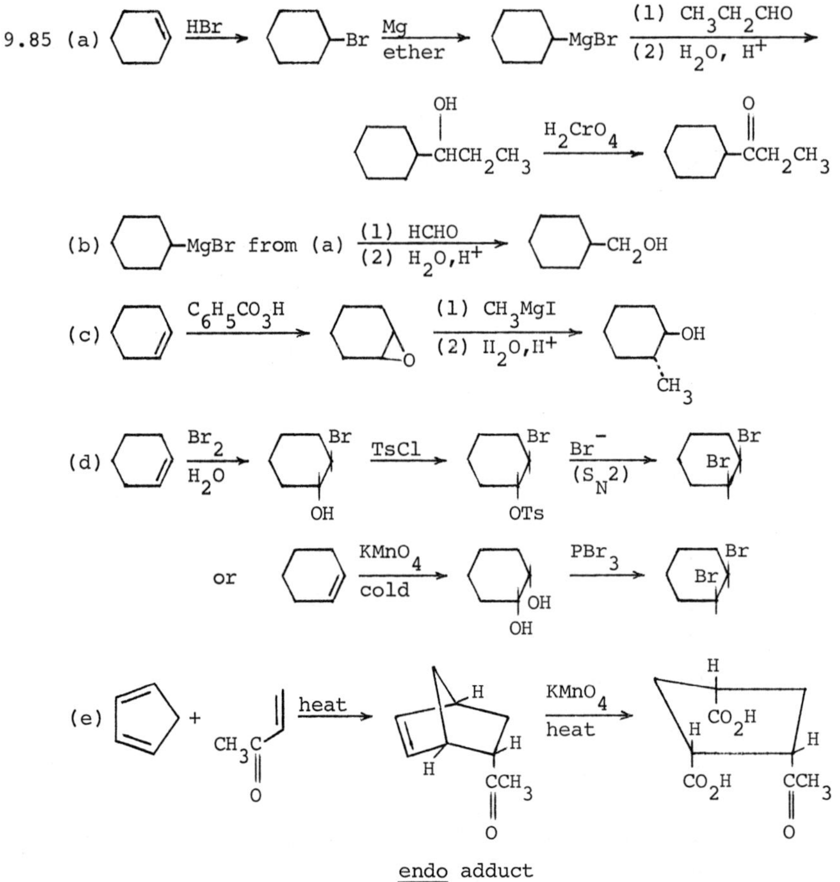

endo adduct

(If the stereochemistry eludes you, make models.)

9.86 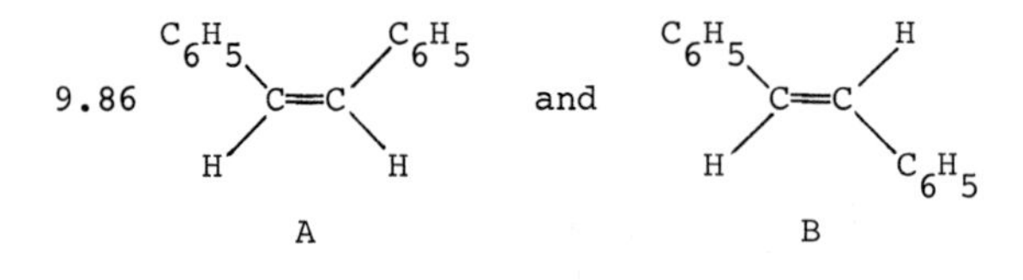

9.87 $(CH_3)_3CC{\equiv}CH \xrightarrow[\text{catalyst}]{H_2} (CH_3)_3CCH{=}CH_2$

A B

Note that the infrared cell was "overloaded." (The CH absorption runs off the bottom of the spectrum.) This fact should warn you that all the absorption in the spectrum is much stronger than would normally be observed. The peak at 1650 cm^{-1} (6.1 μm) is not a C=O peak; this peak is actually weak compared to the CH absorption.

CHAPTER 10

Aromaticity, Benzene, and Substituted Benzenes

Some Important Features

An aromatic compound is one that is substantially stabilized by pi-electron delocalization. To be aromatic, a compound must be cyclic and planar; each ring atom must have a p orbital perpendicular to the plane of the ring; and the number of pi electrons in the ring must fit the formula $4n + 2$, where n is an integer.

Benzene, a typical aromatic compound, undergoes a variety of electrophilic aromatic substitution reactions.

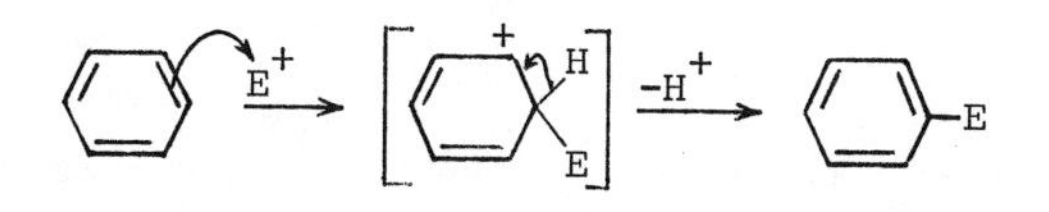

If the first substituent can donate electrons to the ring (by resonance or by the inductive effect), a second substitution occurs ortho and para because of stabilization of the intermediate (see Section 10.10).

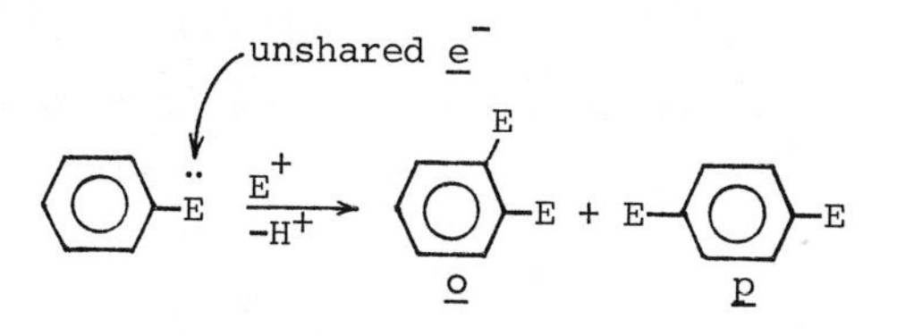

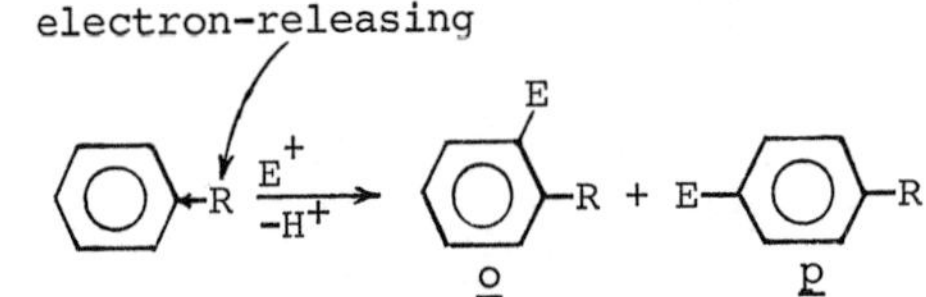

Except for the halogens, the o,p-directors activate the ring toward further electrophilic substitution.

If a first substituent cannot donate electronic charge to the ring, it deactivates the ring toward further electrophilic substitution. In this case, a second substitution occurs meta to the first substituent.

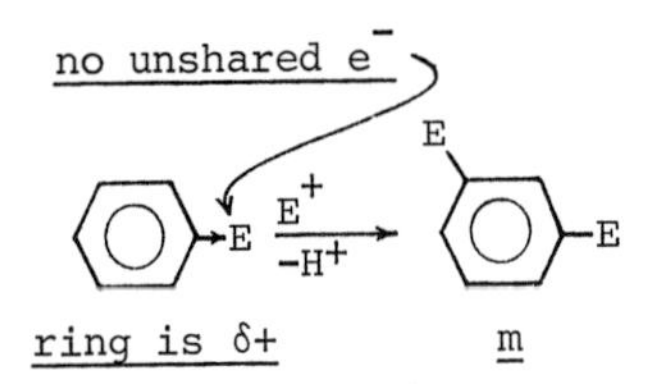

A ring substituted with an electron-withdrawing group is activated toward nucleophilic substitution, however.

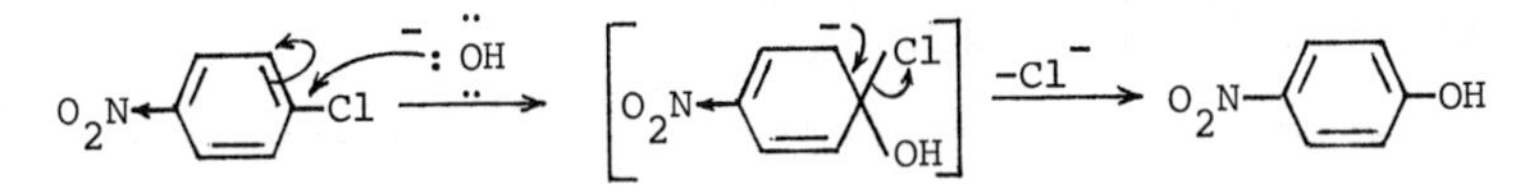

Aromatic nucleophilic substitutions can also occur under "forcing" conditions through a benzyne intermediate (Section 10.15).

The benzylic position of an alkylbenzene has enhanced reactivity toward oxidation and free-radical reactions (Section 10.12). Recall from Chapter 5 that benzylic halides are highly reactive in S_N1 and S_N2 reactions.

Primary arylamines ($ArNH_2$) can be converted to diazonium salts, which can be converted, in turn, to a variety of other compounds (Section 10.14).

Reminders

Except for alkyl and aryl groups, an o,p-director has unshared electrons on the atom adjacent to the ring.

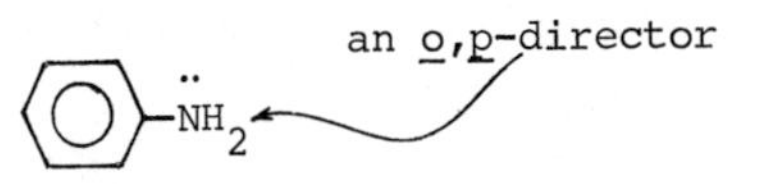

Release of electron density by a substituent activates a ring toward electrophilic substitution, while withdrawal of electrons deactivates a ring.

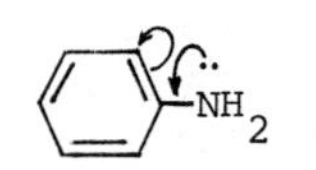

ring is δ- and activated toward E^+

NO_2

ring is δ+ and deactivated toward E^+

Answers to Problems

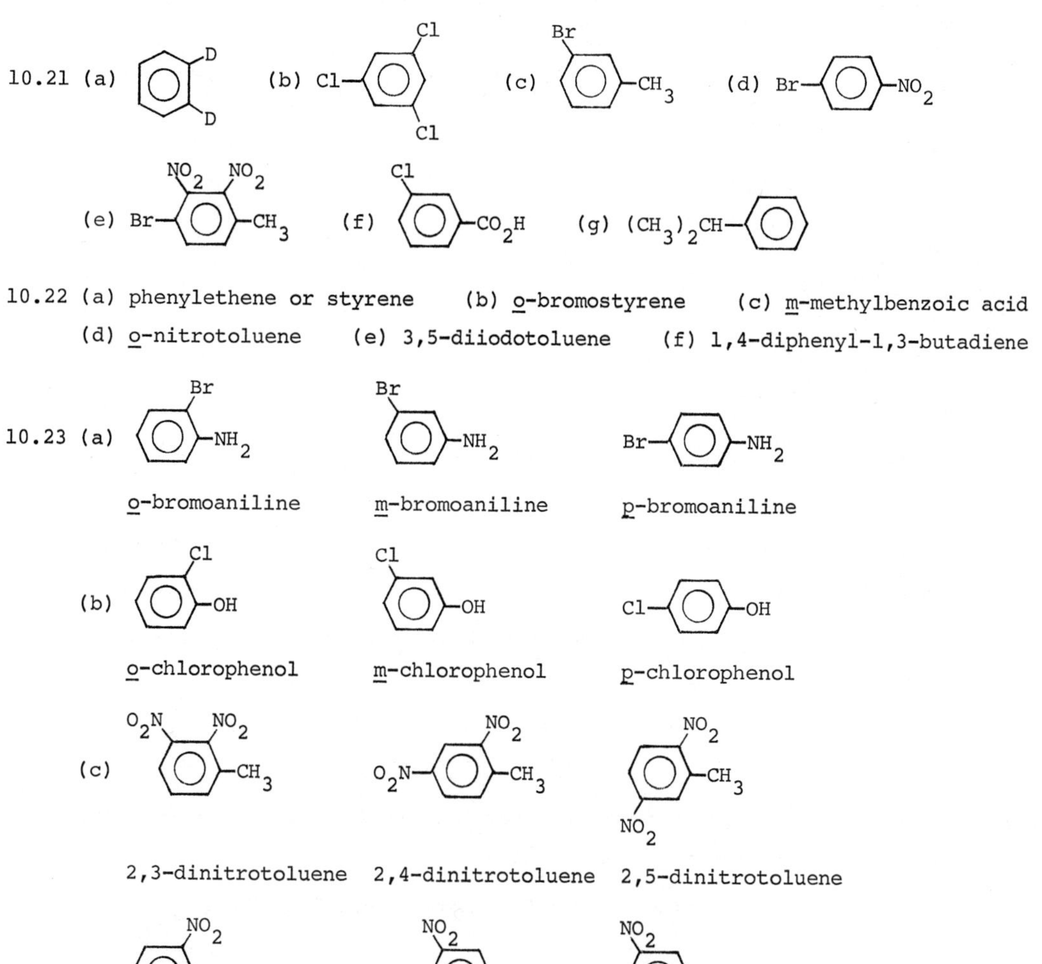

10.21 (a) (b) Cl—⬡—Cl (c) (d) Br—⬡—NO₂

(e) Br—⬡—CH₃ (f) (g) (CH₃)₂CH—⬡

10.22 (a) phenylethene or styrene (b) o-bromostyrene (c) m-methylbenzoic acid

(d) o-nitrotoluene (e) 3,5-diiodotoluene (f) 1,4-diphenyl-1,3-butadiene

10.23 (a)

o-bromoaniline m-bromoaniline p-bromoaniline

(b)

o-chlorophenol m-chlorophenol p-chlorophenol

(c)

2,3-dinitrotoluene 2,4-dinitrotoluene 2,5-dinitrotoluene

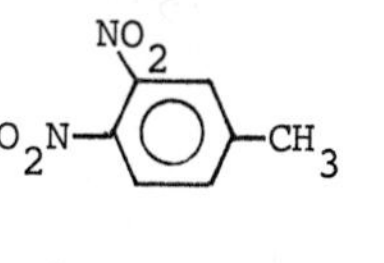

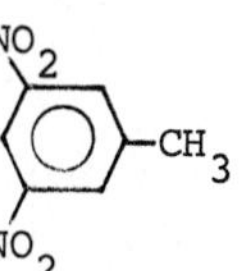

2,6-dinitrotoluene 3,4-dinitrotoluene 3,5-dinitrotoluene

10.24 All the examples shown are aromatic. Compound (a) has 14 pi electrons, compound (b) has ten, and compound (c) has ten.

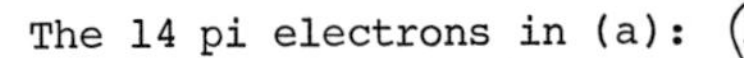

The 14 pi electrons in (a):

The 10 pi electrons in (c):

10.25

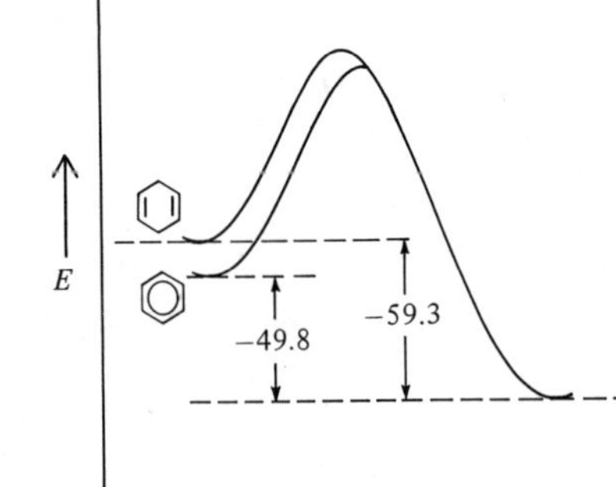

ΔE = -49.8 kcal/mole - (-59.3 kcal/mole) = +9.5 kcal/mole

Therefore, 9.5 kcal/mole must be supplied in the conversion of benzene to 1,4-cyclohexadiene.

10.26 Subtracting the ΔH of hydrogenation of benzene from that of styrene gives the ΔH of hydrogenation of C=C in styrene.

ΔH for C=C in styrene = -76.9 kcal/mole - (-49.8 kcal/mole)
= -27.1 kcal/mole

Subtracting the value for propene from this value shows the added stabilization of C=C in styrene by conjugation.

ΔH of stabilization = -27.1 kcal/mole - (-28.6 kcal/mole)
= 1.5 kcal/mole

Another way to approach this problem is to assume that a calculated ΔH for styrene is equal to the sum of ΔH for benzene and ΔH for propene:

ΔH (theory) for styrene = -49.8 kcal/mole + (-28.6 kcal/mole)
= -78.4 kcal/mole

Then the difference between the actual ΔH and this calculated ΔH is the added stabilization of the double bond in styrene.

-76.9 kcal/mole - (-78.4 kcal/mole) = 1.5 kcal/mole

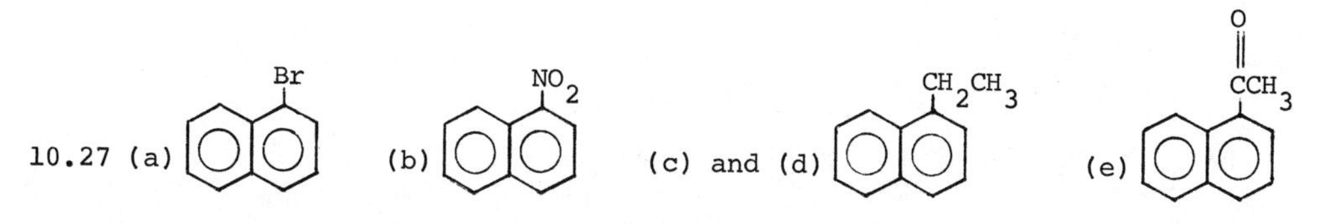

10.28 The rate of reaction for nitrobenzene is much less than the rate for benzene because of electron-withdrawal by the nitro group.

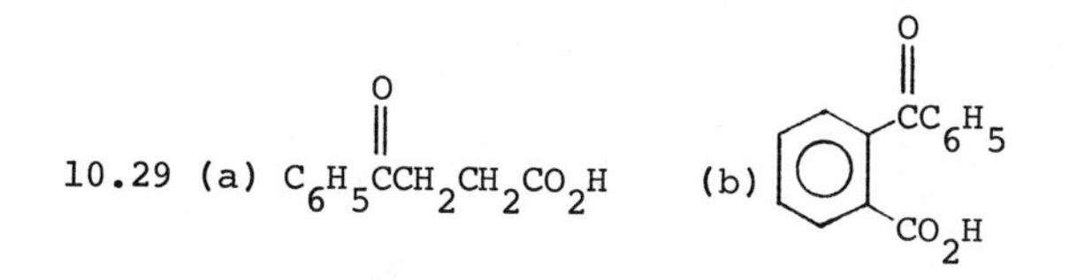

10.30 (a) acetanilide, because the N has unshared electrons that help stabilize the intermediate by resonance.

(b) toluene, because the CH_3 group is electron-releasing by the inductive effect.

(c) p-xylene, because it has two electron-releasing groups on the ring.

(d) m-nitrotoluene, because it has an electron-releasing CH_3 group.

(e) chlorobenzene, because it has only one electron-withdrawing group.

10.31 (a) o,p (b) m (c) o,p (d) m (e) m (f) o,p (g) o,p (h) m (i) o,p

10.32 (a) 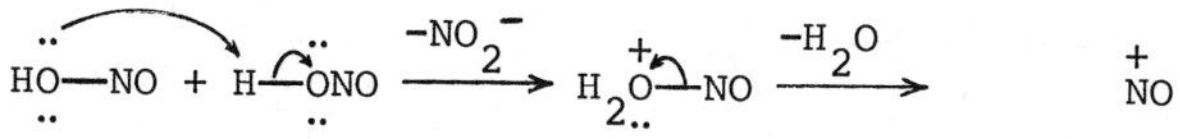 HO—NO + H—ONO $\xrightarrow{-NO_2^-}$ H_2O^+—NO $\xrightarrow{-H_2O}$ $\overset{+}{NO}$

the electrophile

(b) 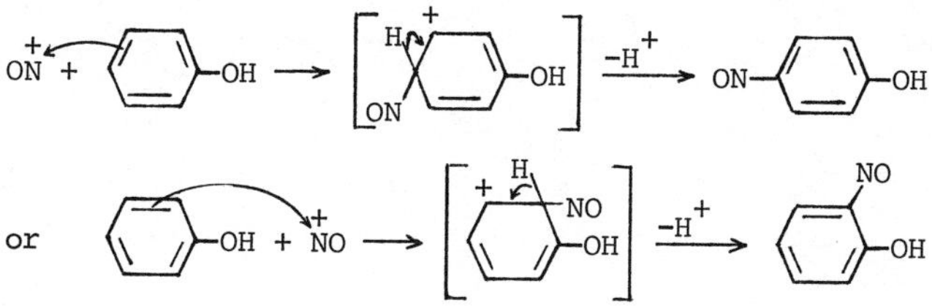 $\overset{+}{ON}$ + C_6H_5—OH → [H, ON, + —OH] $\xrightarrow{-H^+}$ ON—C_6H_4—OH

or C_6H_5—OH + $\overset{+}{NO}$ → [+ H, NO, OH] $\xrightarrow{-H^+}$ NO, C_6H_4—OH

10.33 (a) 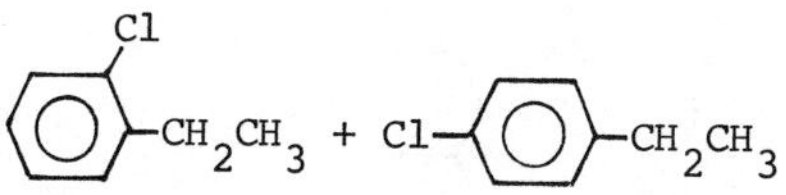 Cl, C_6H_4—CH_2CH_3 + Cl—C_6H_4—CH_2CH_3 (b) $C_6H_5CHBrCH_3$ (c) $C_6H_5CO_2H$

(d) 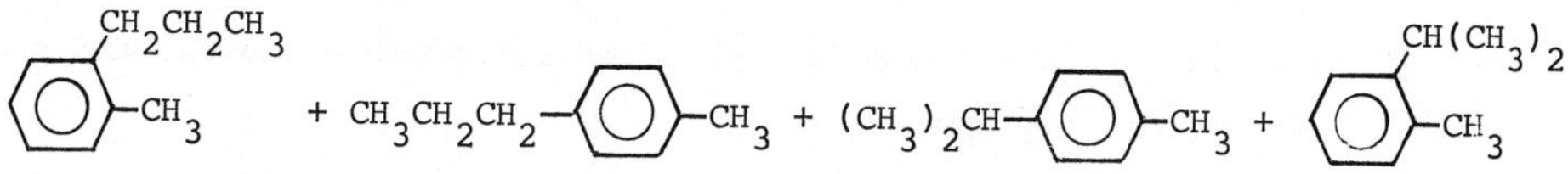 $CH_2CH_2CH_3$, C_6H_4—CH_3 + $CH_3CH_2CH_2$—C_6H_4—CH_3 + $(CH_3)_2CH$—C_6H_4—CH_3 + $CH(CH_3)_2$, C_6H_4—CH_3

(e) 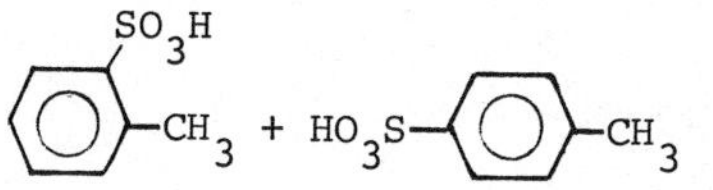 SO_3H, C_6H_4—CH_3 + HO_3S—C_6H_4—CH_3 (f) 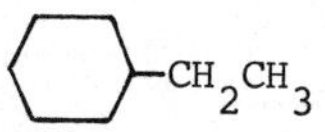 C_6H_{11}—CH_2CH_3

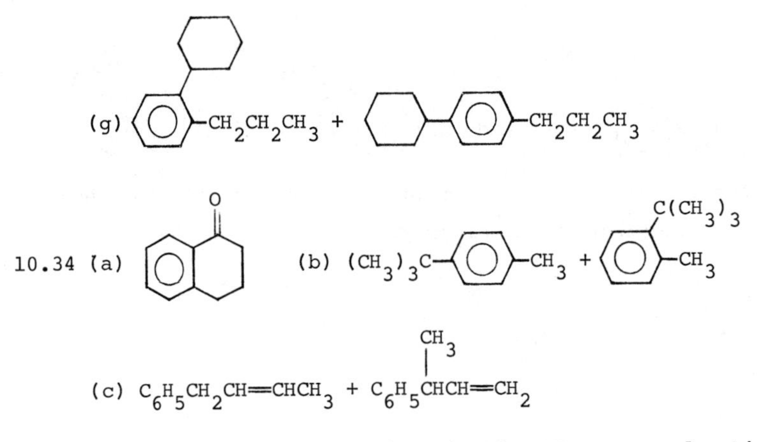

(c) $C_6H_5CH_2CH{=}CHCH_3$ + $C_6H_5CH(CH_3)CH{=}CH_2$

10.35 (a) The presence of two deactivating NO_2 groups deactivates the benzene ring sufficiently that a third substitution does not occur.

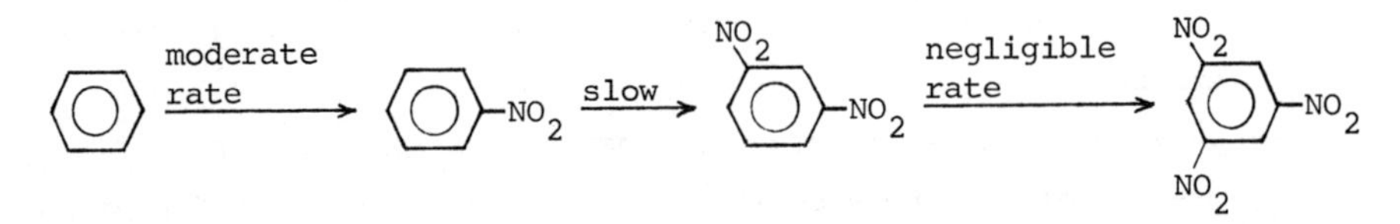

(b) The activating CH_3 group counteracts the deactivation by the two NO_2 groups so that a third substitution can occur.

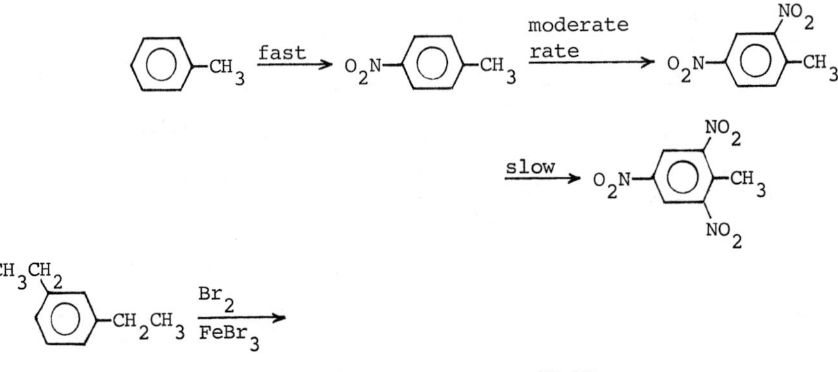

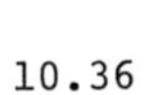

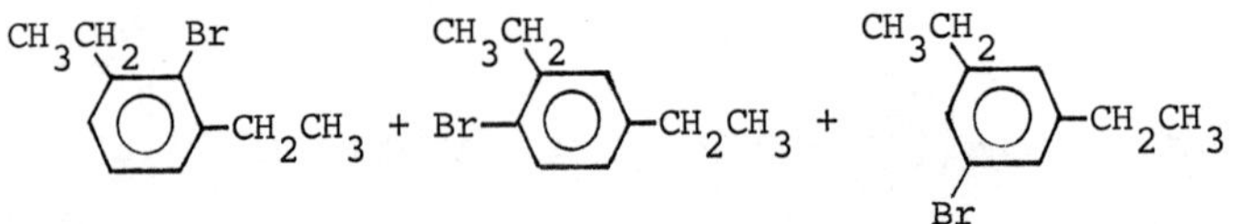

By contrast, o-diethylbenzene yields two monobromo products, and p-diethylbenzene yields only one.

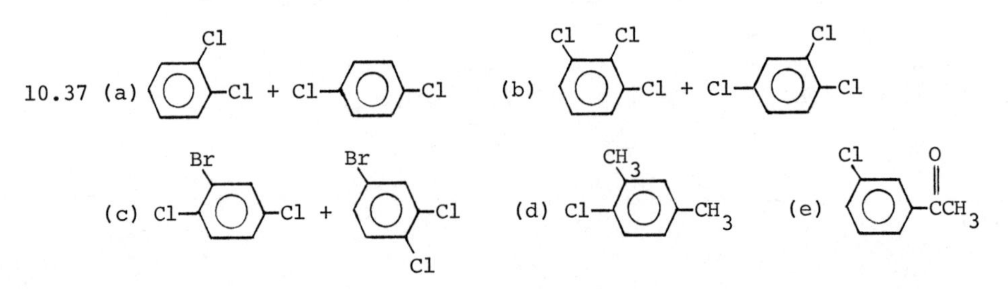

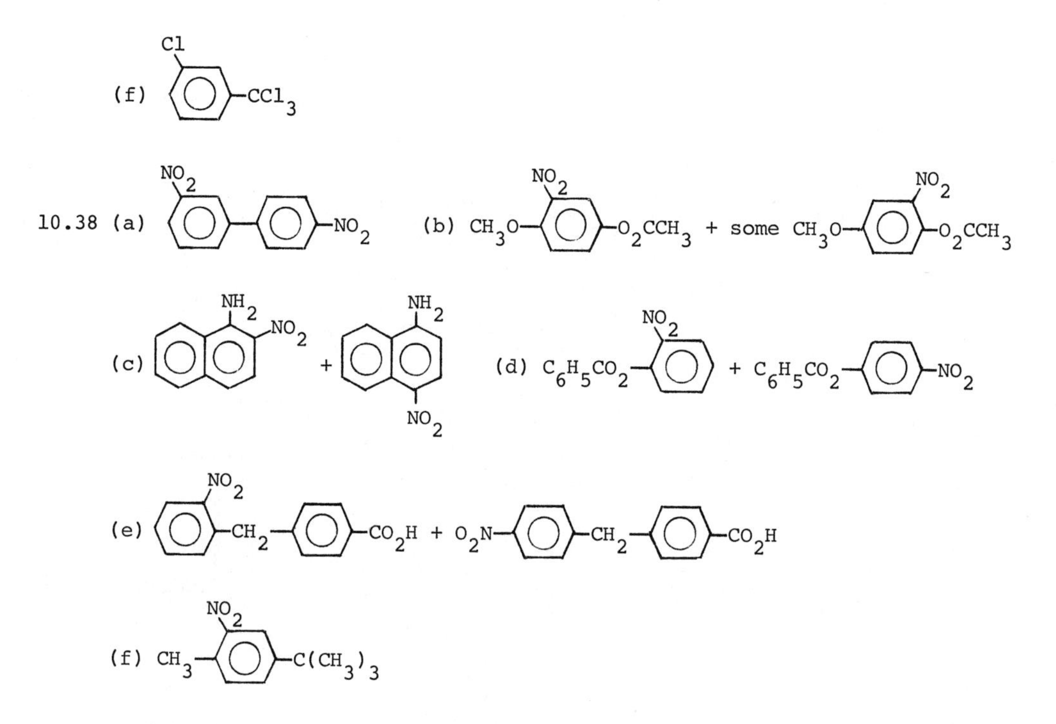

In (a), (c), (d), and (e), the more activated (or less deactivated) ring is attacked. The position of substitution on this ring is determined by the substituent already on the ring. For example,

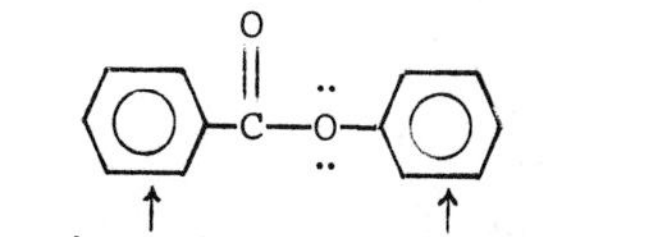

deactivated   activated by o,p-directing RCO_2 group

In (b), attack occurs o to the more strongly directing CH_3O group. In (f), attack occurs o to the CH_3 group (less hindered position).

10.39 (a) $C_6H_6 \xrightarrow[H_2SO_4]{HNO_3} C_6H_5NO_2 \xrightarrow[(2)\ OH^-]{(1)\ Fe, HCl} C_6H_5NH_2 \xrightarrow[0°]{NaNO_2,\ HCl} C_6H_5N_2^+\ Cl^- \xrightarrow[heat]{H_2O, H^+}$

$C_6H_5OH \xrightarrow[(2)\ CH_3I]{(1)\ OH^-} C_6H_5OCH_3$

(b) $C_6H_6 \xrightarrow[AlCl_3]{2\ CH_3I} H_3C-C_6H_4-CH_3 \xrightarrow[heat]{KMnO_4} HO_2C-C_6H_4-CO_2H$

+ o-isomer

(c) $C_6H_5NH_2$ from (a) $\xrightarrow{Br_2}$ Br–(benzene ring, Br)–NH_2 $\xrightarrow[0°]{NaNO_2,\ HCl}$

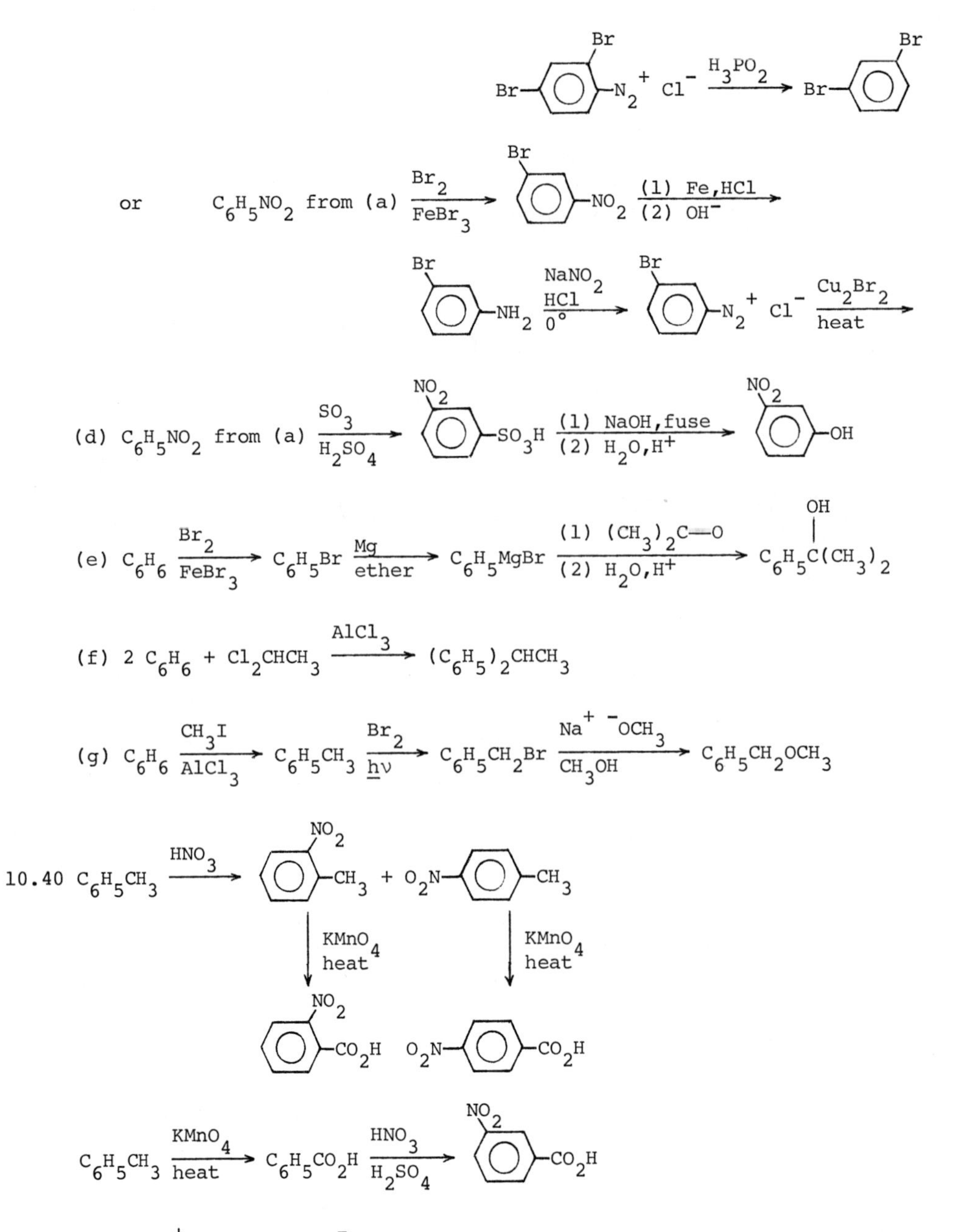

10.41 $(C_6H_5)_3C^+ + H_2O + HSO_4^-$. The triphenylmethyl cation is highly resonance-stabilized because the positive charge is delocalized by the three benzene rings. (See Answer 2.49.)

10.42 (a) $CH_3CH_2-C_6H_4-N_2^+\ Cl^-$ (b) $C_{10}H_7-N_2^+\ Cl^-$ (c) $CH_3CH_2-C_6H_4-CN$

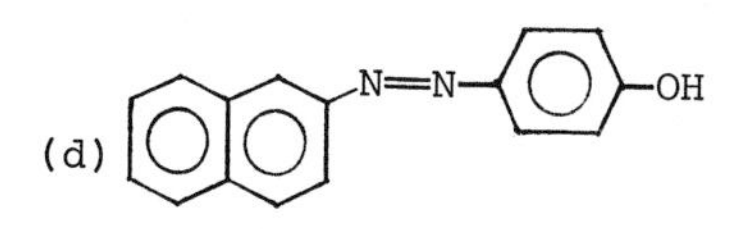

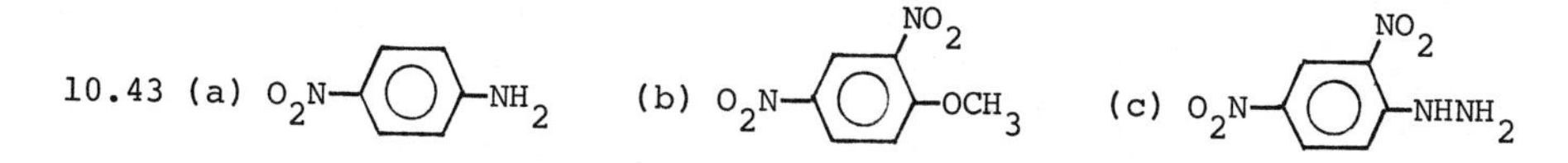

10.44 The reagent ICl must be ionized by the catalyst into positive and negative ions. Because Cl is more electronegative than I, Cl is the negative end of the reagent and I is the positive end. Therefore, I^+ is the attacking electrophile, and the product is C_6H_5I.

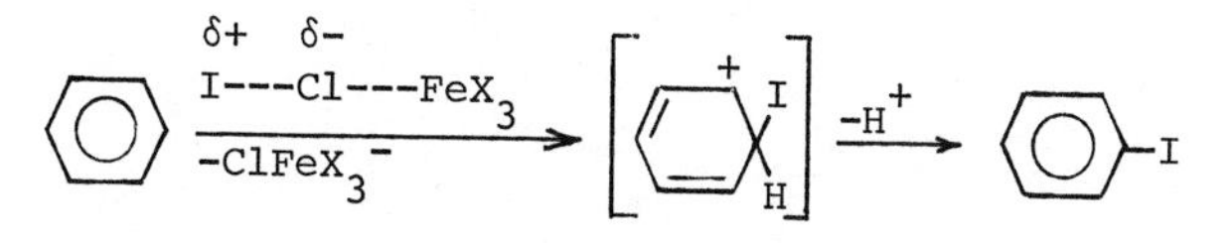

10.45 (a) The unshared electrons on the nitrogen of aniline are delocalized and are less available for donation than those of cyclohexylamine. The positive charge in the product anilinium ion, by contrast, is not delocalized.

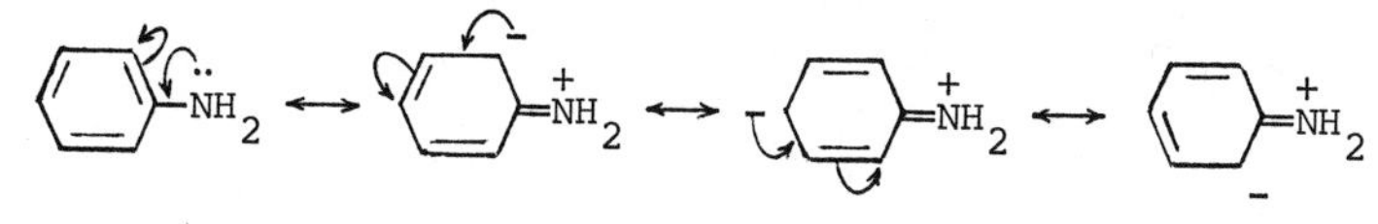

(b) Less basic, because the nitro group helps delocalize the unshared electrons by resonance and by the inductive effect, causing them to be less available for bonding.

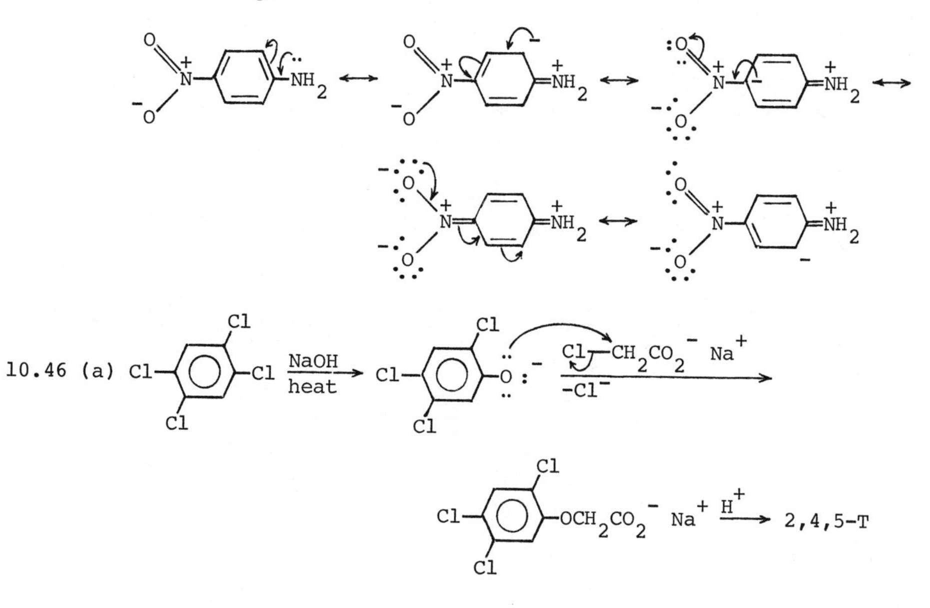

(b)

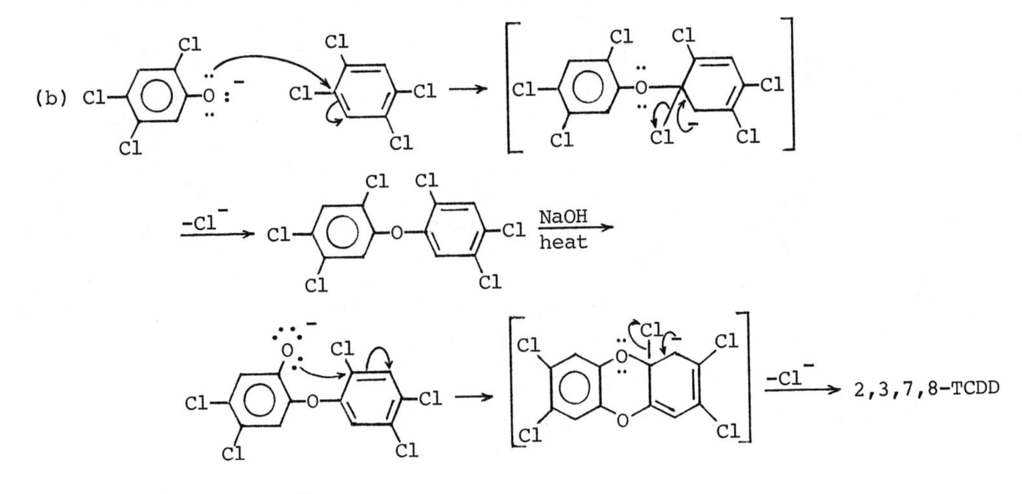

10.47

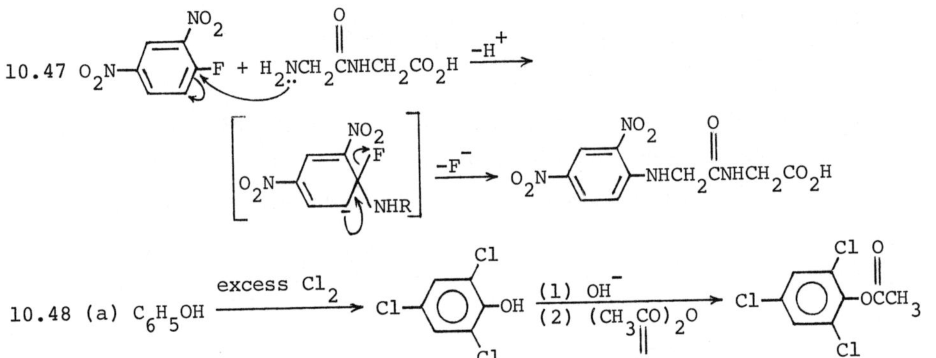

10.48 (a) $C_6H_5OH \xrightarrow{\text{excess } Cl_2}$ 2,4,6-trichlorophenol (Cl–C₆H₂(Cl)₂–OH) $\xrightarrow[(2)\ (CH_3CO)_2O]{(1)\ OH^-}$ Cl–C₆H₂(Cl)₂–$OCCH_3$ (with C=O)

(b) $C_6H_5CH_3 \xrightarrow[h\nu]{Br_2} C_6H_5CH_2Br \xrightarrow[AlBr_3]{C_6H_6} (C_6H_5)_2CH_2 \xrightarrow[h\nu]{Br_2} (C_6H_5)_2CHBr$

or $2\ C_6H_6 \xrightarrow[AlCl_3]{CH_2Cl_2} (C_6H_5)_2CH_2 \xrightarrow[h\nu]{Br_2}$

(c) $C_6H_6 \xrightarrow[AlCl_3]{CH_3CH_2Cl} C_6H_5CH_2CH_3 \xrightarrow[H_2SO_4]{HNO_3} O_2N$–C₆H₄–$CH_2CH_3 \xrightarrow[h\nu]{Br_2}$

O_2N–C₆H₄–$\underset{}{CHCH_3}$ (Br) $\xrightarrow{OH^-} O_2N$–C₆H₄–$CHCH_3$ (OH) $\xrightarrow{H_2CrO_4} O_2N$–C₆H₄–$CCH_3$ (C=O)

(d) $C_6H_5CH_3 \xrightarrow[H_2SO_4]{HNO_3} H_3C$–C₆H₄–$NO_2 \xrightarrow[(2)\ OH^-]{(1)\ Fe,HCl} CH_3$–C₆H₄–$NH_2 \xrightarrow[0°]{NaNO_2,\ HCl}$

CH_3–C₆H₄–$N_2^+\ Cl^- \xrightarrow[heat]{CuCN,\ KCN}$ product

(e) CH_3–C₆H₄–$N_2^+\ Cl^-$ from (d) $\xrightarrow[(2)\ heat]{(1)\ HBF_4} CH_3$–C₆H₄–F

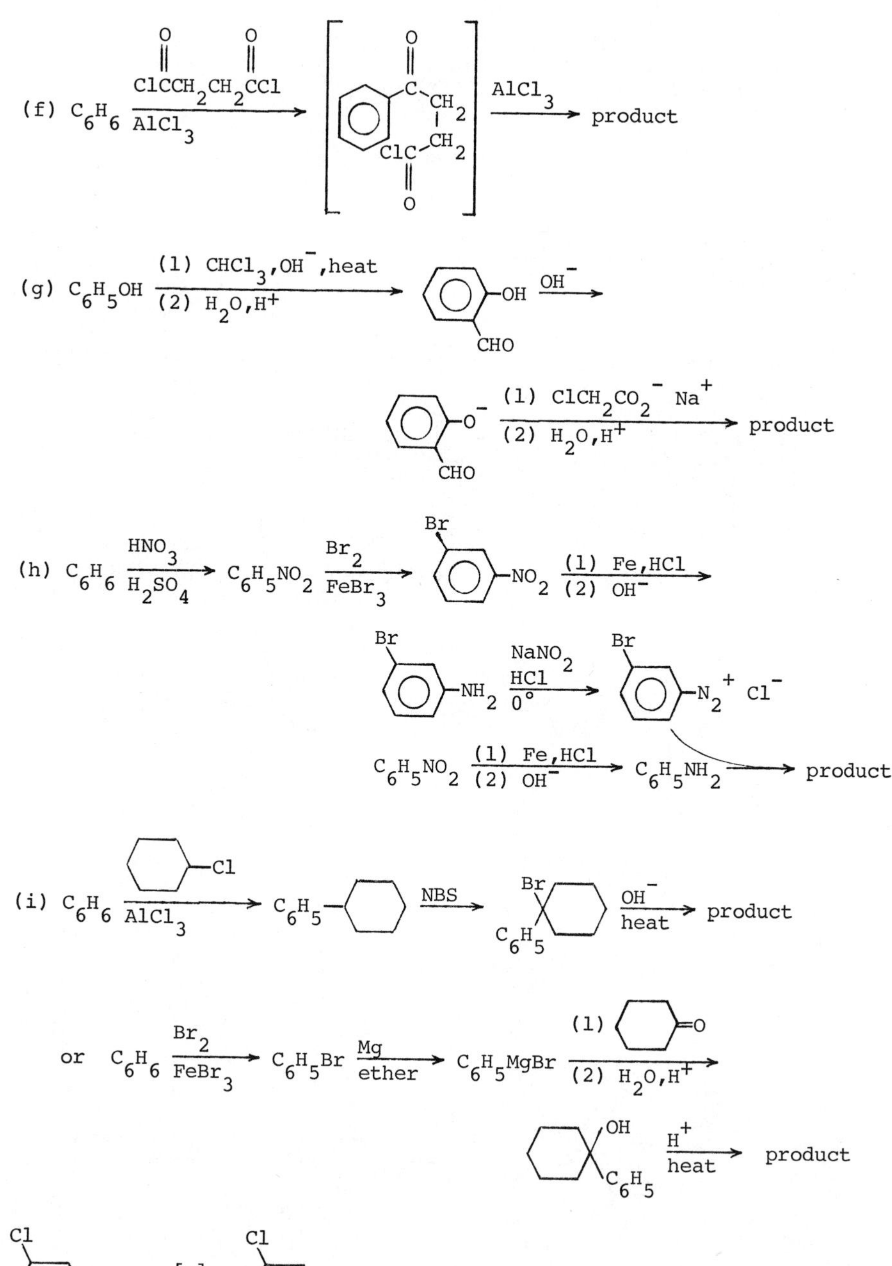

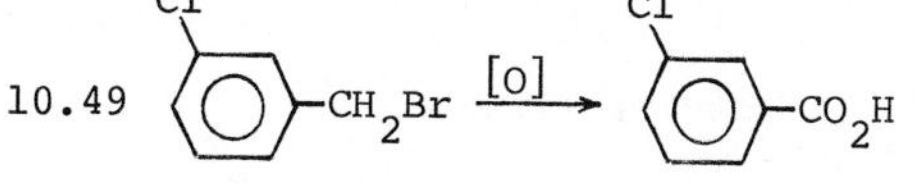

Because Br is lost in the oxidation, it cannot be attached to the ring.

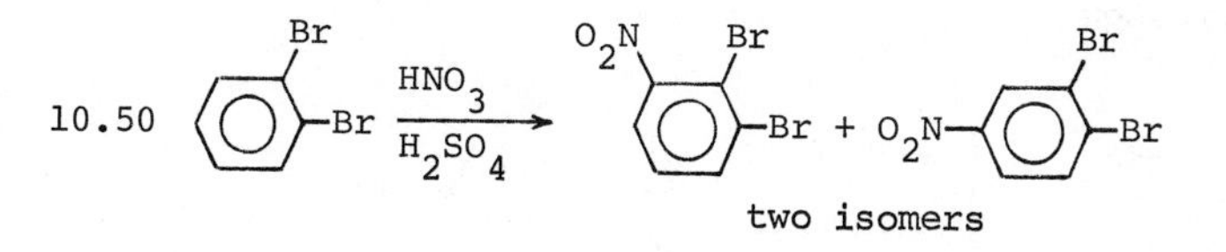

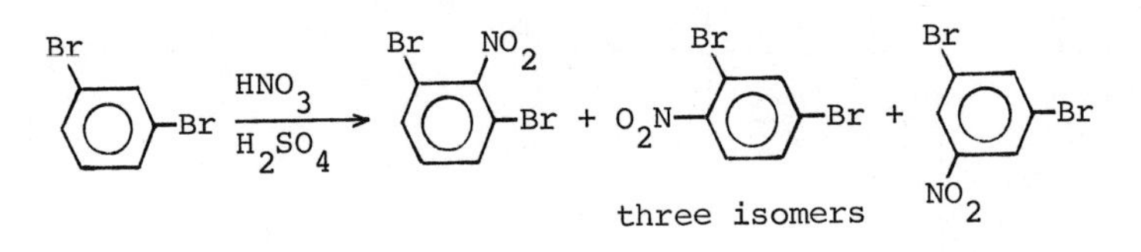

$$\text{Br-}C_6H_4\text{-Br} \xrightarrow[H_2SO_4]{HNO_3} \text{Br-}C_6H_3(NO_2)\text{-Br}$$

one isomer

The melting points of the three isomers are <u>ortho</u>, 6°; <u>meta</u>, 7°; <u>para</u>, 87°.

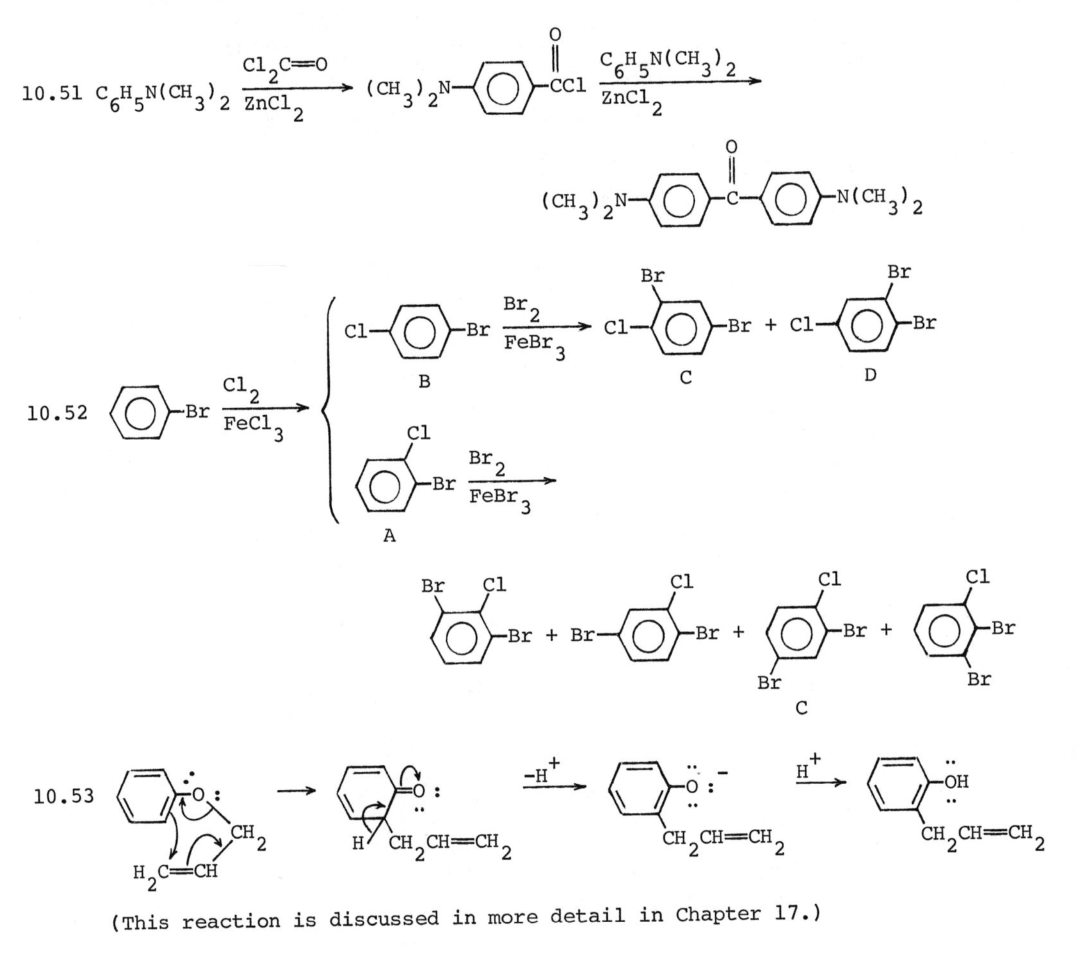

(This reaction is discussed in more detail in Chapter 17.)

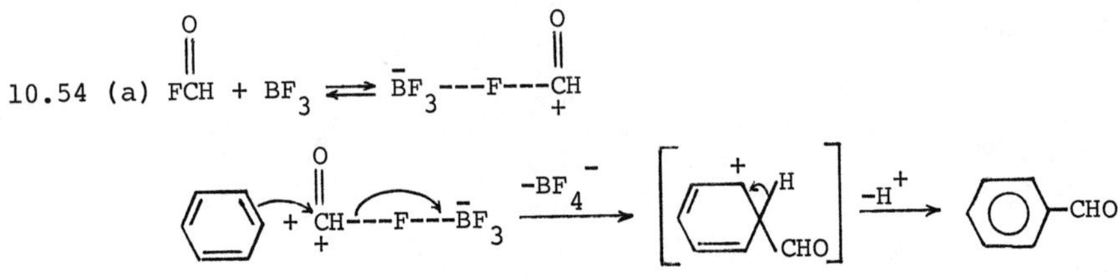

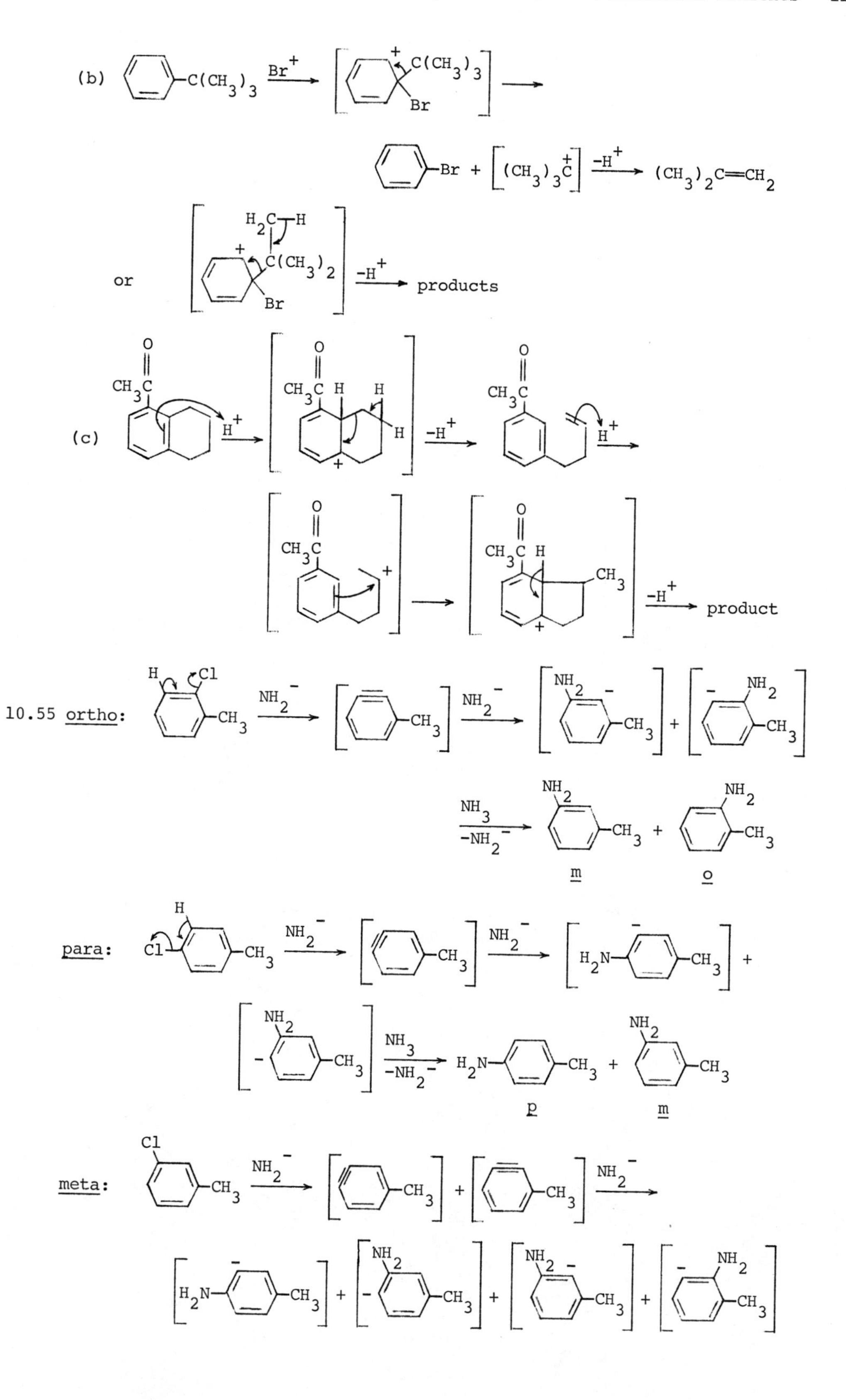
(b)
C(CH3)3
Br+
Br
Br + (CH3)3C+
-H+
(CH3)2C=CH2
or
H2C—H
C(CH3)2
-H+
products
(c)
CH3C
O
H+
H
-H+
H+
CH3
product
10.55 ortho:
H
Cl
CH3
NH2-
NH3
-NH2-
m
o
para:
H2N
p
m
meta:
Cl
NH2-
+

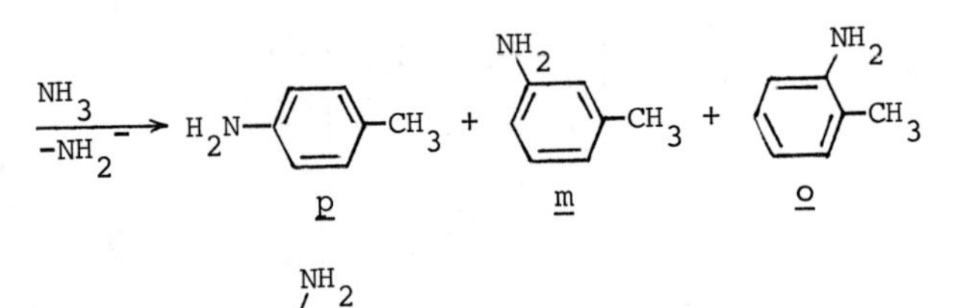

10.56 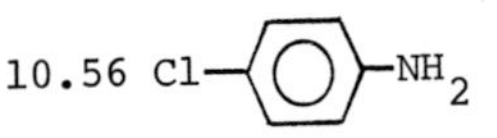
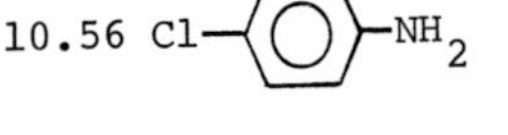

spectrum shown  other component

10.57 $C_6H_5CH_2\overset{O}{\overset{\|}{C}}CH_2C_6H_5$

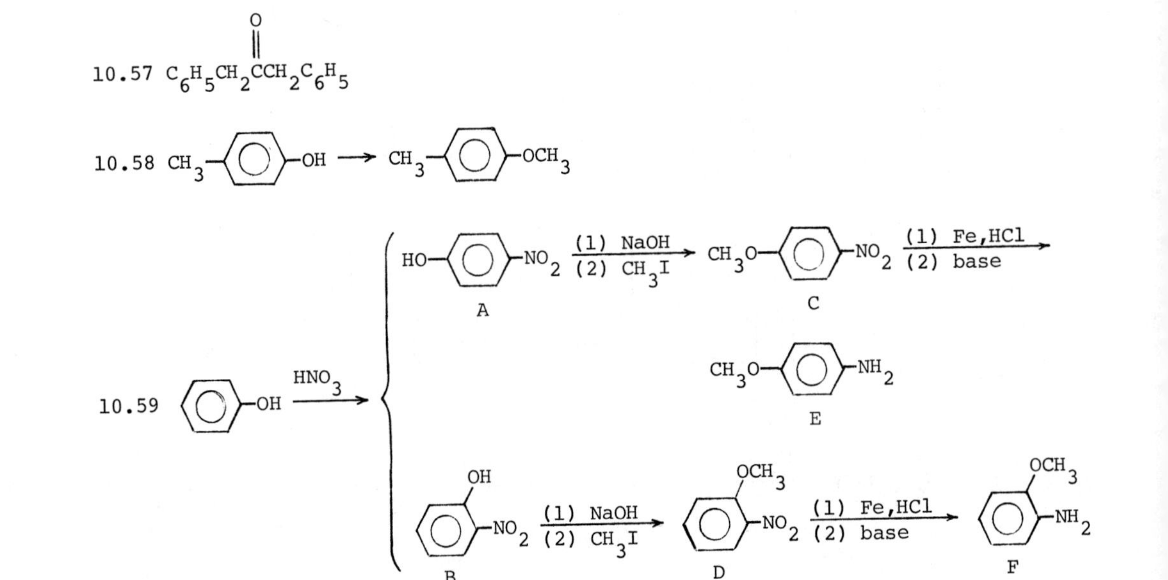

CHAPTER 11

Aldehydes and Ketones

Some Important Features

A carbonyl group (C═O) is polar and can be attacked by an electrophile such as H^+ or by a nucleophile.

In acidic solution, the oxygen is protonated:

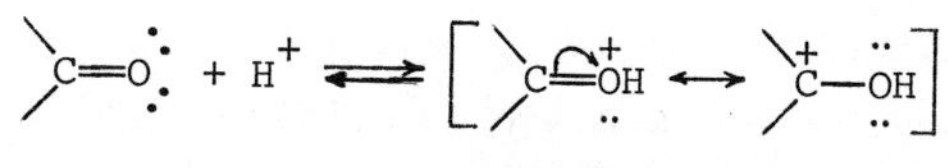

resonance structures for the protonated carbonyl group

In base, the carbon is attacked:

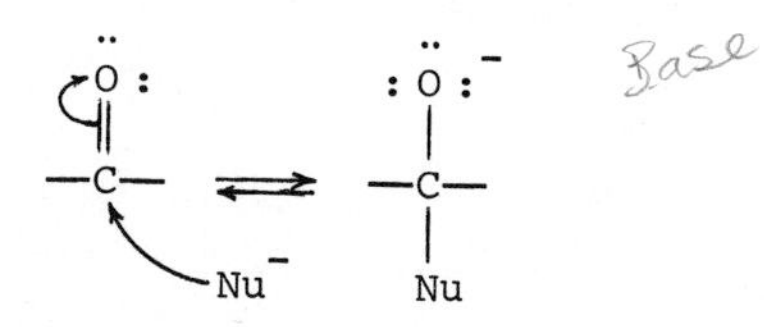

Aldehydes and ketones can undergo many addition reactions. Some of these reactions are reversible; others, such as Grignard reactions, are irreversible.

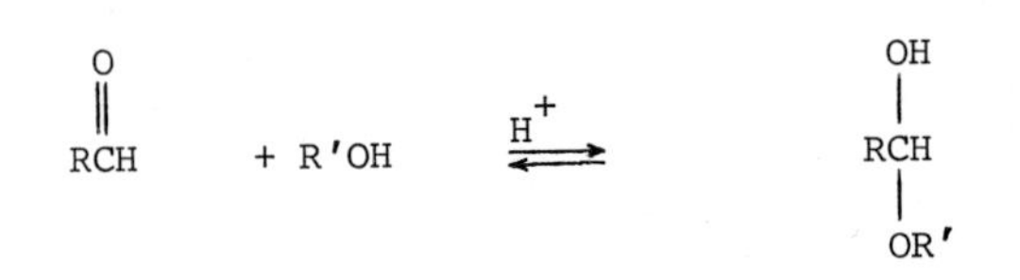

an aldehyde — a hemiacetal

favored by a less positive, more hindered C═O carbon, as in a ketone

favored by a more positive, less hindered C═O carbon, as in an aldehyde

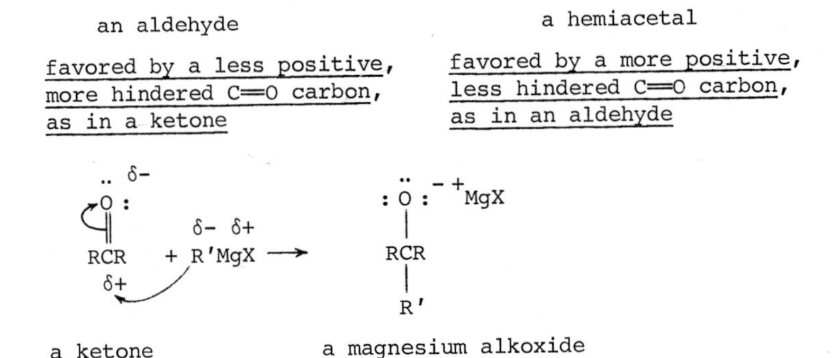

a ketone — a magnesium alkoxide

Aldehydes and ketones can undergo addition-elimination reactions with many nitrogen compounds. The products are favored by resonance-stabilization or by electron-withdrawing groups on the nitrogen. Typical reagents are primary amines (RNH_2) and hydrazine derivatives ($RNHNH_2$).

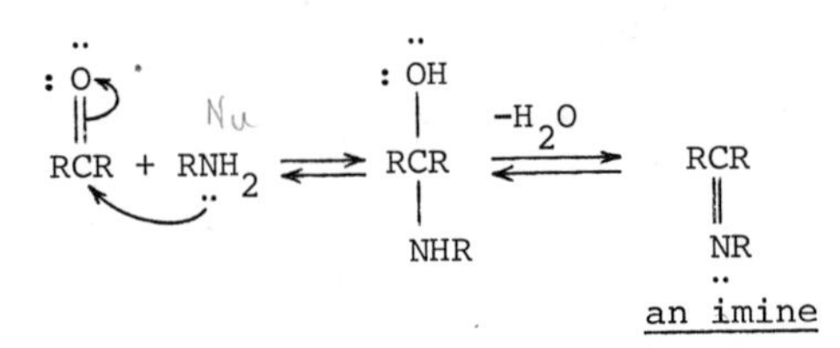

an imine

Aldehydes and ketones undergo catalytic hydrogenation with heat and pressure; however, these conditions also reduce carbon-carbon double bonds. Reduction to an alcohol with a metal hydride does not usually reduce a carbon-carbon double bond. Aldehydes are also very easily oxidized to carboxylic acids.

A hydrogen atom α to a carbonyl group is slightly acidic. (For resonance structures of the product enolate ions, see Section 11.16.)

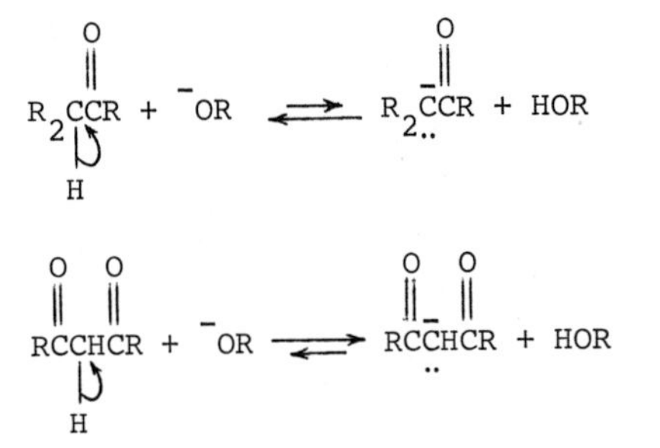

An aldehyde or ketone with an α hydrogen can undergo tautomerism. Unless the enol tautomer is stabilized relative to the keto tautomer (by hydrogen bonding, for example), the keto form is favored.

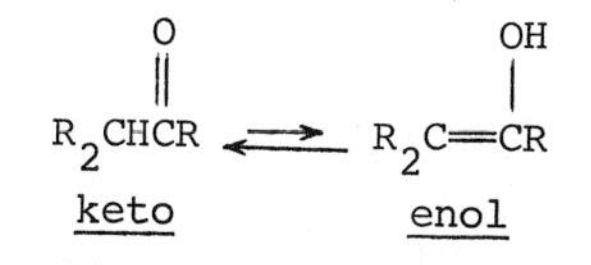

Aldehydes or ketones with the carbonyl group in conjugation with a carbon-carbon double bond can undergo 1,4-addition reactions (either electrophilic or nucleophilic). An isolated carbon-carbon double bond cannot participate in a 1,4-addition reaction.

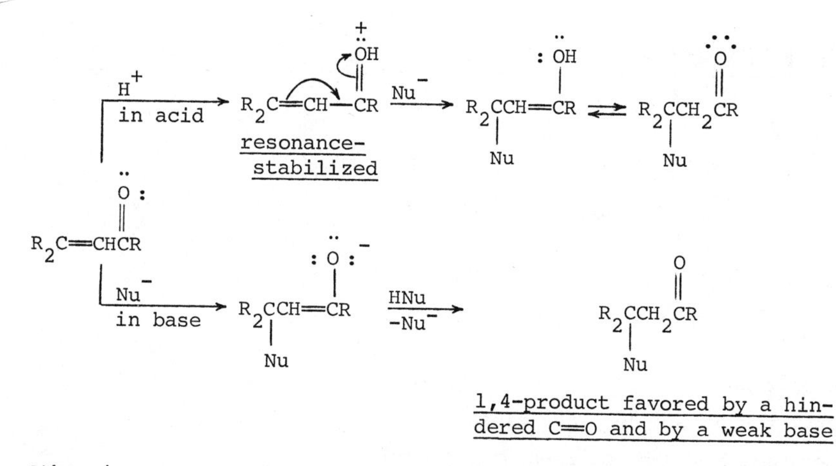

Other important topics covered in this chapter are the synthesis of enamines (Section 11.10B), the Wittig reaction for alkenes (Section 11.12), and α halogenation (Section 11.18).

Reminders

In acidic solution, the C=O oxygen is protonated, and a weak nucleophile can attack the C=O carbon.

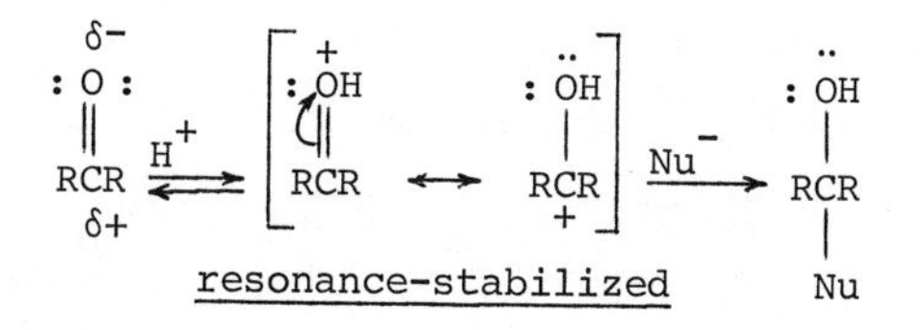

A strong nucleophile, such as CN^-, can attack the C=O carbon without prior protonation of the C=O group.

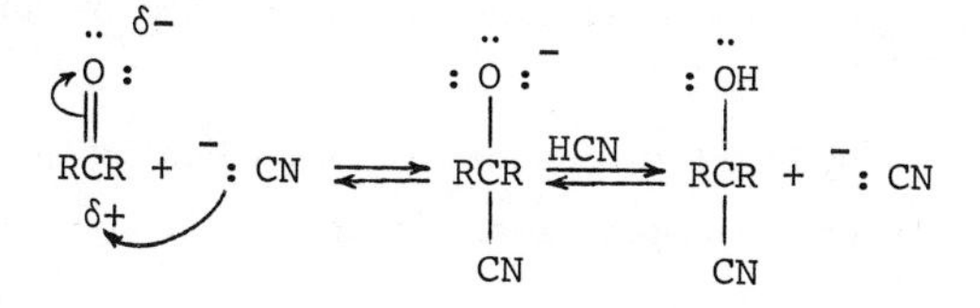

A strong base, such as ^{-}OR, can remove an α hydrogen.

Remember that an sp^2 carbon atom cannot be chiral.

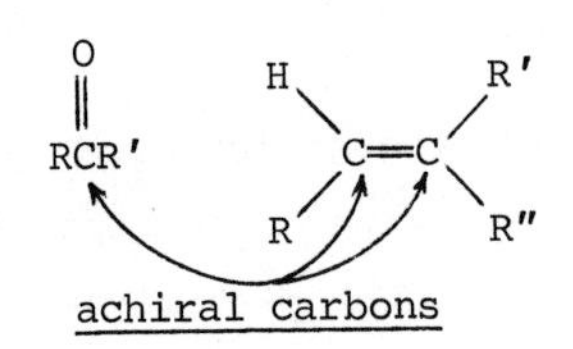

Answers to Problems

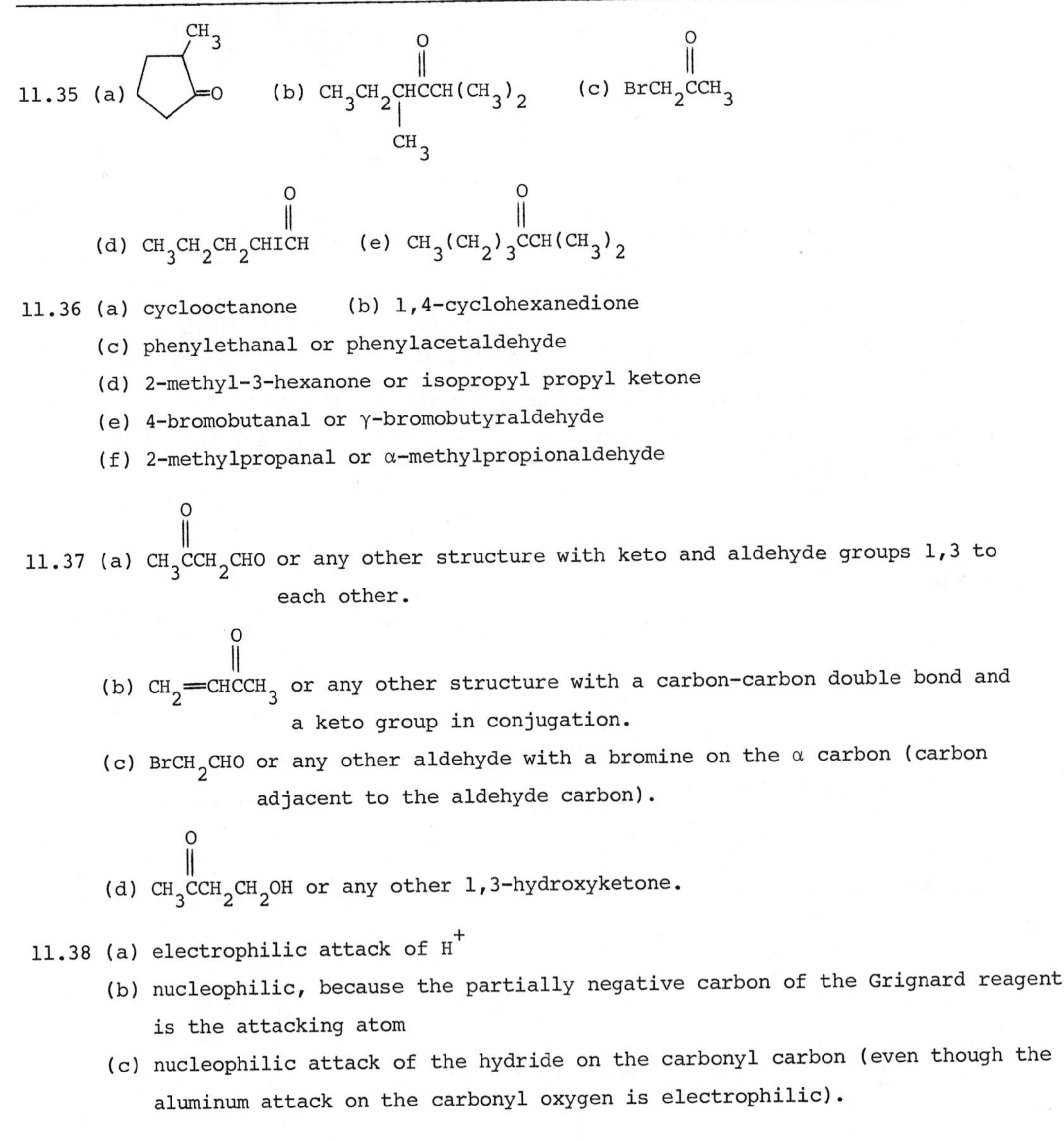

11.35 (a) 2-methylcyclopentanone (ring with CH_3 and =O) (b) $CH_3CH_2CH(CH_3)C(=O)CH(CH_3)_2$ (c) $BrCH_2C(=O)CH_3$

(d) $CH_3CH_2CH_2CHICH(=O)$ (e) $CH_3(CH_2)_3C(=O)CH(CH_3)_2$

11.36 (a) cyclooctanone (b) 1,4-cyclohexanedione

(c) phenylethanal or phenylacetaldehyde

(d) 2-methyl-3-hexanone or isopropyl propyl ketone

(e) 4-bromobutanal or γ-bromobutyraldehyde

(f) 2-methylpropanal or α-methylpropionaldehyde

11.37 (a) $CH_3C(=O)CH_2CHO$ or any other structure with keto and aldehyde groups 1,3 to each other.

(b) $CH_2{=}CHC(=O)CH_3$ or any other structure with a carbon-carbon double bond and a keto group in conjugation.

(c) $BrCH_2CHO$ or any other aldehyde with a bromine on the α carbon (carbon adjacent to the aldehyde carbon).

(d) $CH_3C(=O)CH_2CH_2OH$ or any other 1,3-hydroxyketone.

11.38 (a) electrophilic attack of H^+

(b) nucleophilic, because the partially negative carbon of the Grignard reagent is the attacking atom

(c) nucleophilic attack of the hydride on the carbonyl carbon (even though the aluminum attack on the carbonyl oxygen is electrophilic).

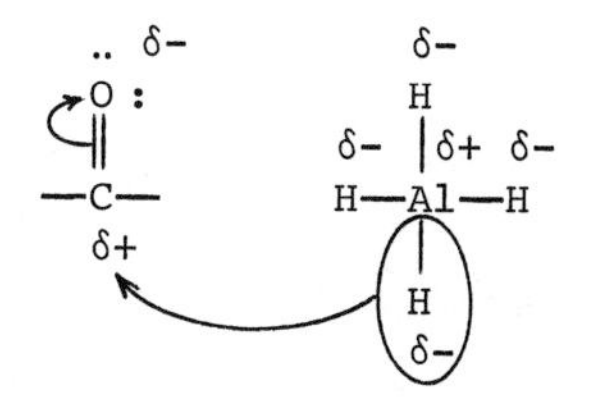

11.39 (c) most stable (a) least stable

Electron withdrawal by Br increases the positive charge on the carbonyl carbon; three bromine atoms are more effective than one.

11.40 $Br_2CHCHO + H_2O \rightleftharpoons Br_2CHCH(OH)_2$

11.41 (a) $CH_3CH_2CHO + CH_3OH \underset{}{\overset{H^+}{\rightleftharpoons}} CH_3CH_2CH(OH)OCH_3 \overset{CH_3OH, H^+}{\rightleftharpoons} CH_3CH_2CH(OCH_3)_2 + H_2O$

(b) $(CH_3)_2C{=}O + HOCH_2CH_2OH \overset{H^+}{\rightleftharpoons} (CH_3)_2C(OH)OCH_2CH_2OH \overset{H^+}{\rightleftharpoons} (CH_3)_2C\langle O\text{-}CH_2CH_2\text{-}O\rangle + H_2O$

Since a five-membered ring can be formed in the hemiacetal-to-acetal step, the cyclization is favored over reaction with another molecule of ethylene glycol.

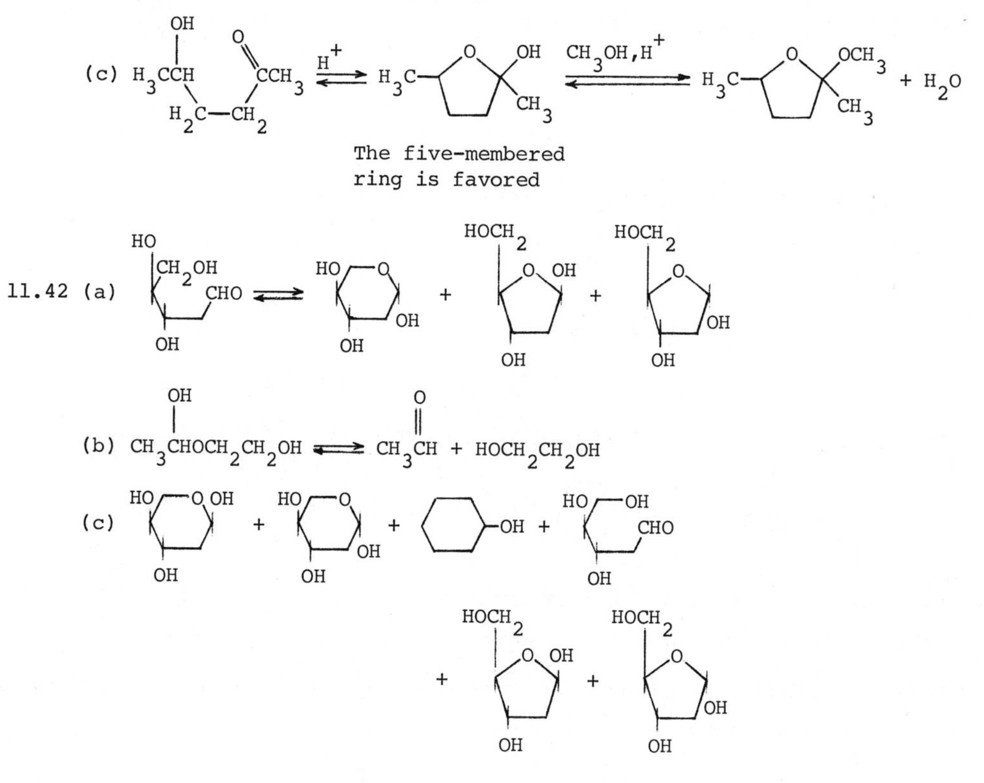

11.43 (a) $(CH_3)_2C{=}O + H_2NNH\text{-}C_6H_3(NO_2)_2 \overset{H^+}{\rightleftharpoons} (CH_3)_2C{=}NNH\text{-}C_6H_3(NO_2)_2 + H_2O$

(b) 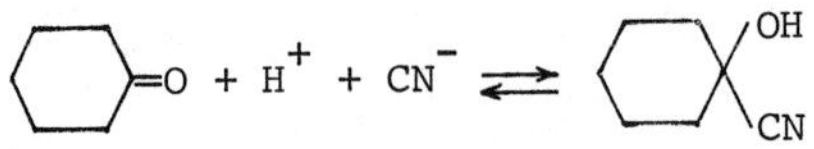

(c) $C_6H_5CHO + H_2NC_6H_5 \overset{H^+}{\rightleftharpoons} C_6H_5CH{=}NC_6H_5 + H_2O$

a 1° amine — an imine

(d) $CH_3CH_2CHO + HN(CH_3)_2 \overset{H^+}{\rightleftharpoons} CH_3CH{=}CHN(CH_3)_2 + H_2O$

a 2° amine — an enamine

11.44 (a) $CH_3CH_2CH_2CHO + HN(CH_2CH_3)_2 \overset{H^+}{\rightleftharpoons}$ (b) 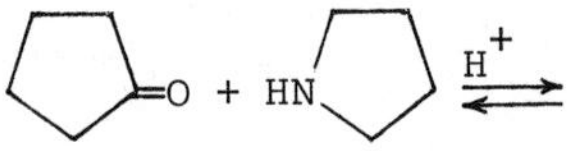

(c) $CH_3CH_2CHO + C_6H_5NH_2 \overset{-H_2O}{\rightleftharpoons}$

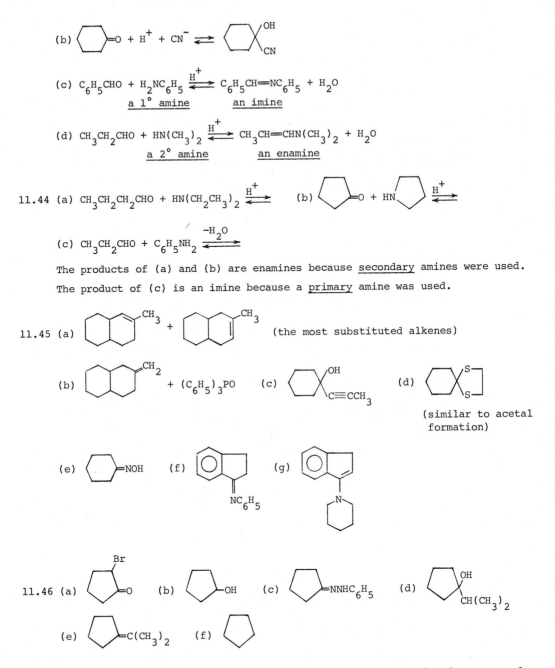

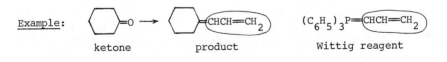

The products of (a) and (b) are enamines because secondary amines were used. The product of (c) is an imine because a primary amine was used.

11.45 (a) 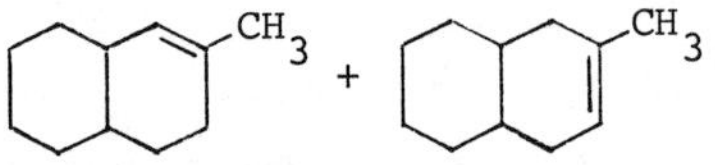 (the most substituted alkenes)

(b) 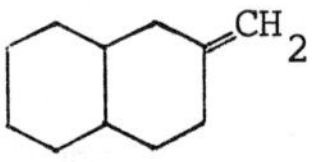 + $(C_6H_5)_3PO$ (c) 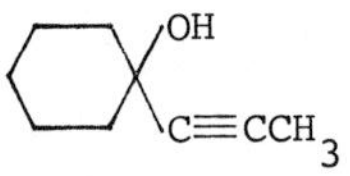 (d) 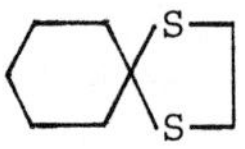

(similar to acetal formation)

(e) 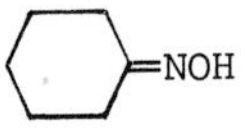 (f) 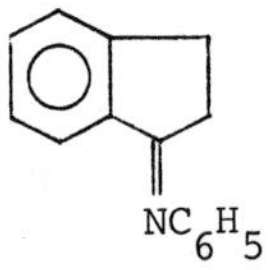 (g) 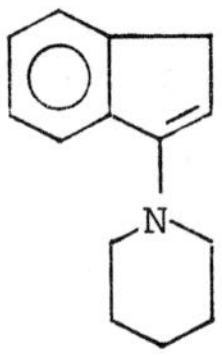

11.46 (a) 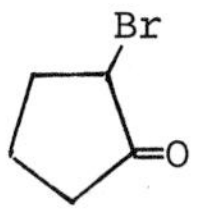 (b) 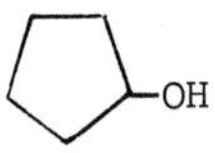 (c) 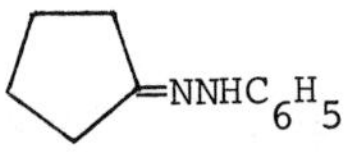 (d) 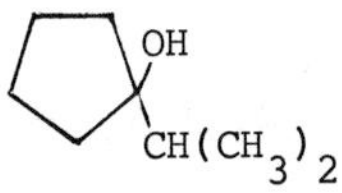

(e) 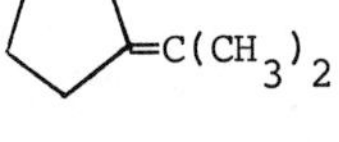 (f) 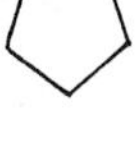

11.47 (b), (a), (c). Compound (b) contains the most hindered carbonyl group, and compound (c) contains the least hindered carbonyl group.

11.48 The structure of the required Wittig reagent may be determined by circling the portion of the product that has replaced the carbonyl oxygen.

Example: 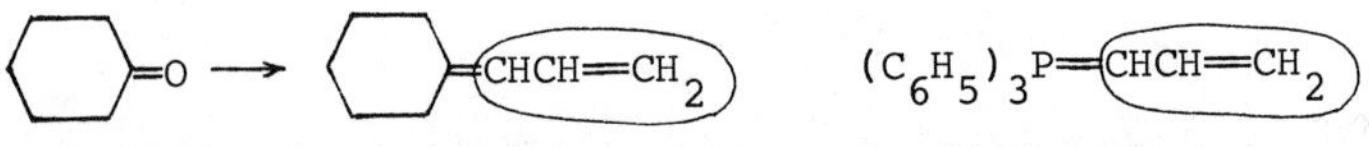

ketone — product — Wittig reagent

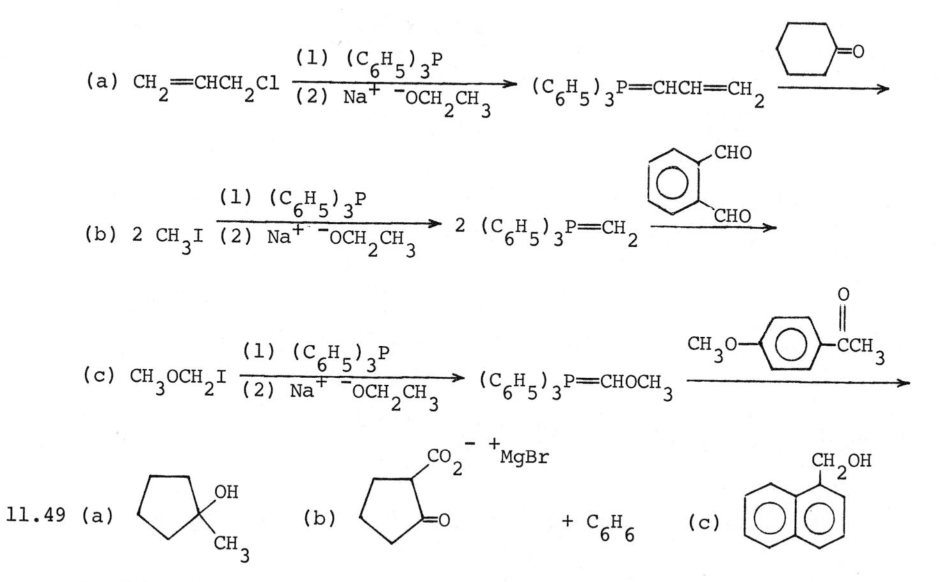

In (b), the starting carbonyl compound contains an acidic proton that reacts with the Grignard reagent at a faster rate than does the keto group.

11.50 (a) $CH_3I \xrightarrow[\text{ether}]{\text{Mg}} CH_3MgI \xrightarrow[(2)\ H_2O, H^+]{(1)\ (CH_3CH_2)_2C{=}O} CH_3CH_2C(OH)(CH_3)CH_2CH_3$

(b) CH_3MgI from (a) $\xrightarrow[(2)\ H_2O, H^+]{(1)\ HCHO} CH_3CH_2OH$

(c) CH_3MgI from (a) $\xrightarrow[(2)\ H_2O, H^+]{(1)\ CH_3CH_2CH_2CHO} CH_3CH_2CH_2CH(OH)CH_3$

11.51 (a) $HOCH_2CH_2CH_2CH_2OH$ (b) no reaction

(c) cyclohexyl-CO_2^- or cyclohexyl-CO_2H after acidification (d) $(CH_3)_2CHNH_2$

(e) $(CH_3)_2CHNHCH_2CH_2OH$ (f) bicyclic amine (N–H) by way of a cyclic imine, bicyclic imine (C=N)

(g) cyclohexyl-CO_2^- + $CHCl_3$

11.52 (a) and (c). Compound (c) is a hemiacetal, which is in equilibrium with the aldehyde in alkaline solution. Compound (d) is an acetal, which is not in equilibrium with the aldehyde under the alkaline conditions of the Tollens test.

11.53 $HOCH_2CH_2CHO + HOCH_2CH_2OH \xrightleftharpoons{H^+} HOCH_2CH_2CH(OCH_2CH_2O)$ (cyclic acetal) $\xrightarrow[OH^-]{KMnO_4}$

$^-O_2CCH_2CH(OCH_2CH_2O) \xrightarrow{H^+} HO_2CCH_2CHO$

Alkaline or neutral conditions must be used for the oxidation to prevent the acetal from reverting to the aldehyde.

11.54 (a) $CH_3\overset{O}{\overset{\|}{C}}CH{=}CH_2 \xrightarrow{HCl} CH_3\overset{O}{\overset{\|}{C}}CH_2CH_2Cl \xrightarrow[-CHCl_3]{Cl_2,OH^-} {}^-O\overset{O}{\overset{\|}{C}}CH_2CH_2Cl \xrightarrow{H^+} HO_2CCH_2CH_2Cl$

(b) $CH_3CH_2CH_2CHO \xrightarrow[(2)\ H_2O,H^+]{(1)\ NaBH_4} CH_3CH_2CH_2CH_2OH \xrightarrow{HBr}$

$CH_3CH_2CH_2CH_2Br \xrightarrow[(2)\ Na^+\ {}^-OCH_2CH_3]{(1)\ (C_6H_5)_3P} CH_3CH_2CH_2CH{=}P(C_6H_5)_3$

$\xrightarrow{CH_3CH_2CH_2CHO} CH_3CH_2CH_2CH{=}CHCH_2CH_2CH_3$

(c) $CH_3CH_2CH_2CH_2Br$ from (b) $\xrightarrow[ether]{Mg} CH_3CH_2CH_2CH_2MgBr \xrightarrow[(2)\ H_2O,H^+]{(1)\ CH_3CH_2\overset{O}{\overset{\|}{C}}CH_3}$

$CH_3CH_2\underset{CH_3}{\underset{|}{\overset{OH}{\overset{|}{C}}}}CH_2CH_2CH_2CH_3$

(d) $CH_3CH_2CH_2CH_2OH$ from (b) $\xrightarrow{K} CH_3CH_2CH_2CH_2O^-\ K^+ \xrightarrow{CH_3CH_2CH_2CH_2Br \text{ from (b)}}$

$(CH_3CH_2CH_2CH_2)_2O$

(e) $CH_3CH_2CH_2CHO \xrightarrow{NH_3,H_2,Pt} CH_3CH_2CH_2CH_2NH_2$

(f) $CH_2{=}CHCH_2CHO \xrightarrow[\text{or other mild ox. agent}]{Ag(NH_3)_2^+} CH_2{=}CHCH_2CO_2^- \xrightarrow{H^+} CH_2{=}CHCH_2CO_2H$

(g) $CH_2{=}CHCH_2CHO \xrightarrow[(2)\ H_2O,H^+]{(1)\ NaBH_4} CH_2{=}CHCH_2CH_2OH$

11.55 (a) $(CH_3CH_2)_2C{=}O \xrightarrow[(2)\ H_2O,H^+]{(1)\ CH_3CH_2MgBr}$

(b) $2\ C_6H_{11}Br \xrightarrow[(2)\ Na^+\ {}^-OCH_2CH_3]{(1)\ 2\ (C_6H_5)_3P} 2\ C_6H_{10}{=}P(C_6H_5)_3 \xrightarrow{OHCCH_2CHO}$

(c) cyclopentanone $+ HOCH_2CH_2OH \xrightarrow{H^+}$ (d) 1-iodocyclohexene $\xrightarrow[\text{ether}]{Mg}$ cyclohexenyl-MgI $\xrightarrow[(2)\ H_2O, H^+]{(1)\ CO_2}$

11.56 (b), (c), (a)

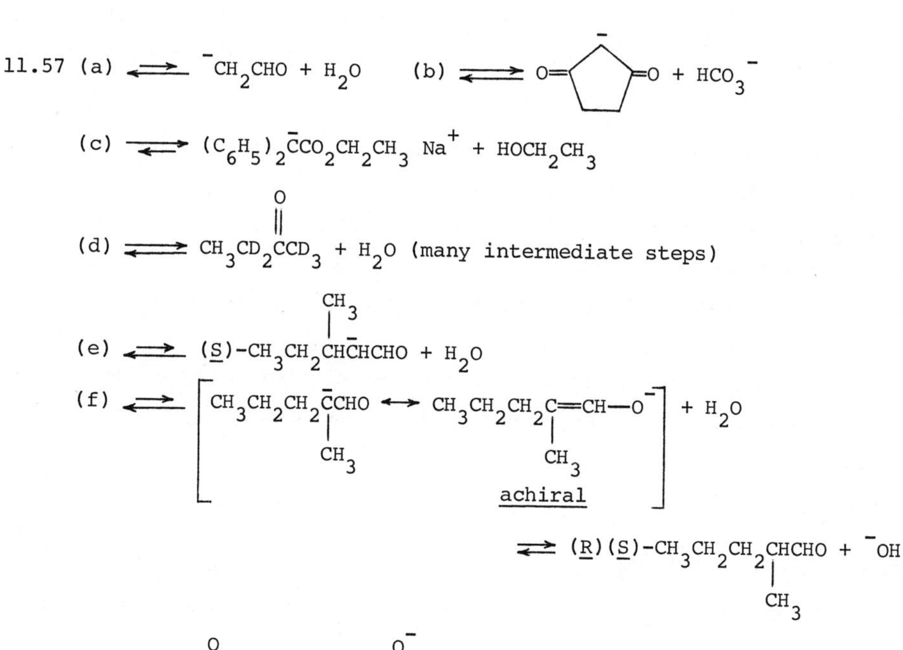

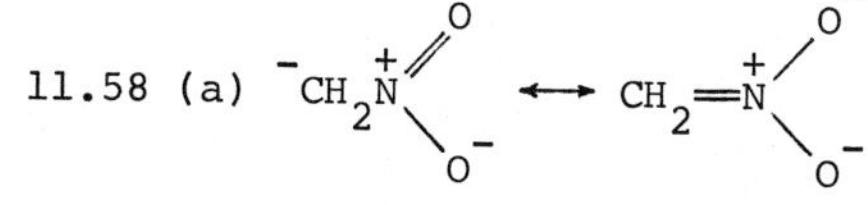

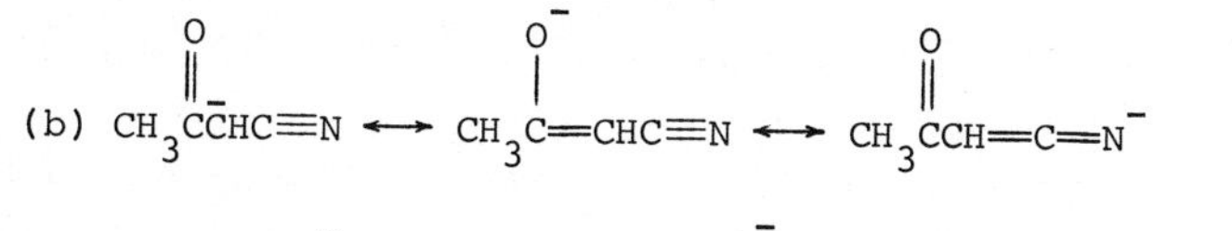

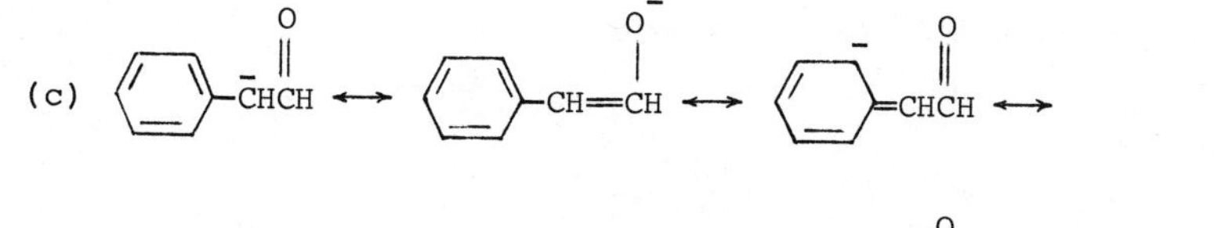

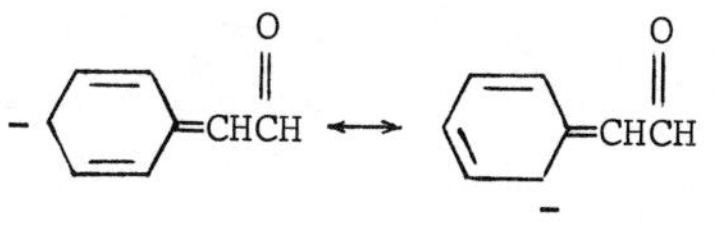

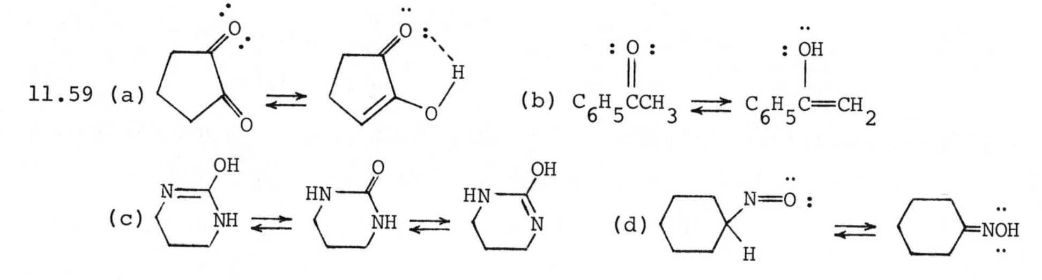

(c) $\bigcirc\!=\!O$ + $HOCH_2CH_2OH$ $\xrightarrow{H^+}$ (d) $\bigcirc\!-I$ $\xrightarrow[ether]{Mg}$ $\bigcirc\!-MgI$ $\xrightarrow[(2) \ H_2O,H^+]{(1) \ CO_2}$

11.56 (b), (c), (a)

11.57 (a) \rightleftharpoons $^-CH_2CHO$ + H_2O (b) \rightleftharpoons $O\!=\!\bigcirc\!=\!O$ + HCO_3^-

(c) \rightleftharpoons $(C_6H_5)_2\bar{C}CO_2CH_2CH_3$ Na^+ + $HOCH_2CH_3$

(d) \rightleftharpoons $CH_3CD_2\overset{\overset{\displaystyle O}{\|}}{C}CD_3$ + H_2O (many intermediate steps)

(e) \rightleftharpoons $(\underline{S})\text{-}CH_3CH_2CH\overset{\overset{\displaystyle CH_3}{|}}{\bar{C}}HCHO$ + H_2O

(f) \rightleftharpoons + H_2O

$\left[CH_3CH_2CH_2\overset{\overset{\displaystyle \|}{C}}{\underset{\underset{\displaystyle CH_3}{}}{}}CHO \leftrightarrow CH_3CH_2CH_2\overset{}{\underset{\underset{\displaystyle CH_3}{}}{C}}\!=\!CH\!-\!O^- \right]$

achiral

\rightleftharpoons $(\underline{R})(\underline{S})\text{-}CH_3CH_2CH_2\overset{}{\underset{\underset{\displaystyle CH_3}{|}}{C}}HCHO$ + ^-OH

11.58 (a) $^-CH_2\overset{+}{N}\!\!\diagdown\!\!\overset{O}{\underset{O^-}{}}$ \leftrightarrow $CH_2\!=\!\overset{+}{N}\!\!\diagdown\!\!\overset{O^-}{\underset{O^-}{}}$

(b) $CH_3\overset{\overset{\displaystyle O}{\|}}{\bar{C}}CHC\!\equiv\!N$ \leftrightarrow $CH_3\overset{O^-}{\underset{}{C}}\!=\!CHC\!\equiv\!N$ \leftrightarrow $CH_3\overset{\overset{\displaystyle O}{\|}}{C}CH\!=\!C\!=\!N^-$

(c)

11.59 (a)

11.60 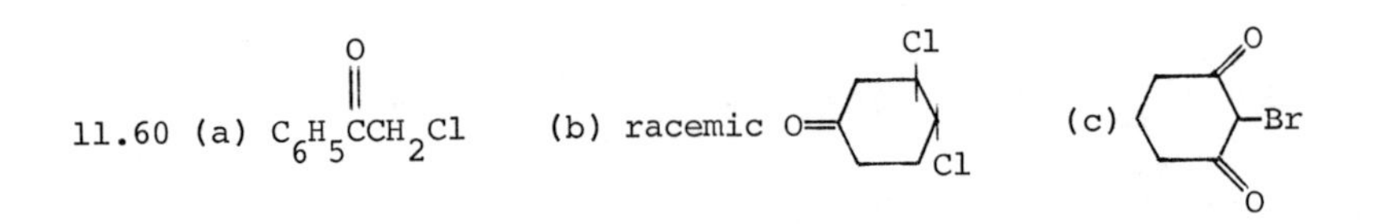

11.61 (a), (d), and (e) would yield a yellow precipitate of CHI_3; (b), (c), and (f) would not show this precipitate because none is a methyl ketone (and none can be oxidized to such a structure).

11.62 (a) Add an acidic solution of 2,4-dinitrophenylhydrazine; cyclohexanone gives a precipitate, while cyclohexanol does not.

(b) Add I_2 in dilute NaOH; 2-pentanone gives a yellow precipitate (CHI_3), while 3-pentanone does not.

(c) Add Tollens reagent; pentanal gives a silver mirror, while 2-pentanone does not.

(d) Add Br_2 in CCl_4; 2-pentene decolorizes the solution very rapidly, while 2-pentanone does not. (2-Pentanone slowly reacts with Br_2 by way of its enol.)

11.63 (a) $BrCH_2CH_2\overset{O}{\overset{\|}{C}}CH_3$ (b) $CH_3CHClCH_2CH_2\overset{O}{\overset{\|}{C}}CH_3$

In (b), the double bond is not in conjugation, and therefore Markovnikov's rule applies.

11.64 (a) $CH_3CHO \xrightarrow[Pt]{NH_3, H_2} CH_3CH_2NH_2 \xrightarrow[\text{1,4-addition}]{CH_2{=}CH{-}\overset{O}{\overset{\|}{C}}CH_2CH_3}$ (b) (cyclohexenyl)$-\overset{O}{\overset{\|}{C}}CH_3 \xrightarrow[HCN]{CN^-}$

11.65 Measure the rate of incorporation of ^{18}O by the water. This rate will be half the rate of hydration and subsequent dehydration.

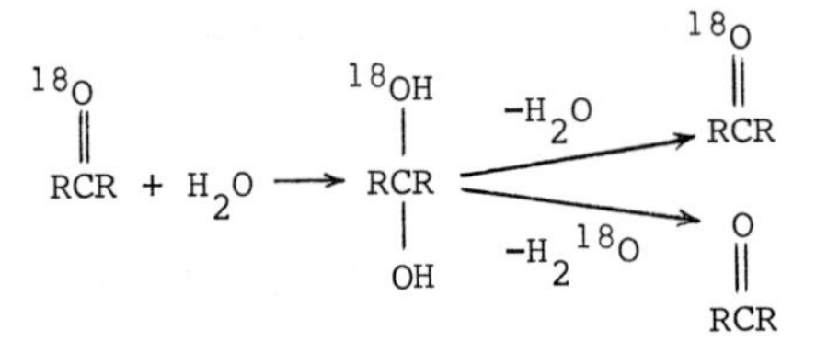

11.66 $RCHO \xrightarrow[H_2, Pt]{NH_3} \underset{\text{a 1° amine}}{RCH_2NH_2} \xrightarrow{RCHO} \underset{\text{an imine}}{RCH_2N{=}CHR} \xrightarrow{H_2, Pt} \underset{\text{a 2° amine}}{RCH_2NHCH_2R}$

The 1° amine product of the reductive amination can react with RCHO to yield an imine. Reduction of this imine yields a 2° amine.

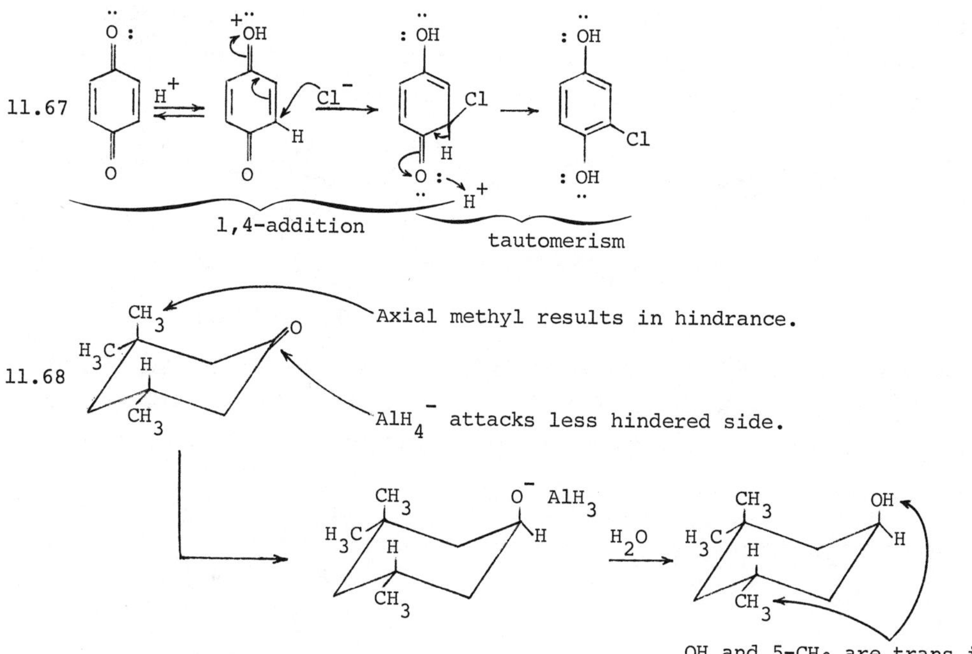

11.67

1,4-addition tautomerism

11.68

Axial methyl results in hindrance.

AlH$_4^-$ attacks less hindered side.

OH and 5-CH$_3$ are <u>trans</u> in predominant product.

11.69 Both electronic and steric effects play a role.

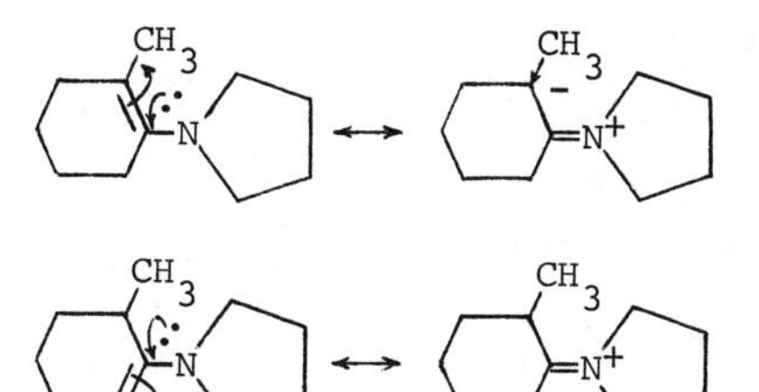

less stable because of electron-releasing CH_3 group and steric hindrance

more stable because the negative carbon is less substituted and less hindered

For resonance, the participating atoms must lie in a plane. The first resonance-stabilized enamine is more hindered than the second.

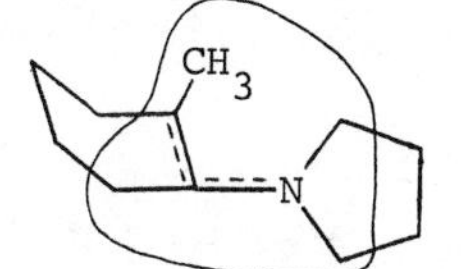

planar and hindered

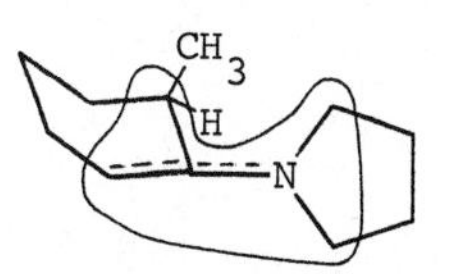

less hindered with CH_3 out of plane

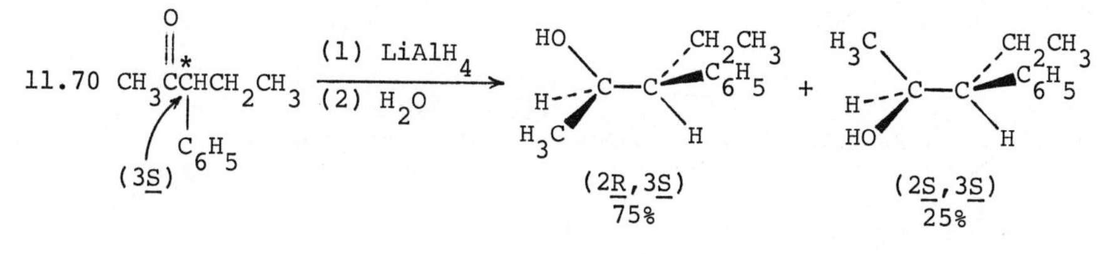

11.70 CH$_3$CCHCH$_2$CH$_3$ $\xrightarrow{(1)\ LiAlH_4}{(2)\ H_2O}$

(2R,3S) 75% (2S,3S) 25%

(a) The starting ketone must be the (3S) enantiomer because that chiral carbon is not affected by the reaction.

(b) The transition state leading to the (2R) diastereomer is less hindered (and of lower energy) than that leading to the (2S) diastereomer. Therefore, the (2R,3S) alcohol is formed at a faster rate.

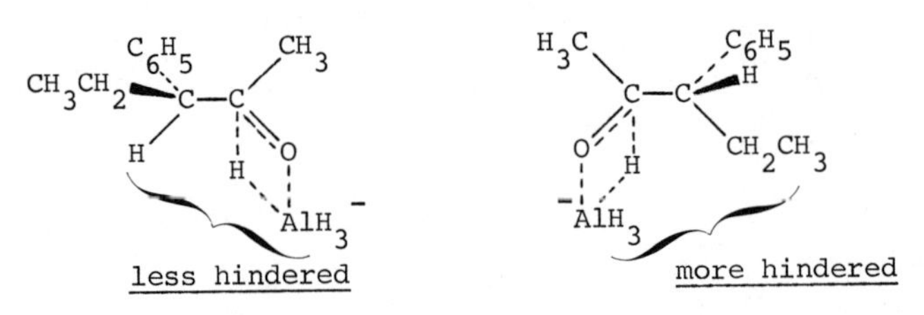

The bulky phenyl group is shown in the rear so that the least hindered transition state can be drawn in each case. (Any other spatial arrangement of the atoms would show greater hindrance.)

11.71 (a)

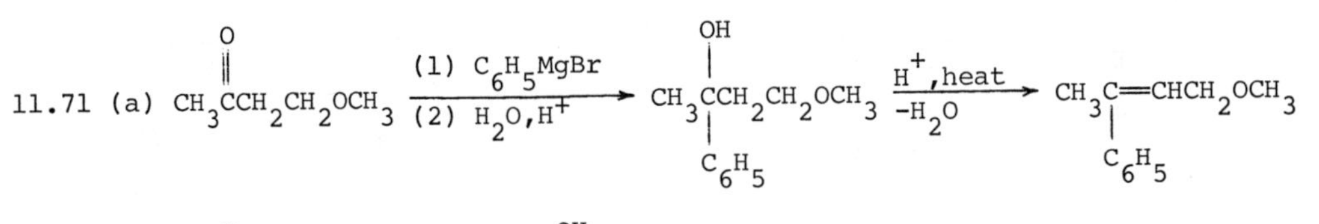

(b) $CH_3\overset{O}{\overset{\|}{C}}CH_3 \xrightarrow[(2)\ H_2O,H^+]{(1)\ NaBH_4} CH_3\overset{OH}{\overset{|}{C}}HCH_3 \xrightarrow[heat]{H_2SO_4} CH_3CH{=}CH_2 \xrightarrow{NBS}$

$BrCH_2CH{=}CH_2 \xrightarrow[(2)\ Na^+\ ^-OCH_2CH_3]{(1)\ (C_6H_5)_3P} (C_6H_5)_3P{=}CHCH{=}CH_2$

$\xrightarrow{(CH_3)_2C{=}O} CH_2{=}CHCH{=}C(CH_3)_2$

(c) $H\overset{O}{\overset{\|}{C}}CH_2\overset{O}{\overset{\|}{C}}H \xrightarrow[(2)\ H_2O,H^+]{(1)\ 2\ CH_3CH_2MgBr} CH_3CH_2\overset{OH}{\overset{|}{C}}HCH_2\overset{OH}{\overset{|}{C}}HCH_2CH_3$

(d) cyclohexanone $\xrightarrow{Cl_2, H^+}$ 2-chlorocyclohexanone (Cl) $\xrightarrow[heat]{KOH,\ CH_3CH_2OH}$ 2-cyclohexenone $\xrightarrow[1,4\text{-addition}]{H_2O, H^+}$

3-hydroxycyclohexanone (HO) $\xrightarrow{H_2CrO_4}$ 1,3-cyclohexanedione

(e)

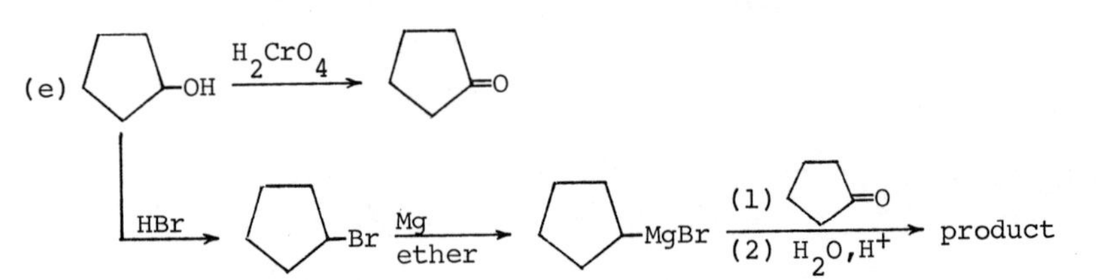

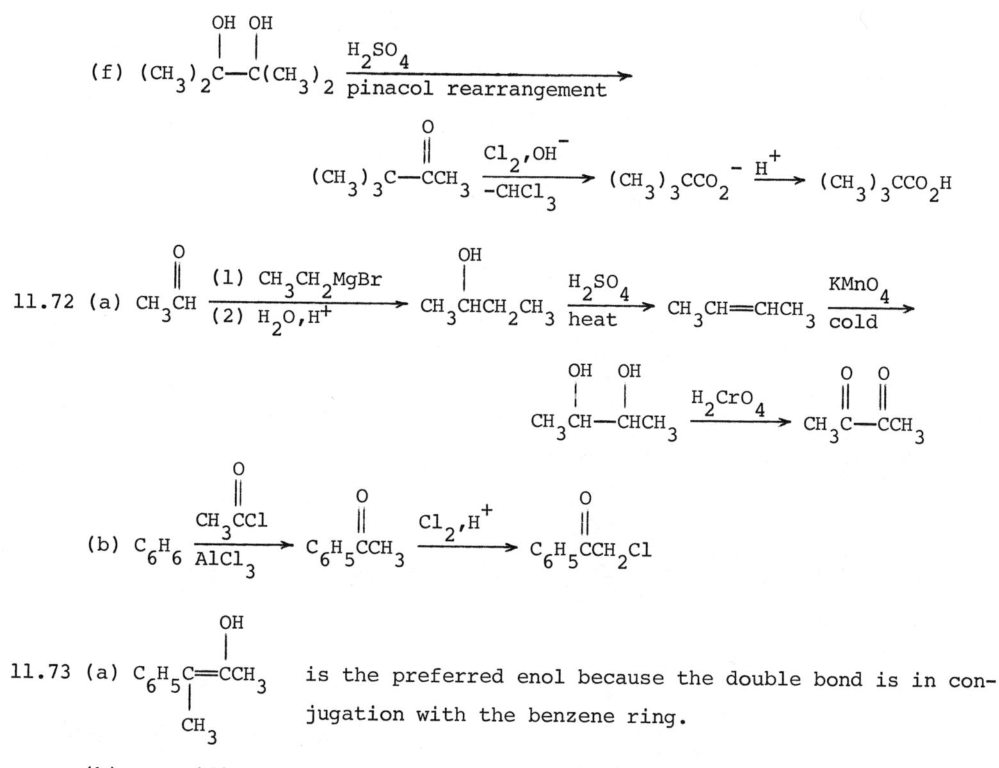

(f) $(CH_3)_2C(OH){-}C(OH)(CH_3)_2 \xrightarrow[\text{pinacol rearrangement}]{H_2SO_4}$

$$(CH_3)_3C{-}\overset{O}{\overset{\|}{C}}{-}CH_3 \xrightarrow[-CHCl_3]{Cl_2,\,OH^-} (CH_3)_3CCO_2^- \xrightarrow{H^+} (CH_3)_3CCO_2H$$

11.72 (a) $CH_3\overset{O}{\overset{\|}{C}}H \xrightarrow[(2)\ H_2O,\,H^+]{(1)\ CH_3CH_2MgBr} CH_3\overset{OH}{\overset{|}{C}}HCH_2CH_3 \xrightarrow[\text{heat}]{H_2SO_4} CH_3CH{=}CHCH_3 \xrightarrow[\text{cold}]{KMnO_4}$

$$CH_3\overset{OH}{\overset{|}{C}}H{-}\overset{OH}{\overset{|}{C}}HCH_3 \xrightarrow{H_2CrO_4} CH_3\overset{O}{\overset{\|}{C}}{-}\overset{O}{\overset{\|}{C}}CH_3$$

(b) $C_6H_6 \xrightarrow[AlCl_3]{CH_3\overset{O}{\overset{\|}{C}}Cl} C_6H_5\overset{O}{\overset{\|}{C}}CH_3 \xrightarrow{Cl_2,\,H^+} C_6H_5\overset{O}{\overset{\|}{C}}CH_2Cl$

11.73 (a) $C_6H_5\underset{CH_3}{\underset{|}{C}}{=}\overset{OH}{\overset{|}{C}}CH_3$ is the preferred enol because the double bond is in conjugation with the benzene ring.

(b) In acidic or basic solution, D_2O reacts with this enol as follows:

$$D\ddot{O}{-}D + C_6H_5\underset{CH_3}{\underset{|}{C}}{=}\overset{:\ddot{O}{-}H}{\overset{|}{C}}CH_3 \rightleftharpoons C_6H_5\underset{CH_3}{\underset{|}{C}}D\overset{:\ddot{O}}{\overset{\|}{C}}CH_3 + D\ddot{O}H$$

D in benzylic position

(c) The ketone undergoes racemization because the chiral carbon is α to the carbonyl group and undergoes tautomerism at that position.

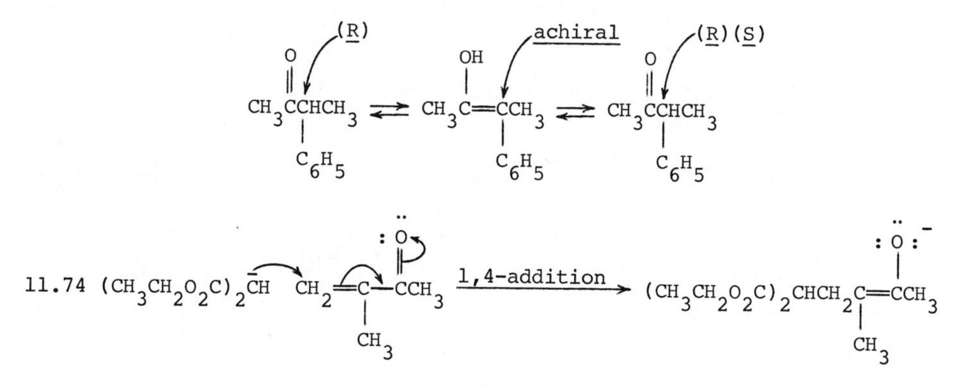

$$CH_3\overset{O}{\overset{\|}{C}}\underset{C_6H_5}{\underset{|}{C}}HCH_3\ (\underline{R}) \rightleftharpoons CH_3\overset{OH}{\overset{|}{C}}{=}\underset{C_6H_5}{\underset{|}{C}}CH_3\ (\text{achiral}) \rightleftharpoons CH_3\overset{O}{\overset{\|}{C}}\underset{C_6H_5}{\underset{|}{C}}HCH_3\ (\underline{R})(\underline{S})$$

11.74 $(CH_3CH_2O_2C)_2\bar{C}H\ \ CH_2{=}\underset{CH_3}{\underset{|}{C}}{-}\overset{:\ddot{O}}{\overset{\|}{C}}CH_3 \xrightarrow{\text{1,4-addition}} (CH_3CH_2O_2C)_2CHCH_2\underset{CH_3}{\underset{|}{C}}{=}\overset{:\ddot{O}:^-}{\overset{|}{C}}CH_3$

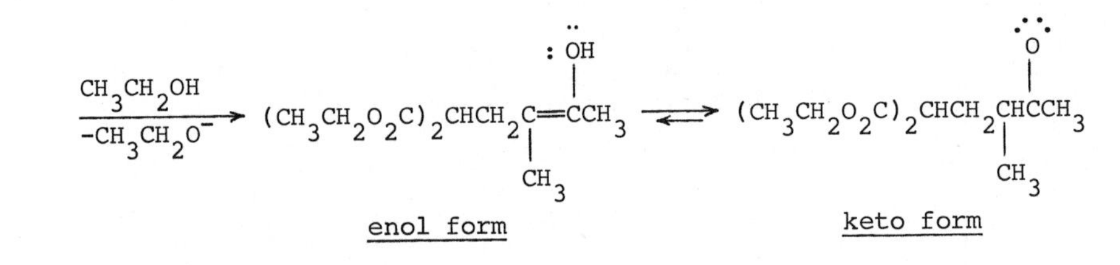

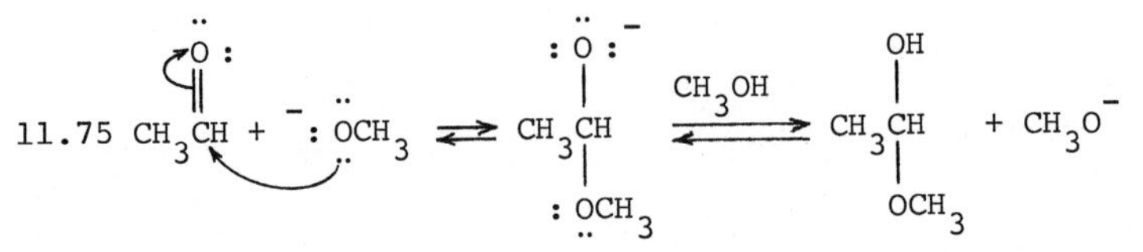

11.76 In acid, the hemiacetal OH can be protonated and forms a leaving group $(-OH_2^+)$. In base, no such leaving group can be formed.

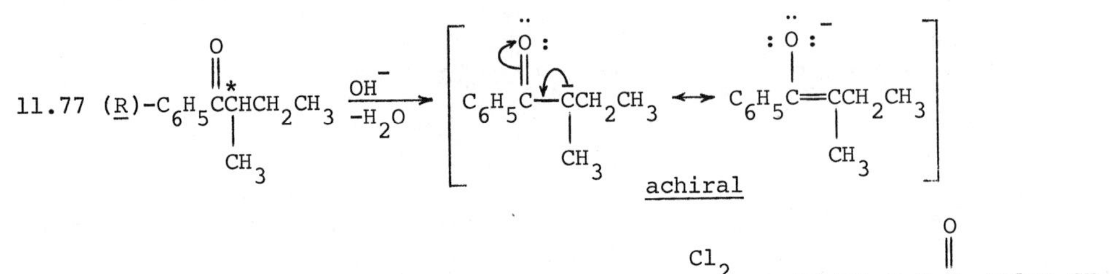

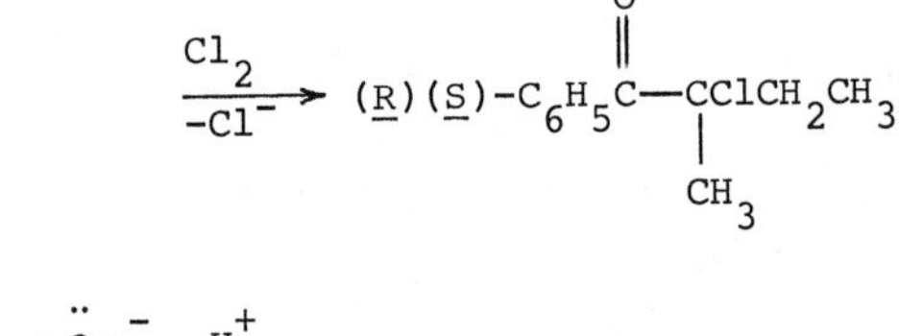

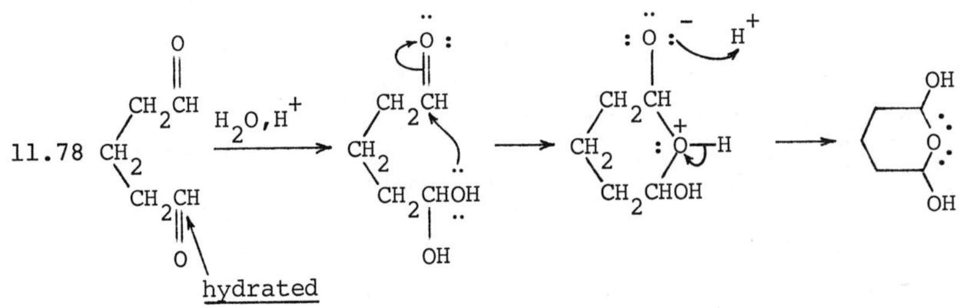

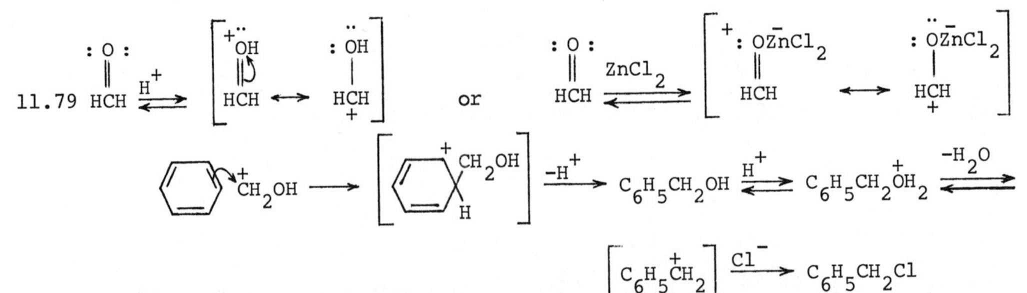

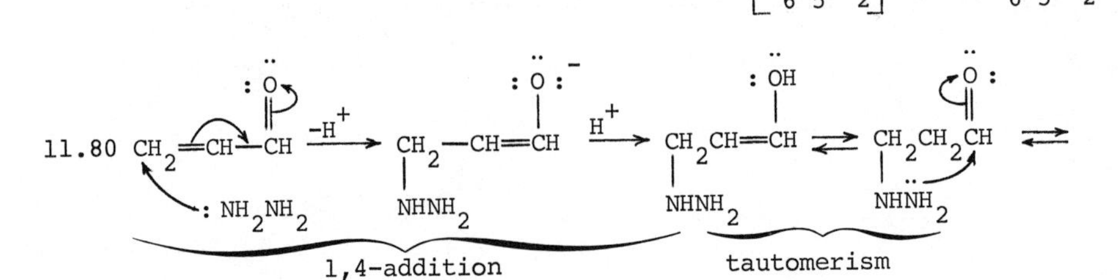

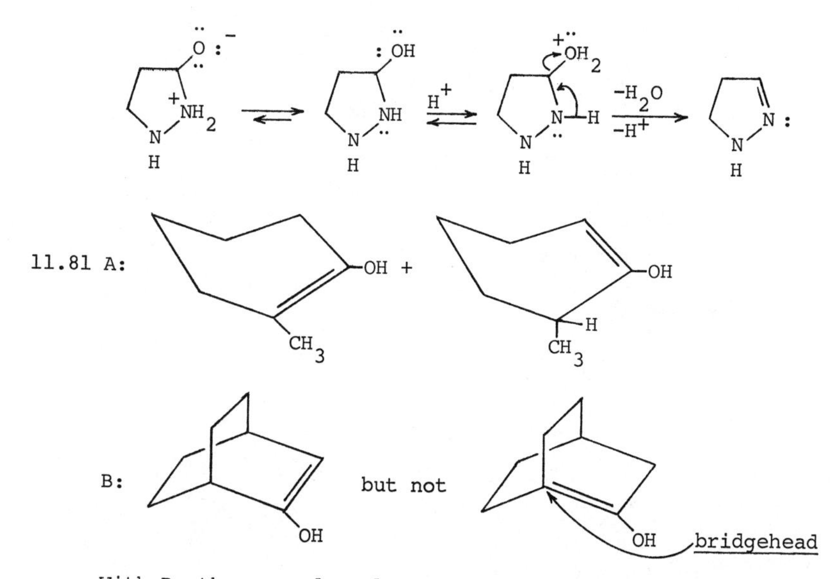

With B, the second enol cannot be formed because the rigid geometry of the ring system does not allow p-orbital overlap between the bridgehead carbon and the carbonyl carbon.

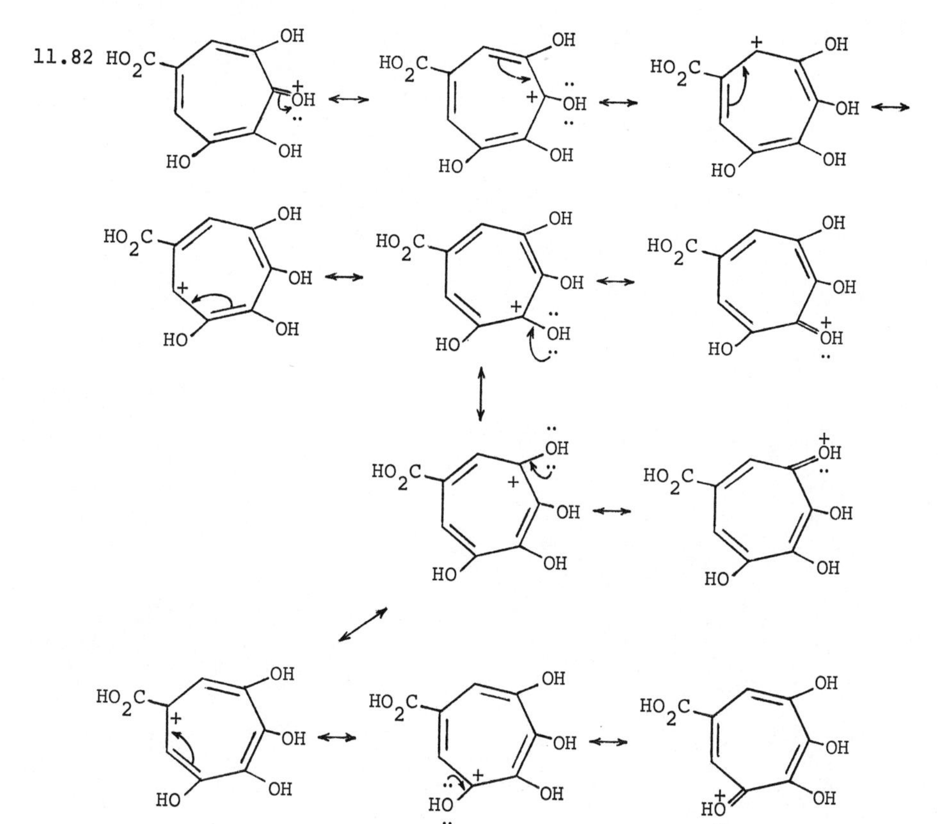

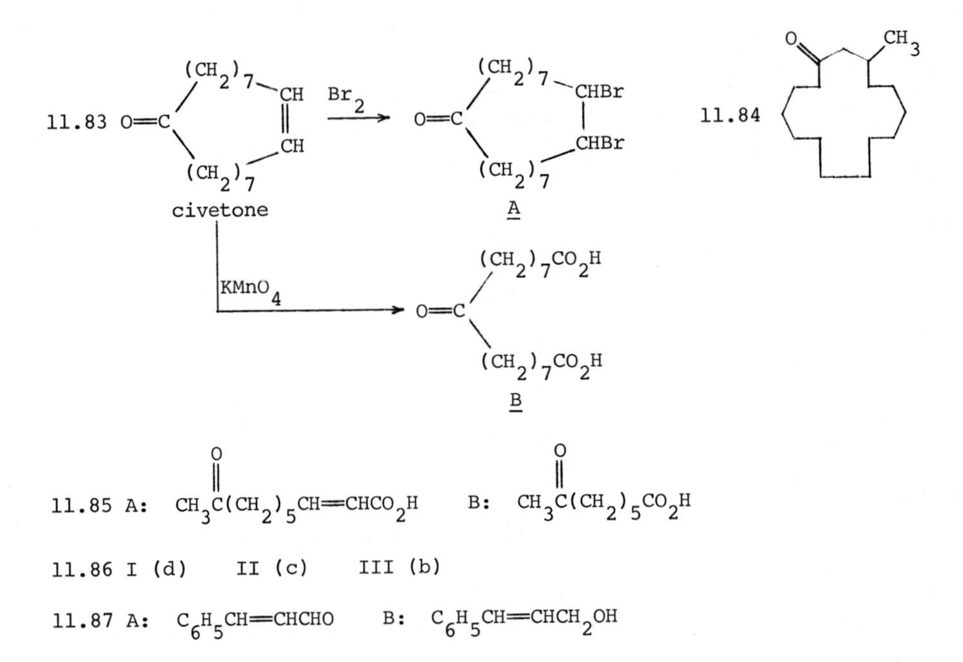

11.85 A: $CH_3C(=O)(CH_2)_5CH{=}CHCO_2H$ B: $CH_3C(=O)(CH_2)_5CO_2H$

11.86 I (d) II (c) III (b)

11.87 A: $C_6H_5CH{=}CHCHO$ B: $C_6H_5CH{=}CHCH_2OH$

CHAPTER 12

Carboxylic Acids

Some Important Features

Carboxylic acids (RCO_2H) are more acidic than alcohols, phenols, or carbonic acid. Carboxylic acids yield carboxylate salts when treated with any base stronger than the carboxylate ion itself.

$$RCO_2H + Na^+ \ HCO_3^- \rightleftharpoons RCO_2^- \ Na^+ + H_2O + CO_2$$

Factors affecting acid strength are discussed in detail in this chapter of the text. Among these factors are the <u>electronegativity</u> of the atom attached to the acidic hydrogen. A greater electronegativity means a stronger acid; therefore, alkanes are extremely weak acids, while amines (R_2NH), alcohols (ROH), and hydrogen halides (HX) are successively stronger acids.

<u>Hybridization</u> of the atom attached to H affects acidity: $RC{\equiv}CH$ is more acidic than $R_2C{=}CHR$ or RH.

The <u>inductive effect</u> increases or decreases acid strength by withdrawing or releasing electron density.

$$ClCH_2CO_2H \rightleftharpoons ClCH_2CO_2^- + H^+$$

$$Cl{-}C_6H_4{-}CO_2H \rightleftharpoons Cl{-}C_6H_4{-}CO_2^- + H^+$$

Cl strengthens the acid by stabilizing the anion relative to the acid through dispersal of the negative charge.

Resonance-stabilization of the anion also increases acid strength. This resonance-stabilization is the primary reason for the acidity of carboxylic acids. Resonance-stabilization of the anion also explains why phenols are more acidic than alcohols.

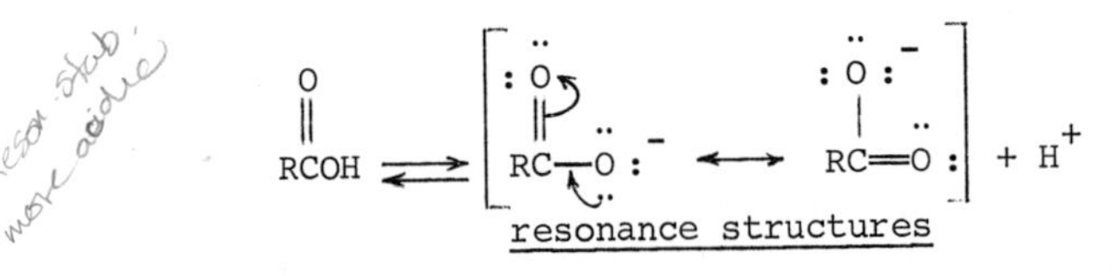

Other factors, such as hydrogen bonding in the anion, can also affect acid strength.

A carboxylic acid may undergo esterification with an alcohol. The rate of esterification decreases with increasing steric hindrance around the carboxyl groups. (Review the mechanism for this reaction in Section 12.9.)

$$\mathrm{RCOH} + \mathrm{R'OH} \underset{}{\overset{H^+}{\rightleftharpoons}} \mathrm{RCOR'} + H_2O$$

an ester

Decarboxylation of β-keto acids is important both in the laboratory and in biological systems.

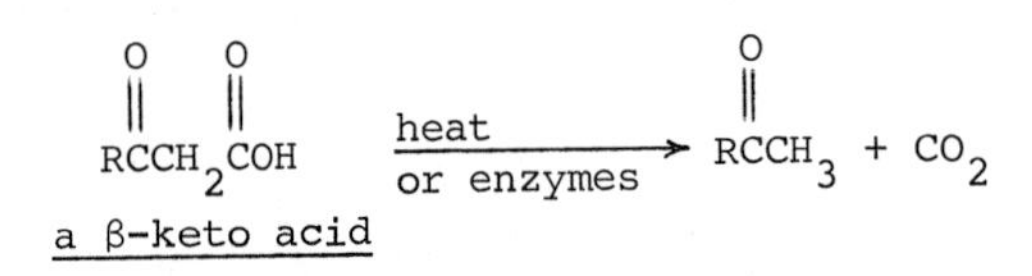

Other reactions of carboxylic acids are reduction by LAH (Section 12.10), 1,4-addition to α,β-unsaturated carboxylic acids (Section 12.11D), and anhydride formation by diacids (Section 12.11B).

Reminders

For calculations of K_a and pK_a values, review Section 1.10.

Remember that the conjugate base of a strong acid is a weak base, while the conjugate base of a weak acid is a strong base.

$$\underset{\text{strong acid}}{RCO_2H} \rightleftharpoons \underset{\text{weak base}}{RCO_2^-} + H^+$$

$$\underset{\text{weak acid}}{ROH} \rightleftharpoons \underset{\text{strong base}}{RO^-} + H^+$$

Because a carboxylic acid contains oxygen atoms with unshared electrons, it can be protonated by a strong acid or it can undergo hydrogen bonding.

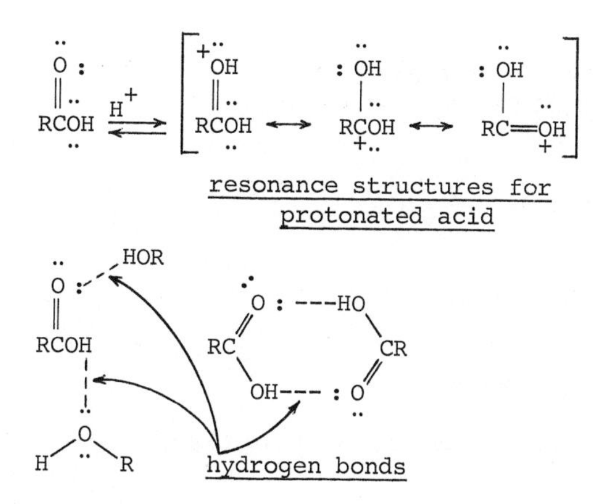

Answers to Problems

12.24 (a) 2,2-dimethylpropanoic acid (b) p-chlorobenzoic acid

(c) 2,3-dibromopentanoic acid or α,β-dibromovaleric acid

(d) magnesium propanoate or magnesium propionate

(e) calcium ethanedioate or calcium oxalate

(f) sodium 2-bromopropanoate or sodium α-bromopropionate

12.25 (a) $CH_2ICH_2CH_2CO_2H$ (b) $HCO_2^-\ K^+$ (c) $CO_2^-\ Na^+$ / $CO_2^-\ Na^+$

(d) $C_6H_5CO_2^-\ Na^+$ (e) CH_3 / CO_2H

12.26 (a) $CH_3CH_2\overset{O}{\overset{\|}{C}}—$ (b) $CH_3CH_2CH_2\overset{O}{\overset{\|}{C}}—$ (c) NO_2 / $\overset{O}{\overset{\|}{C}}—$

12.27 (a)

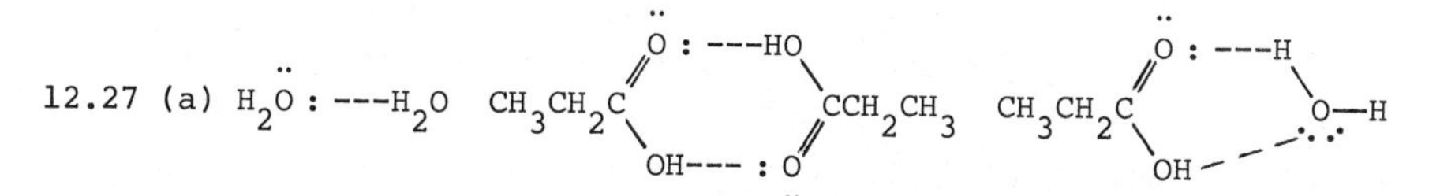

(b)

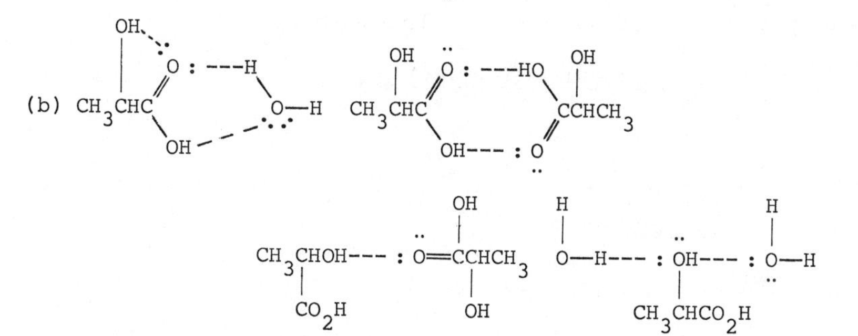

(There are other possibilities.)

12.28 (a) $CH_3CH_2CH_2Br \xrightarrow{KCN} CH_3CH_2CH_2CN \xrightarrow[heat]{H_2O, H^+} CH_3CH_2CH_2CO_2H$

or

$\xrightarrow{Mg, ether} CH_3CH_2CH_2MgBr \xrightarrow{CO_2} CH_3CH_2CH_2CO_2MgBr \xrightarrow{H_2O, H^+} CH_3CH_2CH_2CO_2H$

(b) $CH_3CH_2CH_2CH_2OH \xrightarrow[heat]{KMnO_4}$ (c) $CH_3CH_2CH_2CHO \xrightarrow{KMnO_4}$

(d) $CH_3CH_2CH_2CH{=}CHCH_2CH_2CH_3 \xrightarrow[heat]{KMnO_4}$

In (b), (c), and (d), other oxidizing agents would also be suitable.

(e) $\xrightarrow[heat]{H_2O, H^+}$ (f) $\xrightarrow[heat]{H_2O, H^+}$

12.29 (a) (1) KCN (2) dil. HCl, heat (3) neutralize with NaOH

(b) (1) Mg, diethyl ether (2) CO_2 (3) dil. HCl

(c) (1) CrO_3, 2 pyridine (2) HCN (3) dil. HCl, heat

(d) (1) excess KCN (2) dil. HCl, heat

(e) (1) HBr, H_2O_2 (or BH_3, followed by $Br_2 + OH^-$)

(2) KCN (3) dil. HCl, heat

(f) (1) Mg, diethyl ether (2) CO_2 (3) dil. HCl

(g) dil. HCl, heat

(h) (1) Mg, diethyl ether (2) CH_2CH_2 (epoxide, O bridging) (3) H_2O, H^+ (4) H_2CrO_4

12.30 (a) cyclohexane-1,2-dicarboxylate: ring bearing $CO_2^- \ Na^+$ and $CO_2^- \ Na^+$ (b) $C_6H_5CH_2O_2CCH_2CH_3$ (c) $CH_3CO_2^- + CH_3OH$

(d) $CH_3CO_2^- \ CH_3NH_3^+$ (e) $CH_3CO_2H + ClCH_2CO_2^-$ (f) $HO_2CCH_2CO_2^- \ Na^+$

(g) no reaction (h) $C_6H_5CO_2CH_3$ (i) $C_6H_5CO_2^- \ Li^+$ (j) $C_6H_5O^- \ Li^+$

12.31 $CH_3(CH_2)_3CO_2H \rightleftharpoons CH_3(CH_2)_3CO_2^- + H^+$

$0.200 - 0.00184\ \underline{M}$ $\quad 0.00184\ \underline{M}$ $\quad 0.00184\ \underline{M}$

$$\underline{K}_{\underline{a}} = \frac{(0.00184)(0.00184)}{0.200 - 0.00184} = 1.71 \times 10^{-5}$$

12.32 Mixture $\xrightarrow{NaHCO_3} C_6H_5CO_2^- \ Na^+$
<u>in Solution A</u>

New mixture $\xrightarrow{NaOH} CH_3CH_2-C_6H_4-O^- \ Na^+$
<u>in Solution B</u>

New mixture $\xrightarrow{H_2O}$ no dissolved salts

Solution C

Solution D contains C_6H_5CHO.

12.33 (a) Shake with aqueous $NaHCO_3$; the carboxylic acid dissolves and the ester does not.

(b) Shake with H_2O or aqueous NaOH; phenol dissolves and the ester does not.

(c) Shake with H_2O or aqueous NaOH; phenol dissolves and the alcohol does not.

12.34 (a) 90.08 (same as the molecular weight)

(b) 66.04 (one-half the molecular weight)

12.35 (1) equivalents NaOH = $\underline{NV}$ = (0.307 eq/liter)(0.011 liter) = 3.38×10^{-3}

$$\text{neut. eq. of acid} = \frac{0.250\ g}{3.38 \times 10^{-3}\ \text{equivalents}} = 74.0$$

(2) Calculate the equivalent weights:

(a) 88 (b) 74 (c) 52 (d) 73 (e) 72

Allowing for experimental error, (b), (d), and (e) are possibilities.

12.36 (a) $pK_a = -\log (1.3 \times 10^{-4}) = 4 - \log 1.3 = 3.9$

(b) $pK_a = -\log (3.65 \times 10^{-5}) = 5 - \log 3.65 = 4.44$

12.37 $K_a = \frac{[H^+][A^-]}{[HA]}$ $\qquad$ $pH = -\log [H^+] = 2.50$ $\qquad$ $[H^+] = 10^{-2.50} = 3.20 \times 10^{-3}$

$$K_a = \frac{(10^{-2.50})(10^{-2.50})}{0.010 - (3.20 \times 10^{-3})} = \frac{10^{-5}}{0.00680} = 1.47 \times 10^{-3}$$

12.38 A more electronegative substituent stabilizes the anion relative to the acid to a greater extent; therefore, FCH_2CO_2H is the strongest acid of the three and $BrCH_2CO_2H$ is the weakest.

12.39 (d), (h), (e), (f), (c), (b), (a), (g).

Note the order: alcohol (weakest), water, phenol, carbonic acid (weaker than RCO_2H because HCO_3^- reacts with RCO_2H), carboxylic acids (in order of increasing inductive effect).

12.40 (a) *p*-bromobenzoic acid (b) *m*-bromobenzoic acid

(c) 3,5-dibromobenzoic acid

The reason for these relative acidities is that Br is electron-withdrawing by the inductive effect and stabilizes the conjugate base relative to the acid. In (c), two Br atoms exert a stronger electron-withdrawing effect than one Br atom.

12.41 *p*-methylphenol, phenol, *p*-nitrophenol.

p-Nitrophenol is a stronger acid than phenol because of resonance-stabilization of the anion, which is aided by the nitro group. (See the answer to Problem 7.12 in the text for the resonance structures.) *p*-Methylphenol is a weaker acid than phenol because the methyl group is electron-releasing and destabilizes the anion relative to the conjugate acid.

12.42 (b) is the most acidic and (c) is the least acidic.

12.43 In each case, determine which of the parent acids is the weaker acid (Section 12.7). The anion of that acid is the stronger base.

(a) $CH_3CH{=}CH^-$ (b) $CH_3CO_2^-$ (c) $ClCH_2CO_2^-$ (d) $(CH_3)_3CO^-$

(e) $ClCH_2CH_2CO_2^-$

12.44 *o*-Phthalic acid has a smaller pK_1 (more acidic) because intramolecular hydrogen bonding helps stabilize the anion resulting from the first ionization.

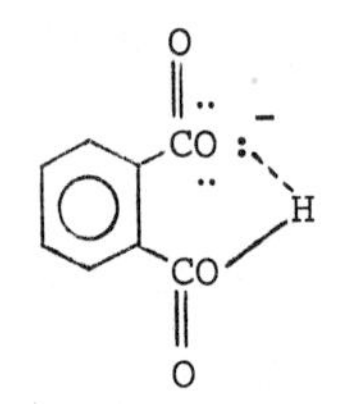

However, the pK_2 of *o*-phthalic acid is larger because of (1) a stabilized reactant, and (2) a dianion that is destabilized by the proximity of two negative charges.

12.45

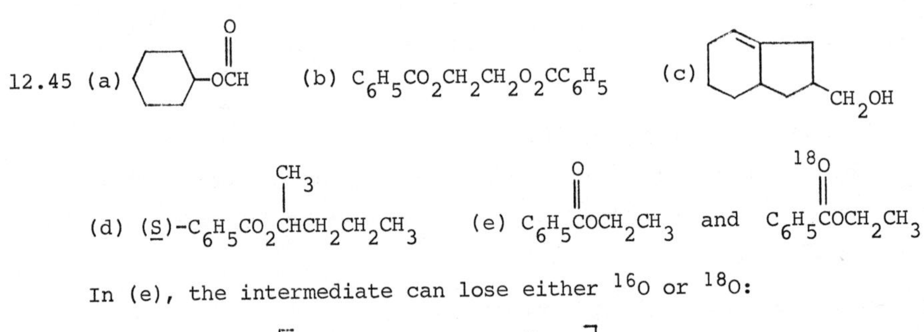

(b) $C_6H_5CO_2CH_2CH_2O_2CC_6H_5$

(d) ($\underline{S}$)-$C_6H_5CO_2CH(CH_3)CH_2CH_2CH_3$

(e) $C_6H_5COCH_2CH_3$ (C=O) and $C_6H_5COCH_2CH_3$ (C=^{18}O)

In (e), the intermediate can lose either ^{16}O or ^{18}O:

$$C_6H_5C(=O){}^{18}OH \xrightleftharpoons{H^+} \left[C_6H_5C(=\overset{+}{O}H){}^{18}OH \longleftrightarrow C_6H_5\overset{+}{C}(OH){}^{18}OH \right] \xrightleftharpoons{CH_3CH_2OH}$$

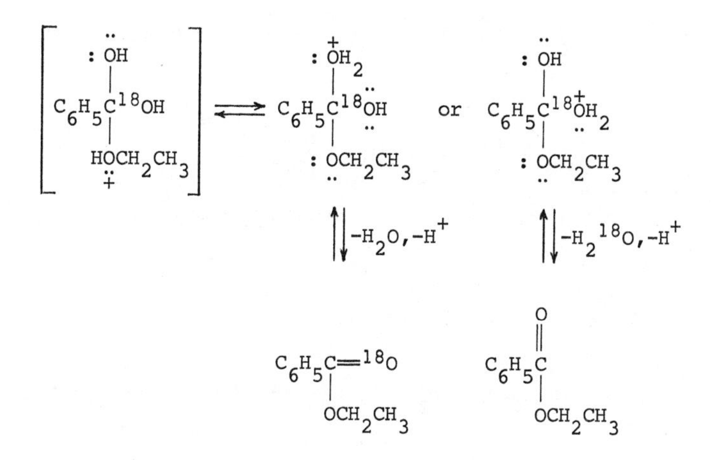

12.46 (b), (c), (a). The order is based upon decreasing steric hindrance.

12.47 (a) $CH_3\overset{*}{C}HBrCO_2\overset{*}{C}H(CH_3)CH_2CH_3$ + $CH_3\overset{*}{C}HBrCO_2\overset{*}{C}H(CH_3)CH_2CH_3$

(R) (R) (S) (R)

(b) These two esters are diastereomers. The enantiomer of the first would be (S,S), while the enantiomer of the second would be (R,S).

12.48 (a) 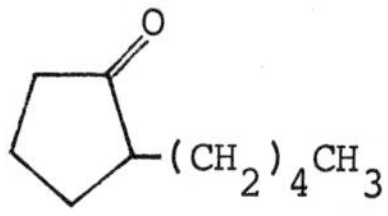 2-(CH$_2$)$_4$CH$_3$-cyclopentanone (b) 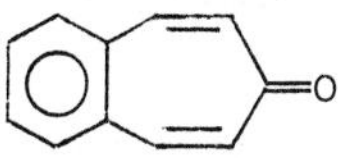 [benzo-fused cycloheptadienone structure]

12.49 $HO_2CCH_2CH_2\overset{O}{\overset{\|}{C}}CO_2H$.

The carboxyl group that is lost is the only one β to a carbonyl group.

12.50 The intermediate enol would have a bridgehead double bond. Such a structure would be extremely strained and cannot be formed under normal conditions.

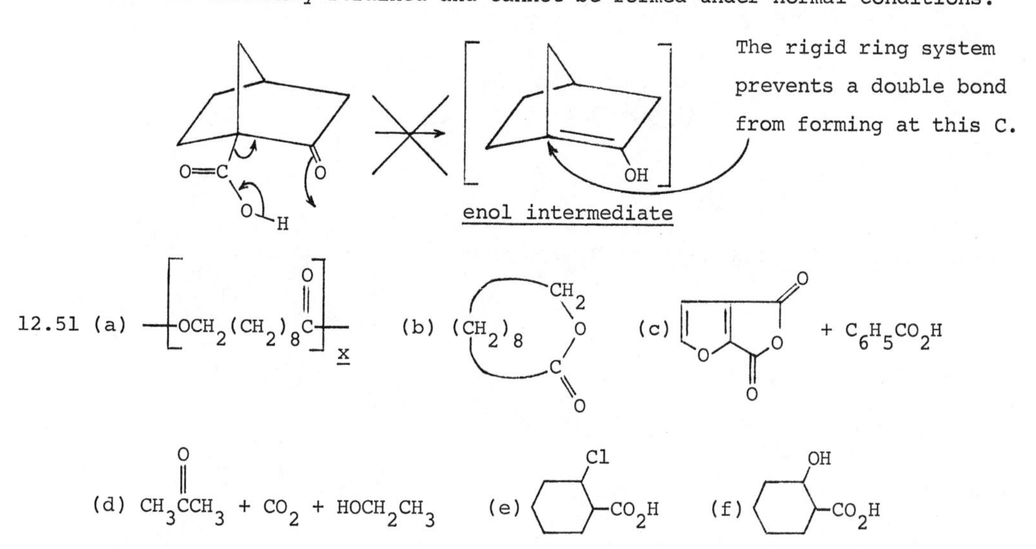

12.52 The oxidation is similar to the oxidation of an alkylbenzene:

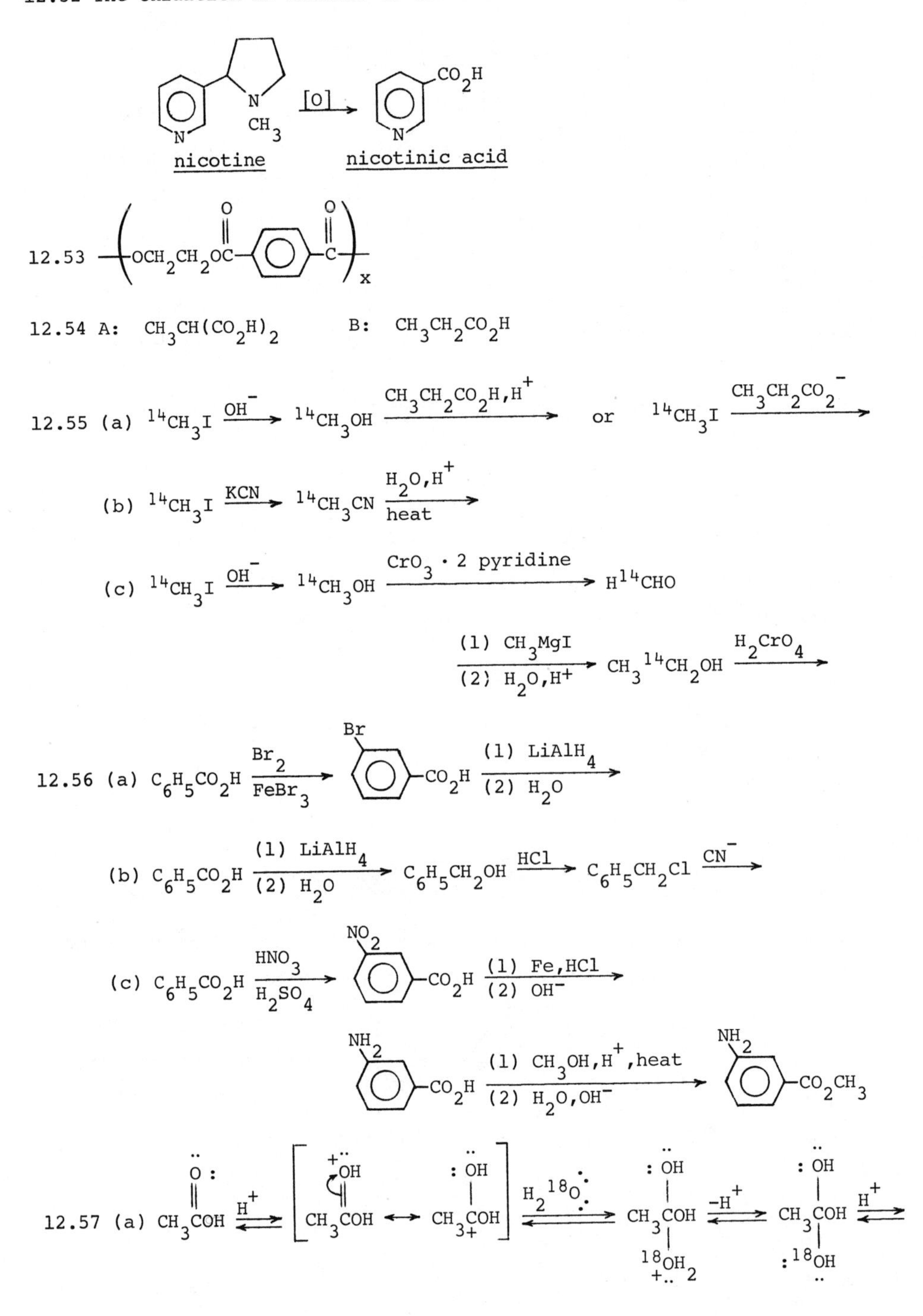

12.53 $\left(-OCH_2CH_2OC-\bigcirc-C-\right)_x$

12.54 A: $CH_3CH(CO_2H)_2$ B: $CH_3CH_2CO_2H$

12.55 (a) $^{14}CH_3I \xrightarrow{OH^-} {}^{14}CH_3OH \xrightarrow{CH_3CH_2CO_2H,H^+}$ or $^{14}CH_3I \xrightarrow{CH_3CH_2CO_2^-}$

(b) $^{14}CH_3I \xrightarrow{KCN} {}^{14}CH_3CN \xrightarrow[\text{heat}]{H_2O,H^+}$

(c) $^{14}CH_3I \xrightarrow{OH^-} {}^{14}CH_3OH \xrightarrow{CrO_3 \cdot 2\ pyridine} H^{14}CHO$

$\xrightarrow[\text{(2) } H_2O,H^+]{\text{(1) } CH_3MgI} CH_3{}^{14}CH_2OH \xrightarrow{H_2CrO_4}$

12.56 (a) $C_6H_5CO_2H \xrightarrow[\text{FeBr}_3]{Br_2}$ (Br-substituted benzene with CO_2H) $\xrightarrow[\text{(2) } H_2O]{\text{(1) } LiAlH_4}$

(b) $C_6H_5CO_2H \xrightarrow[\text{(2) } H_2O]{\text{(1) } LiAlH_4} C_6H_5CH_2OH \xrightarrow{HCl} C_6H_5CH_2Cl \xrightarrow{CN^-}$

(c) $C_6H_5CO_2H \xrightarrow[\text{H}_2SO_4]{HNO_3}$ (NO_2-substituted benzene with CO_2H) $\xrightarrow[\text{(2) } OH^-]{\text{(1) } Fe,HCl}$

(NH_2-substituted benzene with CO_2H) $\xrightarrow[\text{(2) } H_2O,OH^-]{\text{(1) } CH_3OH,H^+,heat}$ (NH_2-substituted benzene with CO_2CH_3)

12.57 (a) $CH_3\overset{\overset{\ddots}{O}:}{\overset{\|}{C}}OH \overset{H^+}{\rightleftharpoons} \left[CH_3\overset{\overset{+}{\ddots}{OH}}{\overset{\|}{C}}OH \longleftrightarrow CH_3\overset{\overset{\ddots}{OH}}{\overset{|}{C}}OH \atop 3+ \right] \xrightarrow{H_2{}^{18}O:} CH_3\overset{\overset{\ddots}{OH}}{\overset{|}{C}}OH \atop {}^{18}OH_2 \overset{-H^+}{\rightleftharpoons} CH_3\overset{\overset{\ddots}{OH}}{\overset{|}{C}}OH \atop :{}^{18}OH \overset{H^+}{\rightleftharpoons}$

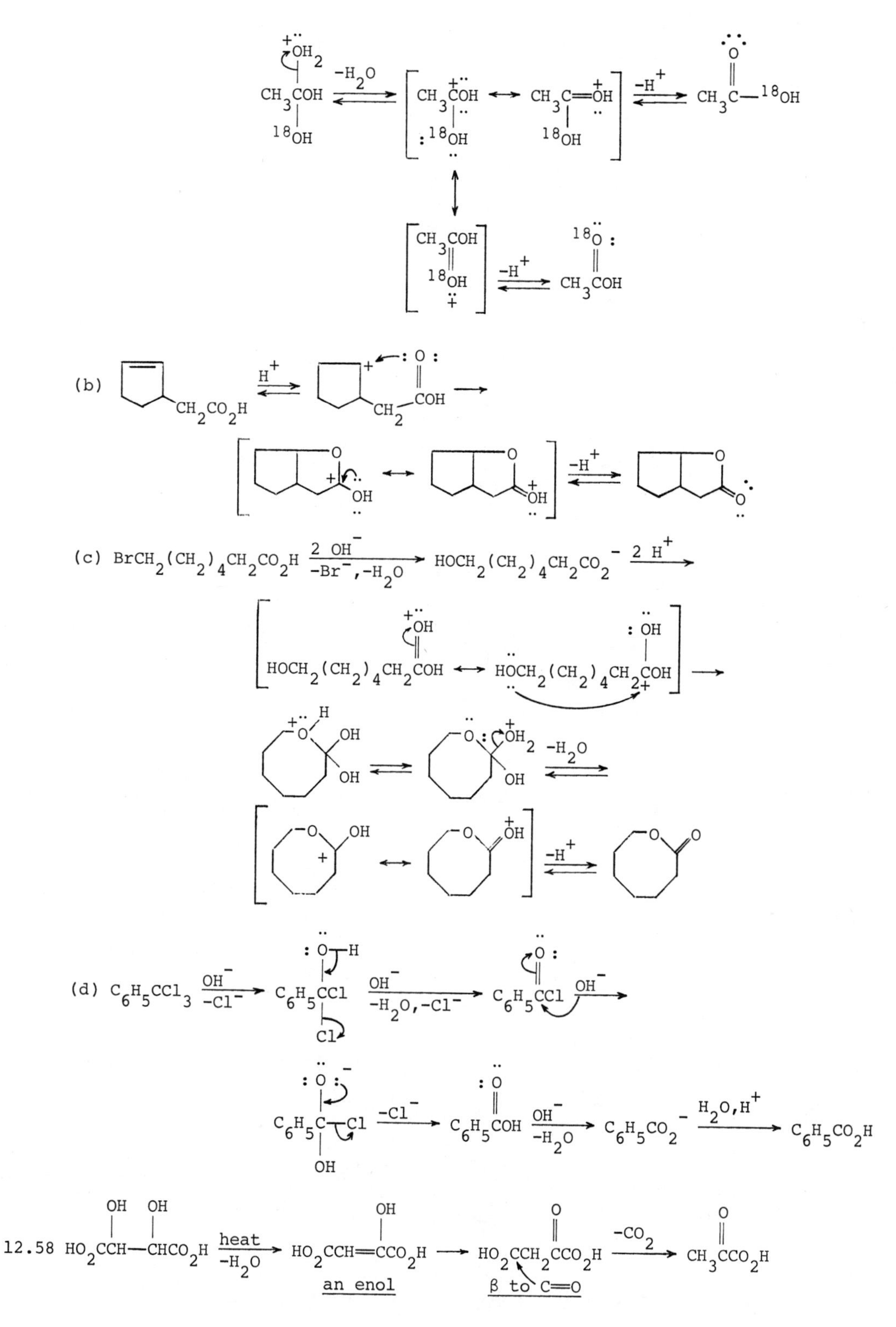

12.58 $HO_2CCH(OH)—CH(OH)CO_2H \xrightarrow[-H_2O]{heat} HO_2CCH{=}C(OH)CO_2H \longrightarrow HO_2CCH_2C(=O)CO_2H \xrightarrow{-CO_2} CH_3C(=O)CO_2H$

an enol; β to C=O

12.59 A carboxylate does not contain a true carbonyl group; it is a composite of two resonance structures.

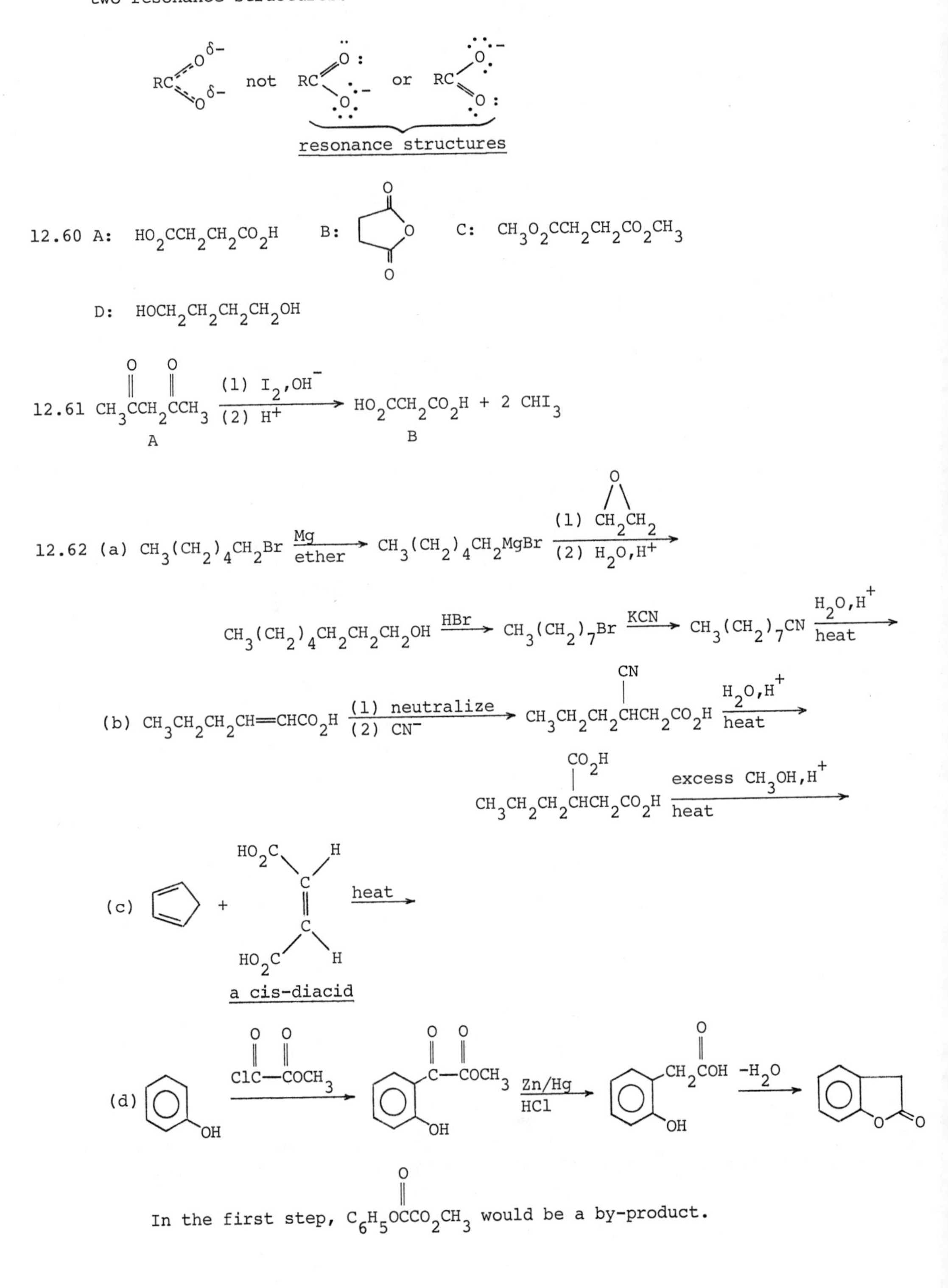

12.60 A: $HO_2CCH_2CH_2CO_2H$ B: C: $CH_3O_2CCH_2CH_2CO_2CH_3$

 D: $HOCH_2CH_2CH_2CH_2OH$

12.61 $CH_3\overset{O}{\overset{||}{C}}CH_2\overset{O}{\overset{||}{C}}CH_3$ $\xrightarrow[\text{(2) H}^+]{\text{(1) I}_2,\text{OH}^-}$ $HO_2CCH_2CO_2H + 2\ CHI_3$
 A B

12.62 (a) $CH_3(CH_2)_4CH_2Br \xrightarrow[\text{ether}]{\text{Mg}} CH_3(CH_2)_4CH_2MgBr \xrightarrow[\text{(2) H}_2O,\text{H}^+]{\text{(1) CH}_2\text{CH}_2}$

 $CH_3(CH_2)_4CH_2CH_2CH_2OH \xrightarrow{\text{HBr}} CH_3(CH_2)_7Br \xrightarrow{\text{KCN}} CH_3(CH_2)_7CN \xrightarrow[\text{heat}]{\text{H}_2O,\text{H}^+}$

 (b) $CH_3CH_2CH_2CH=CHCO_2H \xrightarrow[\text{(2) CN}^-]{\text{(1) neutralize}} CH_3CH_2CH_2\overset{CN}{\overset{|}{C}}HCH_2CO_2H \xrightarrow[\text{heat}]{\text{H}_2O,\text{H}^+}$

 $CH_3CH_2CH_2\overset{CO_2H}{\overset{|}{C}}HCH_2CO_2H \xrightarrow[\text{heat}]{\text{excess CH}_3\text{OH,H}^+}$

 (c) + [cis-diacid structure] $\xrightarrow{\text{heat}}$
 a cis-diacid

 (d) $\xrightarrow{\overset{O\ \ O}{\overset{||\ ||}{ClC-COCH_3}}}$ $\xrightarrow[\text{HCl}]{\text{Zn/Hg}}$ $\xrightarrow{-H_2O}$

 In the first step, $C_6H_5\overset{O}{\overset{||}{O}}CCO_2CH_3$ would be a by-product.

12.63 (a) $CH_3CH_2Br \xrightarrow{KCN} CH_3CH_2CN \xrightarrow[heat]{H_2O,H^+} CH_3CH_2CO_2H$

$CH_3CH_2Br \xrightarrow{OH^-} CH_3CH_2OH \xrightarrow[heat]{CH_3CH_2CO_2H,H^+} CH_3CH_2CO_2CH_2CH_3$

A Grignard reaction could also be used to prepare the carboxylic acid.

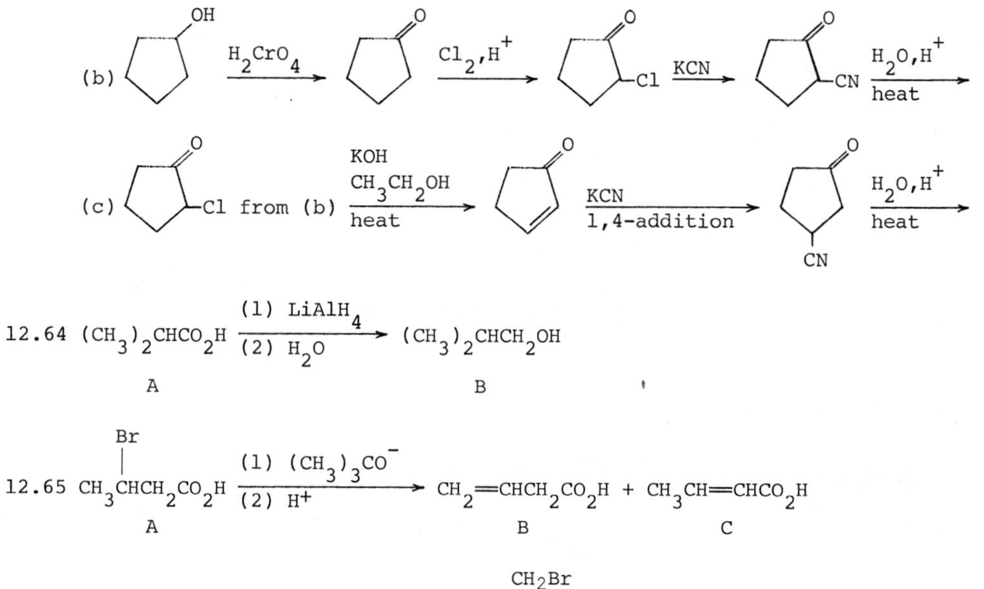

12.64 $(CH_3)_2CHCO_2H \xrightarrow[(2)\ H_2O]{(1)\ LiAlH_4} (CH_3)_2CHCH_2OH$

A B

12.65 $CH_3CH(Br)CH_2CO_2H \xrightarrow[(2)\ H^+]{(1)\ (CH_3)_3CO^-} CH_2{=}CHCH_2CO_2H + CH_3CH{=}CHCO_2H$

A B C

A could not be $(CH_3)_2CBrCO_2H$ or $CH_3CH(CH_2Br)CO_2H$ because only one alkenyl acid could result from the elimination reaction. For B, note the $—CH_2$ bending mode at about 920 cm^{-1} (10.8 μm) in the infrared spectrum. The nmr spectrum of C shows CH_3 absorption (the doublet at about 1.9) and $C\underline{H}{=}C\underline{H}$ downfield (besides the offset $—CO_2H$). From these data, you cannot tell if C is the _cis_ or _trans_ acid.

CHAPTER 13

Derivatives of Carboxylic Acids

Some Important Features

The common derivatives of carboxylic acids are esters, amides, acid halides, acid anhydrides, and nitriles. Except for nitriles, these compounds all undergo nucleophilic acyl substitution (addition-elimination) reactions with reagents such as water, alcohols, ammonia, or amines. Nitriles undergo similar reactions. (See the chapter-end summary in the text.)

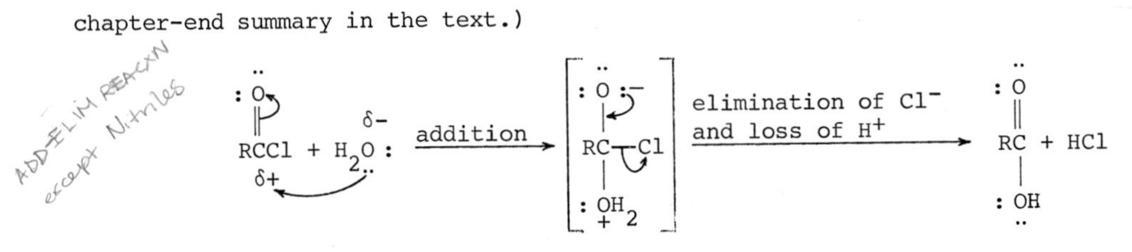

The reactivity of the carboxylic acid derivatives depends partly on the leaving group (Section 13.1).

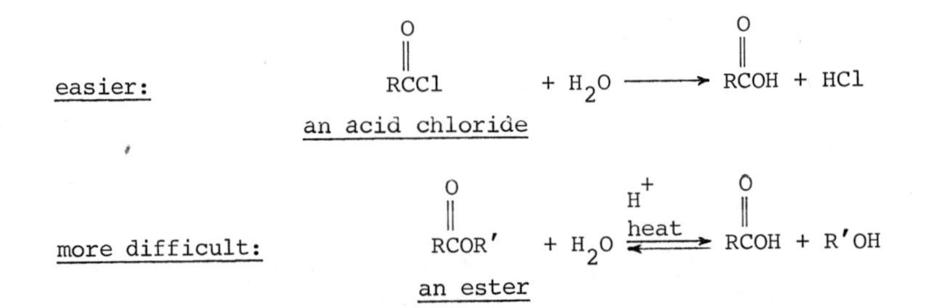

In your study of the reactions of the carboxylic acid derivatives, pay particular attention to the similarities of the reactions and their mechanisms. The reactions of esters and amides, for example, are quite similar.

Simplified mechanisms:

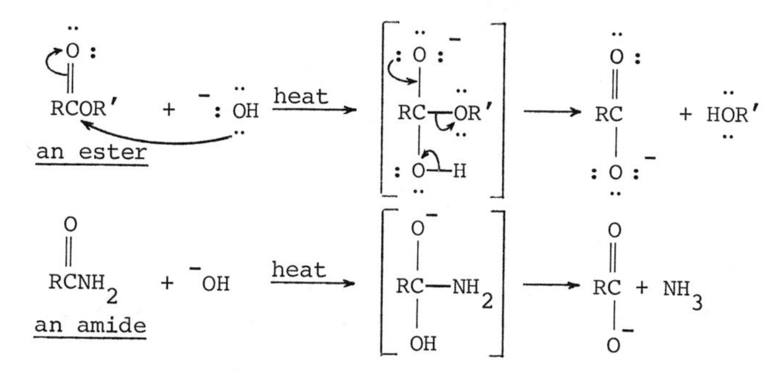

All carboxylic acid derivatives contain unsaturation and therefore can be reduced by catalytic hydrogenation or by LAH.

$$\underset{\text{an ester}}{\mathrm{RCOR'}}\ (\mathrm{C{=}O})\ \xrightarrow[\text{heat,pressure}]{H_2,\mathrm{Ni}}\ \mathrm{RCH_2OH + HOR'}$$

Other topics covered in this chapter are α halogenation of acid halides, some other types of esters (lactones, polyesters, thioesters), some other types of amides (lactams, imides, carbamates, and others), and the reactions of nitriles.

Reminders

In writing mechanisms for the reactions of carboxylic acid derivatives, use electron dots for unshared electrons. They will help you see where protonation can occur and the direction of the electron shifts. (Remember that protonation can occur in acidic solution, but not in alkaline solution.)

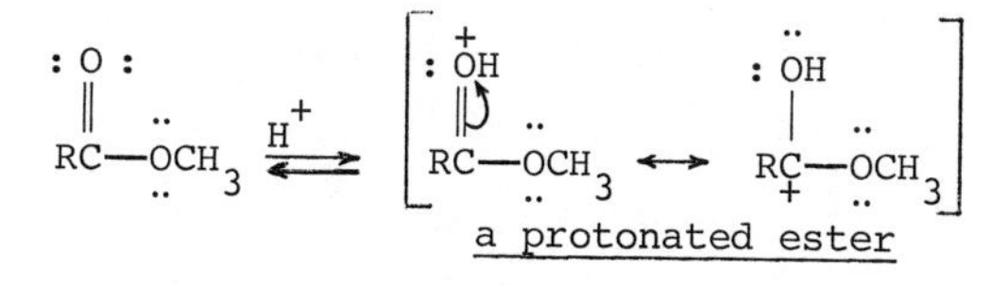

You have probably noticed that we are using more-complex structures now than we did earlier in the text. Do not be intimidated by a complex structure. For example, under saponification conditions, the only reactive functional group in the following compound is an ester group.

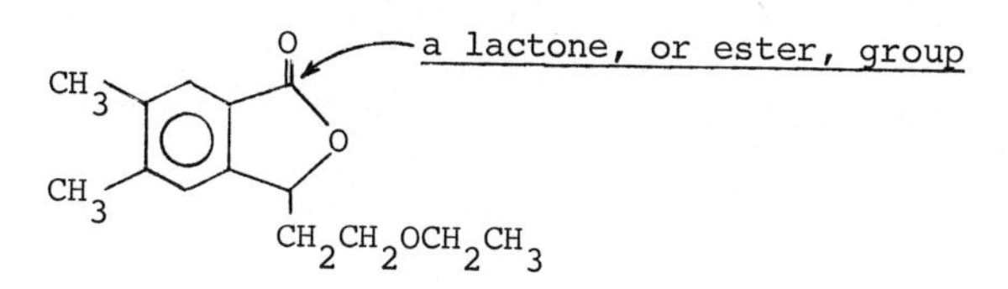

Answers to Problems

13.32 (a) p-bromobenzoyl bromide (b) pentanoic propanoic anhydride

(c) N,N-diethylbenzamide (d) hexanenitrile

(e) p-nitrophenyl acetate or p-nitrophenyl ethanoate

(f) methyl p-nitrobenzoate

13.33 (a) 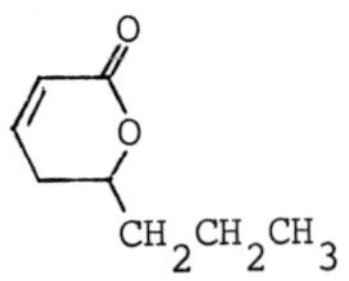 (b) $CH_3CH_2\overset{O}{\overset{\|}{C}}N(CH_2CH_3)_2$ (c) $CH_3NH\overset{O}{\overset{\|}{C}}NHCH_3$

(d) $CH_3CH_2CH_2\underset{}{\overset{CH_3}{\overset{|}{C}}}HCN$ (e) $CH_3\overset{NH_2}{\overset{|}{C}}HCO_2CH_2CH_3$ (f) $CH_3CHClCH_2CN$ (g) $C_6H_5\overset{O}{\overset{\|}{C}}O\overset{O}{\overset{\|}{C}}H$

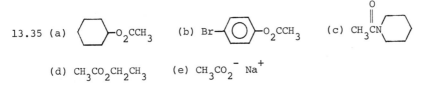

13.34 (b), (a), (c), based primarily upon the difference in electronegativities between the two atoms of the bond.

13.35 (a) 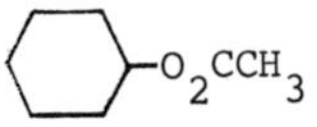 (b) 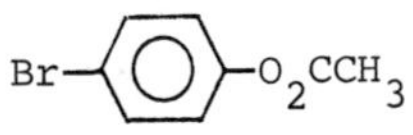 (c) 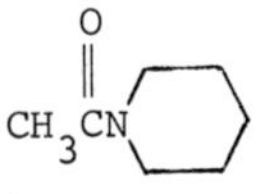

(d) $CH_3CO_2CH_2CH_3$ (e) $CH_3CO_2^-\ Na^+$

In each reaction except (e), CH_3CO_2H (or $CH_3CO_2^-$) would also be an organic product.

13.36 (a) 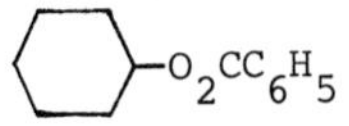 (b) 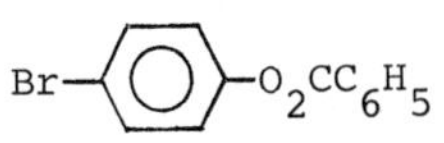 (c) 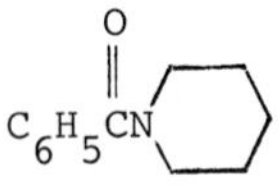

(d) $C_6H_5CO_2CH_2CH_3$ (e) $C_6H_5CO_2^-\ Na^+$

13.37 (a) $C_6H_5CO_2^- + CH_3CH_2OH$ (b) $C_6H_5CO_2^- + CH_3CH_2NH_2$

(c) $C_6H_5CO_2^- + CH_3CH_2CO_2^-$ relative rates: (b) > (c) > (a)

13.38 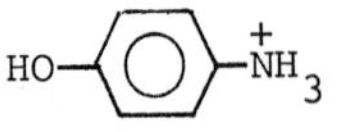 $+ CH_3CO_2H$

13.39 (a) 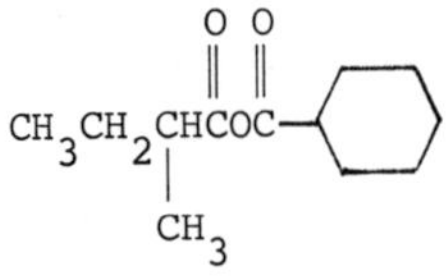 (b) 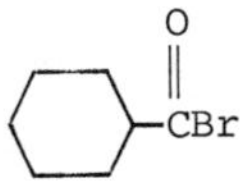 (c) 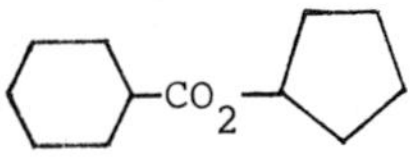

(d) 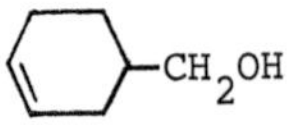 $+ CH_3CH_2OH$ (e) $CH_3CH_2CH_2\overset{O}{\overset{\|}{C}}Cl$ (f) $2\ CH_3CH_2OD + D_2O + CO_2$

(g) $CH_3CH_2\overset{O}{\overset{\|}{C}}CH_2CH{=}CH_2$ (h) $CH_3CH_2\overset{O}{\overset{\|}{C}}CH(CH_3)_2$

13.40 (a) $CH_3\overset{O}{\overset{\|}{C}}O^- + \left[C_6H_5\overset{OH}{\overset{|}{C}}HCl\right] \xrightarrow[-Cl^-]{OH^-} \left[C_6H_5\overset{OH}{\overset{|}{C}}HOH\right] \xrightarrow{-H_2O} C_6H_5\overset{O}{\overset{\|}{C}}H$

a hydrated aldehyde

(b) 1,2,4-benzenetricarboxylate (^-O_2C, ^-O_2C, CO_2^- on the benzene ring) + $HOCH_3$

(c) cyclopentane bearing $\overset{CH_3}{\overset{|}{C}}{=}CHOH$, CO_2^- and CH_3 $\rightleftharpoons$ cyclopentane bearing $\overset{CH_3}{\overset{|}{C}}HCHO$, CO_2^- and CH_3

an enol

13.41 (a) $CH_3CO_2CH_2CH_3 \xrightarrow[(2)\ H^+]{(1)\ H_2O,OH^-,heat} CH_3CO_2H + CH_3CH_2OH$

(b) $CH_3CO_2CH_2CH_3 \xrightarrow[(2)\ H_2O]{(1)\ LiAlH_4} 2\ CH_3CH_2OH$

(c) $CH_3CO_2CH_2CH_3 \xrightarrow[(2)\ H_2O,H^+]{(1)\ 2\ CH_3MgI} (CH_3)_3COH$

(d) $CH_3CO_2CH_2CH_3 \xrightarrow[(2)\ SOCl_2]{(1)\ H_2O,H^+,\ heat} CH_3\overset{O}{\overset{\|}{C}}Cl \xrightarrow[AlCl_3]{C_6H_6} CH_3\overset{O}{\overset{\|}{C}}C_6H_5$

(e) $CH_3CO_2CH_2CH_3 \xrightarrow[heat]{NaOH,H_2O} CH_3CO_2^-\ Na^+ + HOCH_2CH_3$

(f) $CH_3CO_2CH_2CH_3 \xrightarrow{CH_3NH_2} CH_3\overset{O}{\overset{\|}{C}}NHCH_3 + HOCH_2CH_3$

13.42 Saponification is the reaction of choice because heating in dilute HCl would lead to addition of HCl and H_2O to the carbon-carbon double bond and allylic rearrangement.

$$CH_3CH{=}CHCH_2O\overset{O}{\overset{\|}{C}}CH_3 \xrightarrow[(2)\ \text{neutralize with}\ H^+]{(1)\ H_2O,OH^-,heat} CH_3CH{=}CHCH_2OH + CH_3CO_2H$$

13.43 (a) $(C_6H_5CH_2)_2Cd$ or $(C_6H_5CH_2)_2CuLi + Cl\overset{O}{\overset{\|}{C}}CH_2CH_3$

$C_6H_5CH_2\overset{O}{\overset{\|}{C}}Cl + Cd(CH_2CH_3)_2$ or $LiCu(CH_2CH_3)_2$

(b) $[(CH_3)_2CHCH_2C(CH_3)_2]_2CuLi + ClC(=O)C_6H_5$

$(CH_3)_2CHCH_2C(CH_3)_2C(=O)Cl + Cd(C_6H_5)_2$ or $LiCu(C_6H_5)_2$

In (b), $[(CH_3)_2CHCH_2C(CH_3)_2]_2Cd$ is unsuitable because 3° (as well as 2°) cadmium reagents are unstable.

13.44 (a) $[HOC(=O)OH] \longrightarrow H_2O + CO_2$ (b) $CH_3CH_2OC(=O)OCH_2CH_3$

(c) $ClC(=O)OCH_2CH_3$ (d) $H_2NC(=O)OCH_2CH_3 + NH_4Cl$

In (d), note that the more reactive group is the first to leave; continued treatment with NH_3 would yield $H_2NC(=O)NH_2$.

13.45 (a) ($\underline{R}$)-$CH_3C(=O)NHCH(CH_3)(CH_2)_5CH_3$ (b) $C_6H_5C(=O)—^{18}OCH(CH_3)_2$

13.46 The effect of the electron-releasing methyl groups on the benzene ring makes the carbonyl carbon atom less positive and less susceptible to attack. Also, there would be considerable steric hindrance in the intermediate leading to the ester.

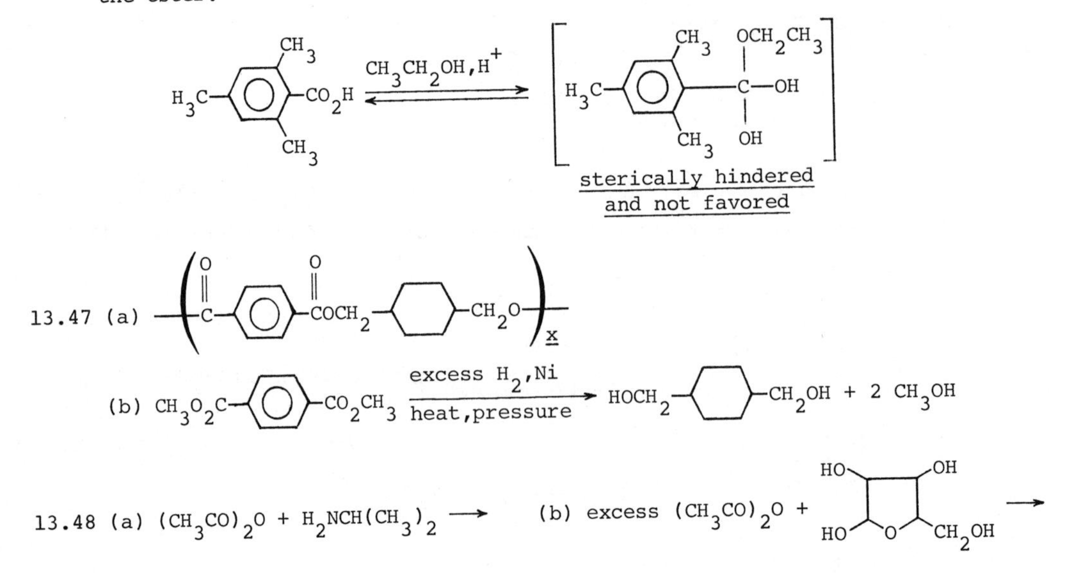

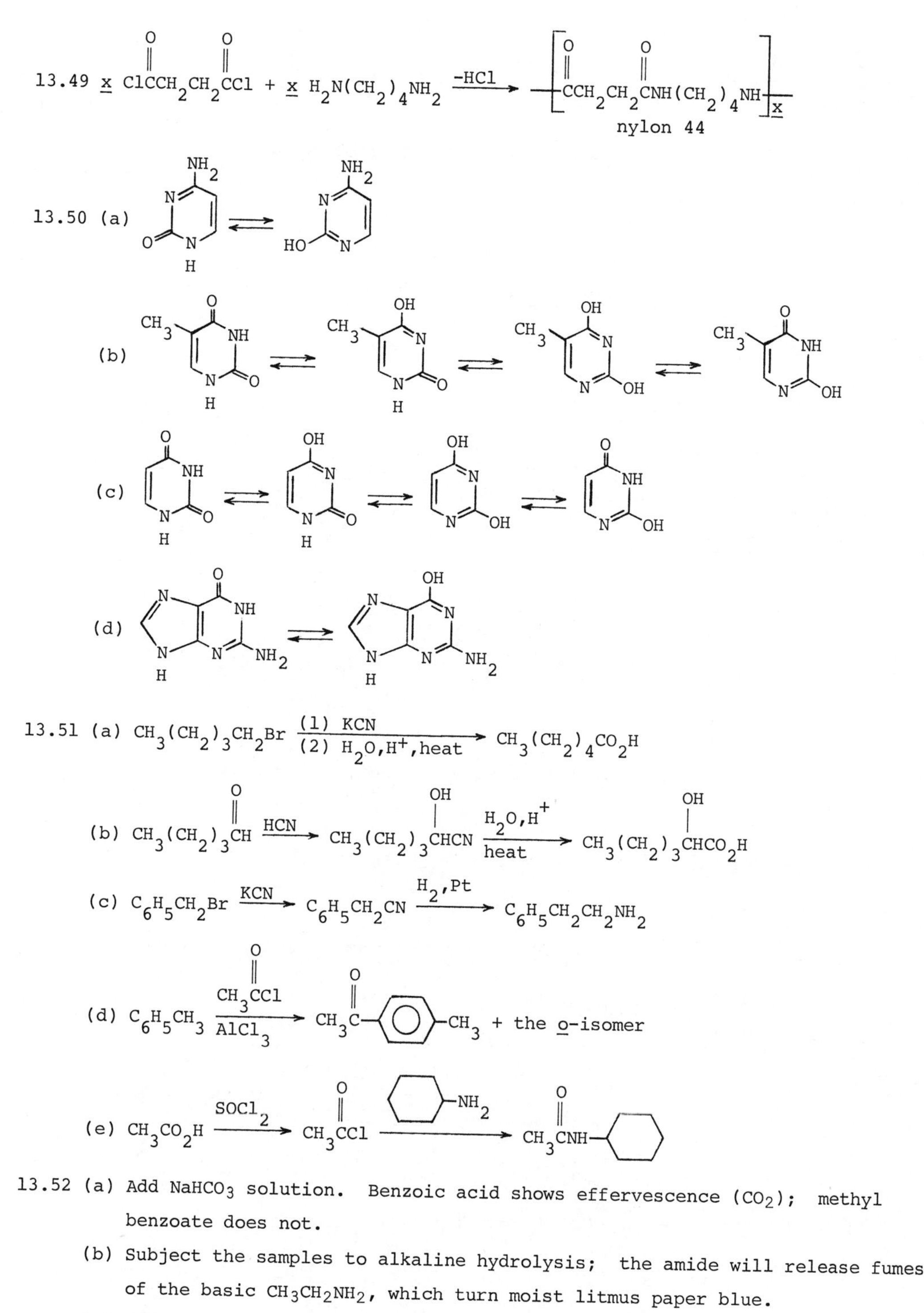

13.49 \underline{x} ClCCH$_2$CH$_2$CCl + \underline{x} H$_2$N(CH$_2$)$_4$NH$_2$ $\xrightarrow{-HCl}$ nylon 44

13.50 (a)

(b)

(c)

(d)

13.51 (a) CH$_3$(CH$_2$)$_3$CH$_2$Br $\xrightarrow[\text{(2) H}_2\text{O,H}^+\text{,heat}]{\text{(1) KCN}}$ CH$_3$(CH$_2$)$_4$CO$_2$H

(b) CH$_3$(CH$_2$)$_3$CH $\xrightarrow{\text{HCN}}$ CH$_3$(CH$_2$)$_3$CHCN $\xrightarrow[\text{heat}]{\text{H}_2\text{O,H}^+}$ CH$_3$(CH$_2$)$_3$CHCO$_2$H

(c) C$_6$H$_5$CH$_2$Br $\xrightarrow{\text{KCN}}$ C$_6$H$_5$CH$_2$CN $\xrightarrow{\text{H}_2\text{,Pt}}$ C$_6$H$_5$CH$_2$CH$_2$NH$_2$

(d) C$_6$H$_5$CH$_3$ $\xrightarrow[\text{AlCl}_3]{\text{CH}_3\text{CCl}}$ CH$_3$C— ⬡ —CH$_3$ + the \underline{o}-isomer

(e) CH$_3$CO$_2$H $\xrightarrow{\text{SOCl}_2}$ CH$_3$CCl → CH$_3$CNH—

13.52 (a) Add $NaHCO_3$ solution. Benzoic acid shows effervescence (CO_2); methyl benzoate does not.

(b) Subject the samples to alkaline hydrolysis; the amide will release fumes of the basic $CH_3CH_2NH_2$, which turn moist litmus paper blue.

(c) Add $AgNO_3$ solution; C_6H_5COCl gives a AgCl precipitate.

13.53 (a) C_6H_{11}–CH_2OH + CH_3OH (b) C_6H_{11}–CO_2H

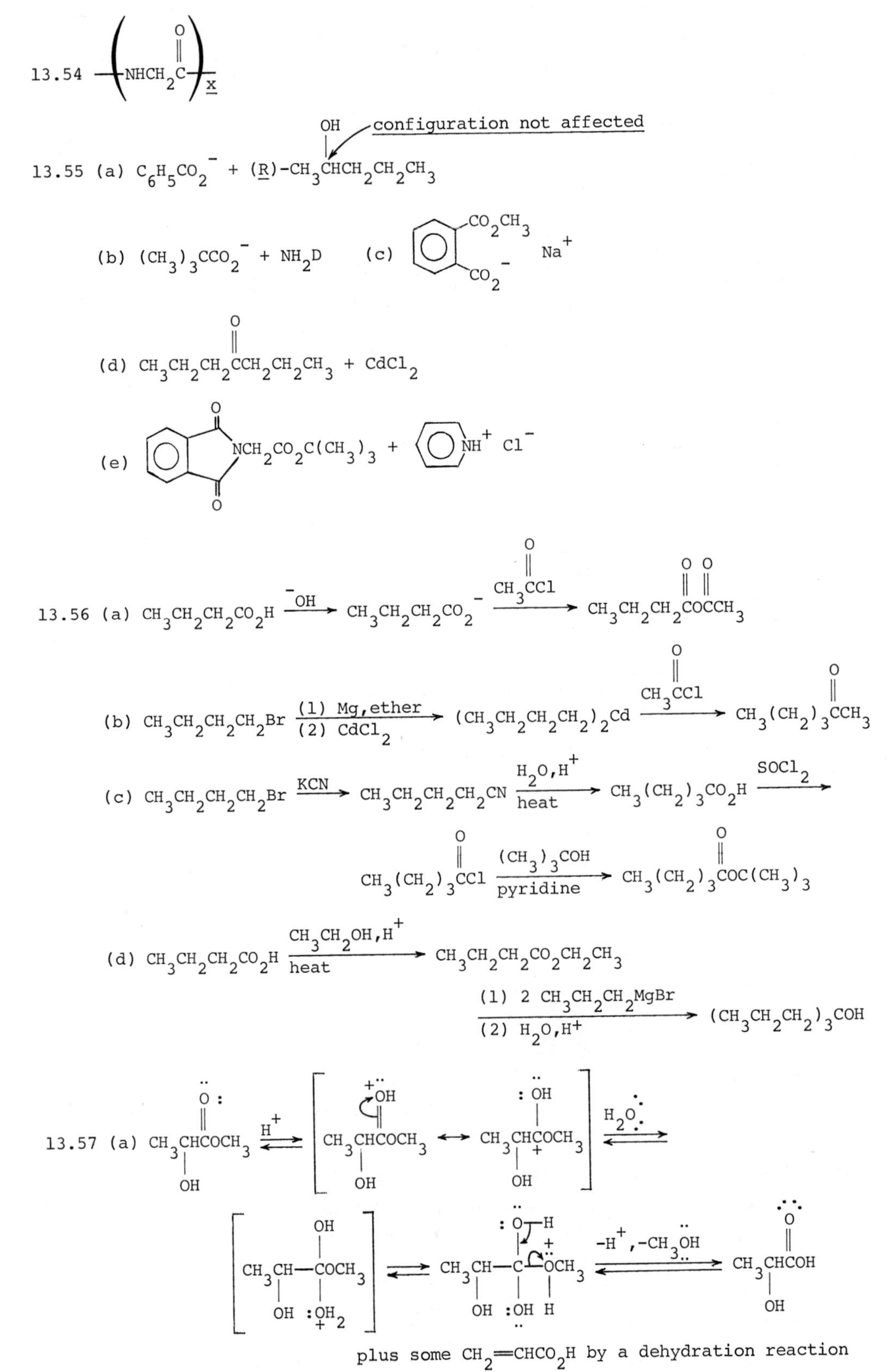
13.54
NHCH2C
13.55 (a) C6H5CO2⁻ + (R)-CH3CHCH2CH2CH3
configuration not affected
(b) (CH3)3CCO2⁻ + NH2D
(c) CO2CH3 CO2⁻ Na+
(d) CH3CH2CH2CCH2CH2CH3 + CdCl2
(e) NCH2CO2C(CH3)3 + NH+ Cl⁻
13.56 (a) CH3CH2CH2CO2H ⁻OH CH3CH2CH2CO2⁻ CH3CCl CH3CH2CH2COCCH3
(b) CH3CH2CH2CH2Br (1) Mg,ether (2) CdCl2 (CH3CH2CH2CH2)2Cd CH3CCl CH3(CH2)3CCH3
(c) CH3CH2CH2CH2Br KCN CH3CH2CH2CH2CN H2O,H+ heat CH3(CH2)3CO2H SOCl2
CH3(CH2)3CCl (CH3)3COH pyridine CH3(CH2)3COC(CH3)3
(d) CH3CH2CH2CO2H CH3CH2OH,H+ heat CH3CH2CH2CO2CH2CH3
(1) 2 CH3CH2CH2MgBr (2) H2O,H+ (CH3CH2CH2)3COH
13.57 (a) CH3CHCOCH3 H+ CH3CHCOCH3 CH3CHCOCH3 H2O
CH3CH—COCH3 CH3CH—C—OCH3 -H+,-CH3OH CH3CHCOH
plus some CH2=CHCO2H by a dehydration reaction

(b) 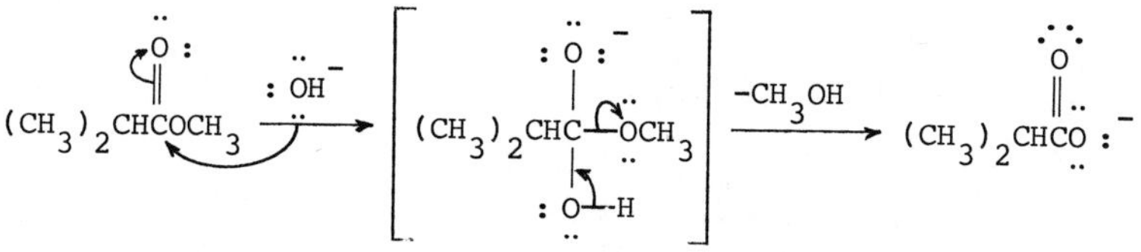

13.58 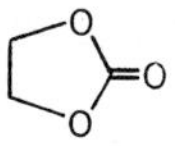 and 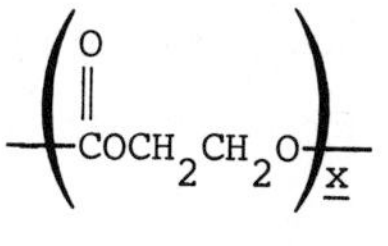

13.59 A, B, C, and D could all be found in solution:

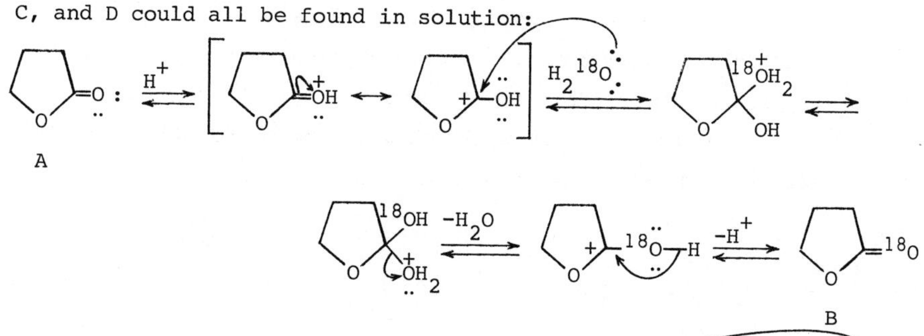

and

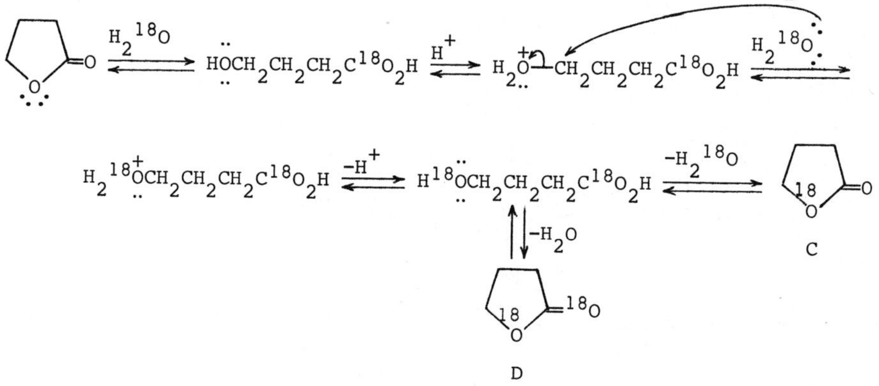

13.60 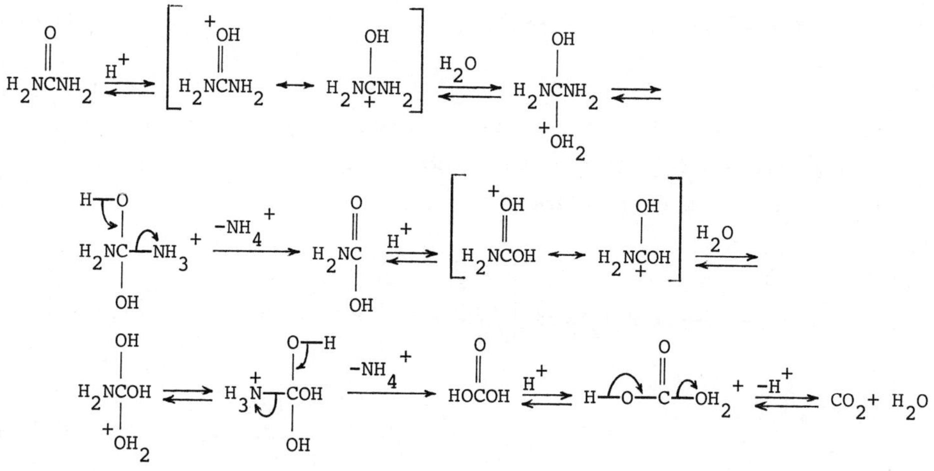

13.61 (a) $\underset{\text{A}}{CH_2{=}CH\overset{\overset{O}{\|}}{C}NH_2} \xrightarrow{Br_2} CH_2BrCHBr\overset{\overset{O}{\|}}{C}NH_2$

$\xrightarrow[\text{heat}]{H_2O,OH^-} CH_2{=}CHCO_2^- + NH_3$

(b) The infrared spectrum would show double NH stretching absorption at about 3125-3570 cm^{-1} (2.8-3.2 μm) plus amide NH (bending) and C=O absorption at about 1515-1670 cm^{-1} (6.0-6.6 μm) and 1650-1700 cm^{-1} (5.88 μm), respectively. Also, $=CH_2$ absorption would be observed at 3135 cm^{-1} (3.2 μm) and 887 cm^{-1} (11.3 μm). C=C absorption would appear in the same general region as the amide I and II bands, about 1620-1680 cm^{-1} (5.95-6.17 μm).

The nmr spectrum would show a singlet for NH_2 (area 2) and the characteristic 8-12 peaks (total area 3) at around $\delta = 5$ ppm for $-CH{=}CH_2$.

13.62

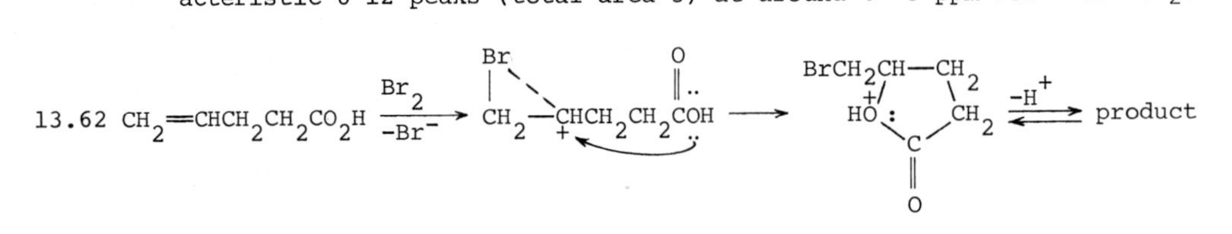

13.63

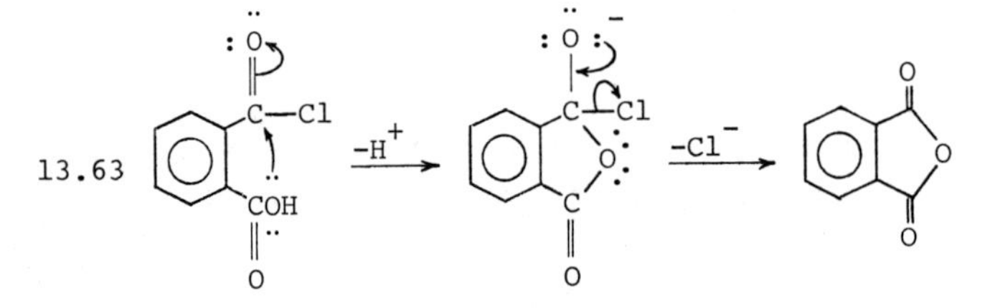

13.64 When A is hydrolyzed, the $^{18}O{-}R$ bond is not broken.

$$CH_3\overset{\overset{O}{\|}}{C}{-}^{18}OCH_2CH_3 \underset{}{\overset{H_2O,H^+}{\rightleftharpoons}} \left[CH_3\underset{OH}{\overset{OH}{C}}{-}^{18}OCH_2CH_3 \right] \rightleftharpoons CH_3\overset{\overset{O}{\|}}{C}OH + H^{18}OCH_2CH_3$$

When B is subjected to hydrolysis, however, the t-butyl group can break off as a 3° carbocation, leading to a mixture of products. One possibility for a carbocation reaction follows:

$$CH_3\overset{\overset{O}{\|}}{C}{-}^{18}O{-}C(CH_3)_3 \rightleftharpoons \left[CH_3\overset{\overset{O}{\|}}{C}{-}^{18}O^- \longleftrightarrow CH_3\overset{\overset{O^-}{|}}{C}{=}^{18}O \right] + \left[\overset{+}{C}(CH_3)_3 \right]$$

$$\overset{H_2O,H^+}{\rightleftharpoons} CH_3\overset{\overset{^{18}O}{\|}}{C}OH + HOC(CH_3)_3$$

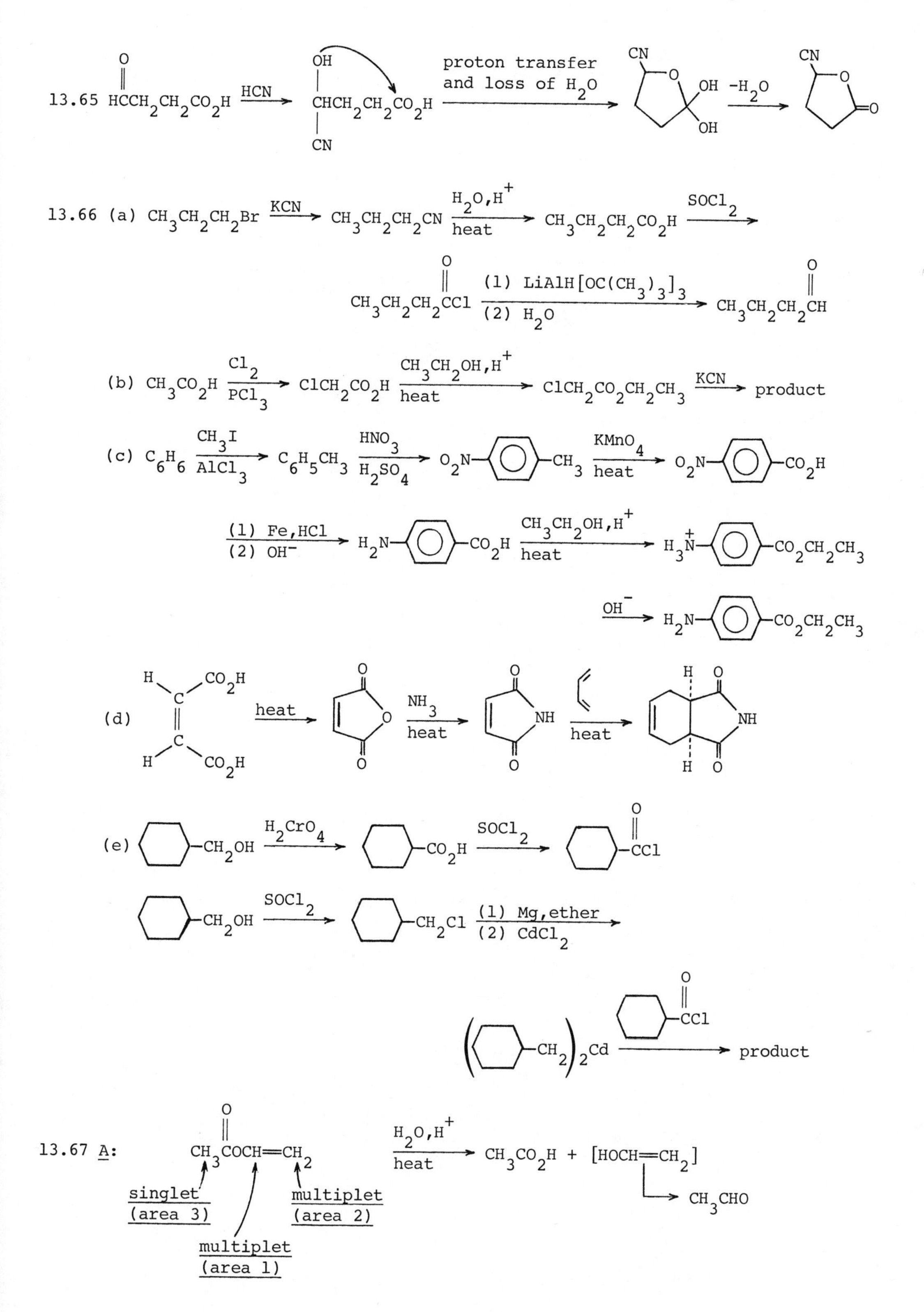
13.65
HCN
proton transfer and loss of H_2O
$-H_2O$
13.66 (a)
KCN
H_2O, H^+ heat
$SOCl_2$
(1) $LiAlH[OC(CH_3)_3]_3$
(2) H_2O
(b)
Cl_2 PCl_3
CH_3CH_2OH, H^+ heat
KCN
product
(c)
CH_3I $AlCl_3$
HNO_3 H_2SO_4
$KMnO_4$ heat
(1) Fe, HCl
(2) OH^-
CH_3CH_2OH, H^+ heat
OH^-
(d)
heat
NH_3 heat
heat
(e)
H_2CrO_4
$SOCl_2$
$SOCl_2$
(1) Mg, ether
(2) $CdCl_2$
product
13.67 A:
H_2O, H^+ heat
CH_3CO_2H + $[HOCH{=}CH_2]$
CH_3CHO
singlet (area 3)
multiplet (area 1)
multiplet (area 2)

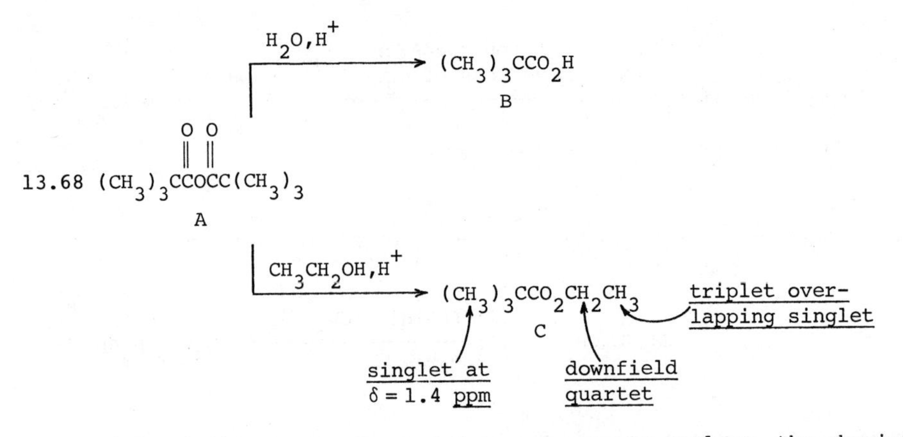

Although the nmr spectrum of C is not easy to analyze, the chemistry implies an ethyl ester.

CHAPTER 14

Enolates and Carbanions: Building Blocks for Organic Synthesis

Some Important Features

A hydrogen atom α to a carbonyl group is slightly acidic and may be removed by a base. The reason for this acidity of carbonyl compounds is the resonance-stabilization of the enolate ion. The extent of enolate formation is determined by the acidity of the carbonyl compound and by the strength of the base.

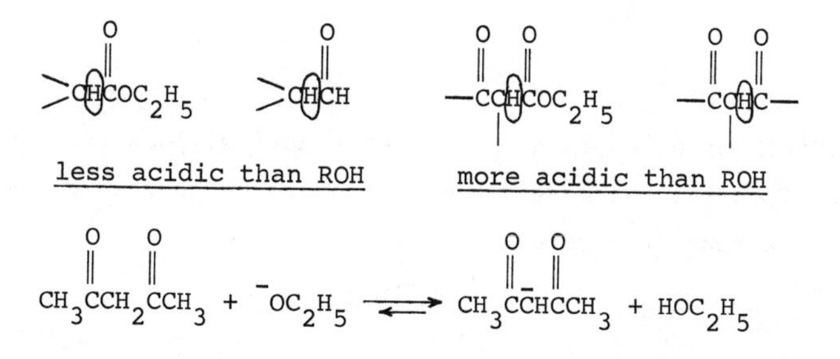

Enolate ions are useful as synthetic intermediates. For example, enolate ions can act as nucleophiles in substitution reactions with alkyl halides.

An alkylation reaction:

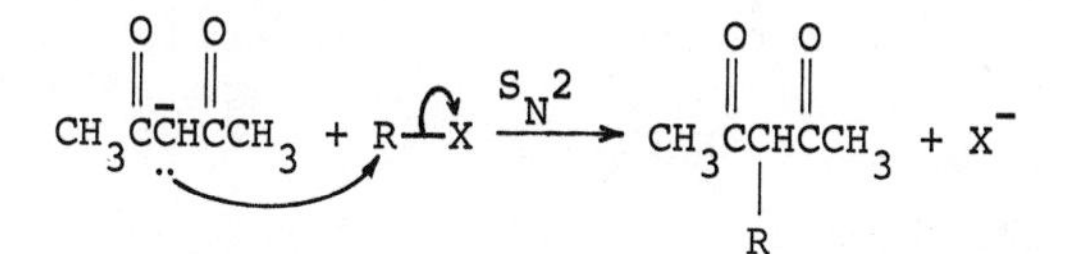

Enolate ions can also attack carbonyl groups.

An aldol condensation:

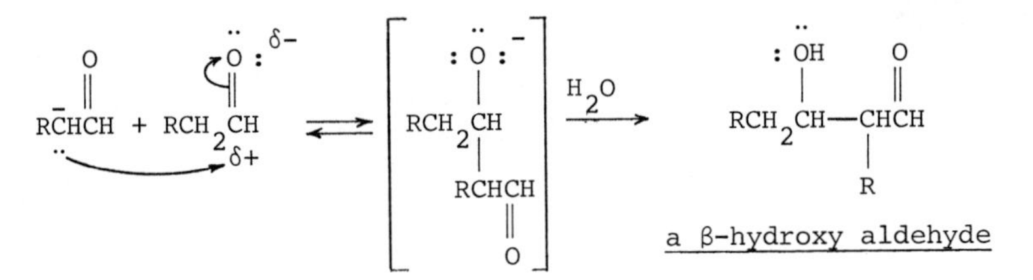

An ester condensation:

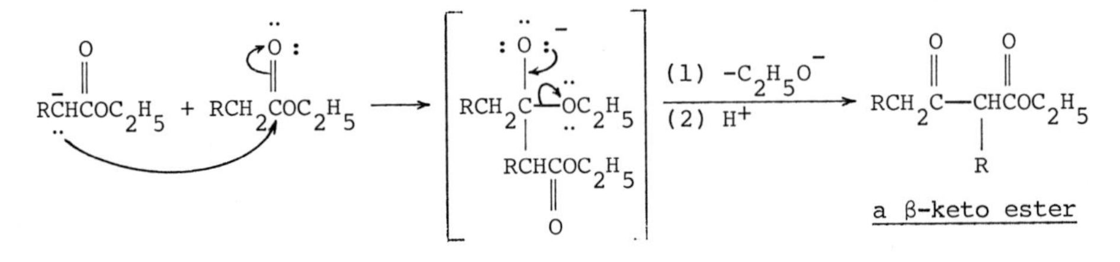

Enolate ions can also attack an α,β-unsaturated carbonyl compound in a 1,4-addition reaction (Michael addition).

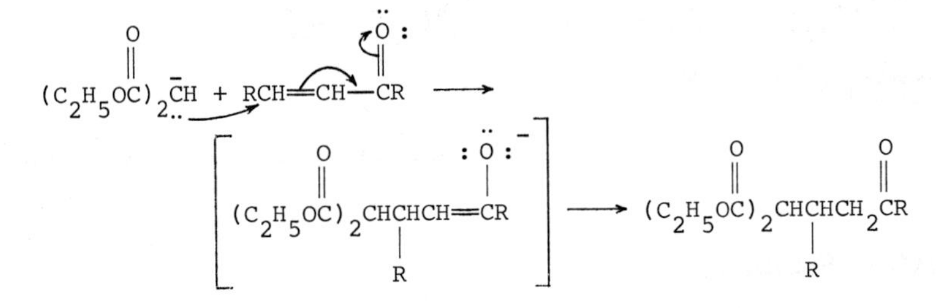

The products of enolate reactions may be subjected to further reactions, such as a second enolate reaction, hydrolysis, decarboxylation, or dehydration. These topics are discussed in the text.

Reminders

In predicting the products of an enolate or related reaction, look first for the most acidic hydrogen. Then look for the most likely target of nucleophilic attack – for example, a carbon-halogen bond or a carbonyl group.

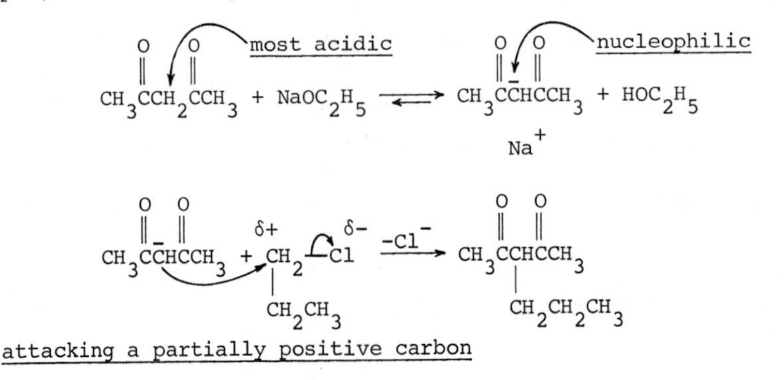

In solving problems asking for a synthetic route, first determine the type of product. Is the product a substituted acetic acid? (Try a malonic ester alkylation.) A substituted acetone? (Try an acetoacetic ester alkylation.) An α,β-unsaturated aldehyde? (Try an aldol condensation.) A β-keto ester? (Try an ester condensation.) After you deduce a likely route, use your pencil to divide the structure into its pieces.

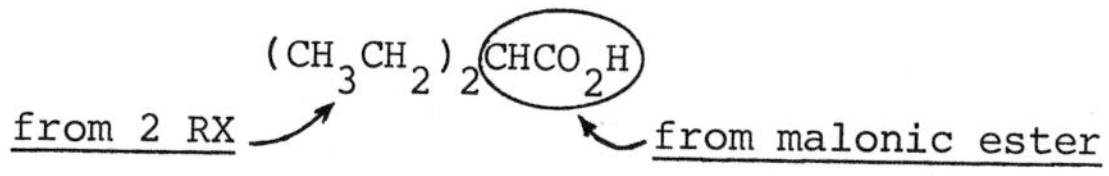

$(CH_3CH_2)_2CHCO_2H$ — $(CH_3CH_2)_2$ from 2 RX; $CHCO_2H$ (circled) from malonic ester

At this point, you can work the problem backwards.

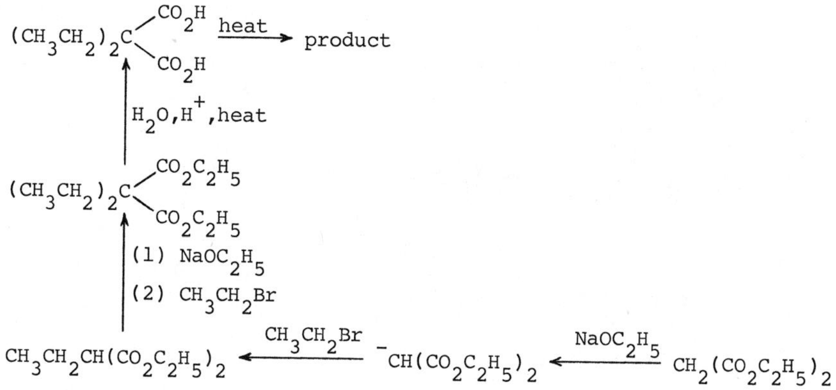

$$(CH_3CH_2)_2C(CO_2H)_2 \xrightarrow{\text{heat}} \text{product}$$

$$\uparrow \; H_2O, H^+, \text{heat}$$

$$(CH_3CH_2)_2C(CO_2C_2H_5)_2$$

$$\uparrow \; (1)\ NaOC_2H_5 \quad (2)\ CH_3CH_2Br$$

$$CH_3CH_2CH(CO_2C_2H_5)_2 \xleftarrow{CH_3CH_2Br} {}^-CH(CO_2C_2H_5)_2 \xleftarrow{NaOC_2H_5} CH_2(CO_2C_2H_5)_2$$

Answers to Problems

14.26 (a) $C_6H_5CH(CH_3)CO_2C_2H_5$ (H on CH circled) (b) $CH_3C(=O)CH_2CN$ (H_2 circled) (c) $OHCC(CH_3)HCH{=}CHCHO$ (H circled)

(d) $(CH_3)_3CCH_2CO_2H$ (H circled) (e) $CH_3CH_2NO_2$ (H_2 circled) (f) $CH_3CCl_2CHCl_2$ (H circled)

In (f), the circled proton is acidic because of the cumulative inductive effect of the chlorine atoms.

14.27 (a) $C_2H_5O_2CCH_2CN + C_2H_5O^- \rightleftharpoons C_2H_5O_2C\bar{C}HCN + C_2H_5OH$

(b) 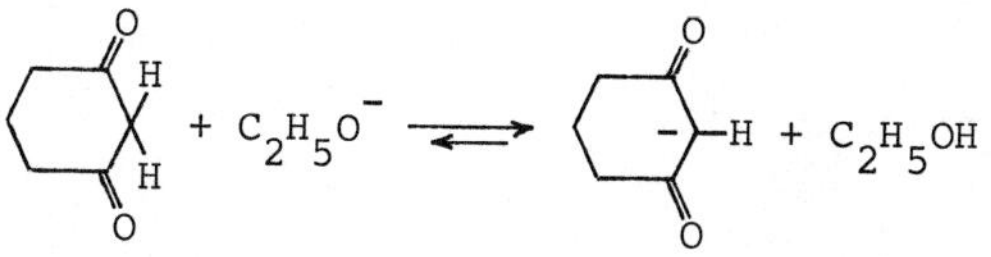 cyclohexane-1,3-dione (C-2 bearing two H) $+ C_2H_5O^- \rightleftharpoons$ carbanion at C-2 (–H) $+ C_2H_5OH$

(c) 2-cyanocyclohexanone (C-2 bearing CN and H) $+ C_2H_5O^- \rightleftharpoons$ carbanion at C-2 (–CN) $+ C_2H_5OH$

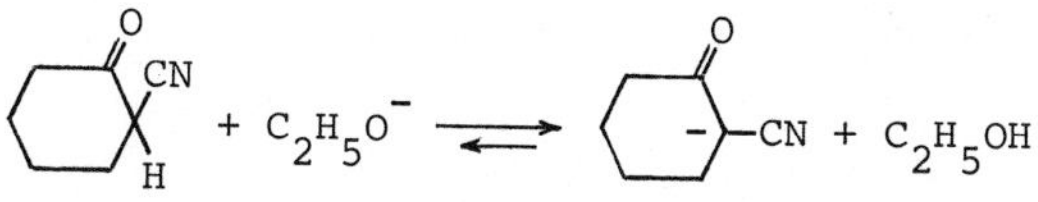

14.28 (a) 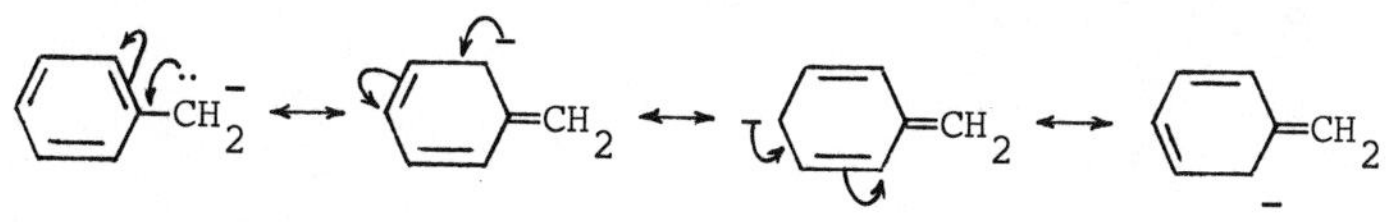 $C_6H_5\ddot{C}H_2^- \longleftrightarrow$ ring anion $=CH_2$ $\longleftrightarrow$ ring anion $=CH_2$ $\longleftrightarrow$ ring anion $=CH_2$

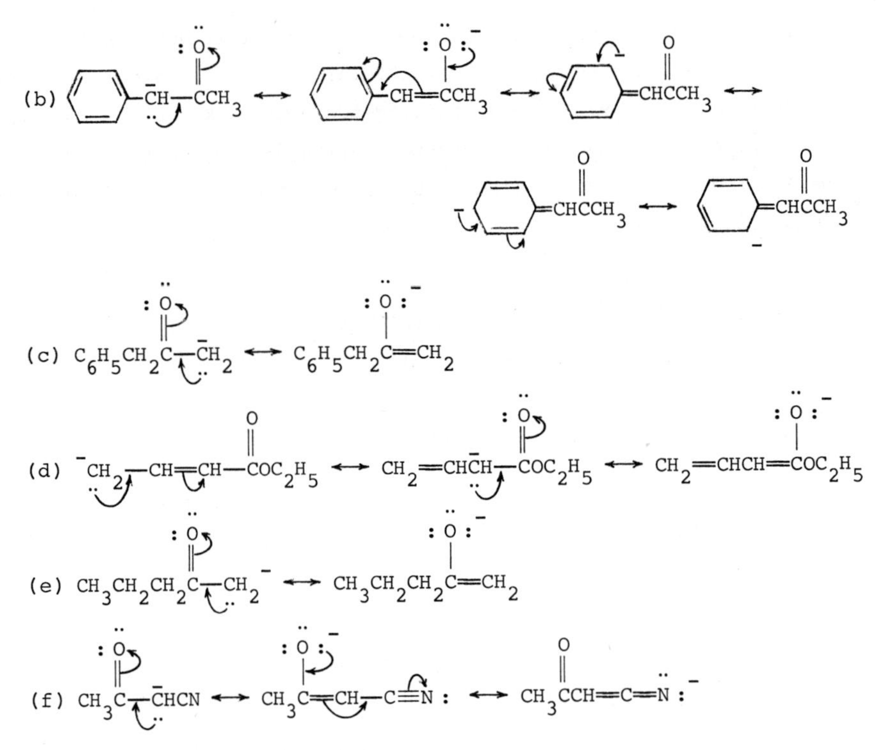

14.29 (b), (a), (e), (d), (c), (f)

The compounds with hydrogens α to only one carbonyl group, (b) and (a), are less acidic than ethanol (e), while those with hydrogens α to two carbonyl groups are more acidic. None of the other compounds is as acidic as a carboxylic acid (f). Also note that a hydrogen α to an aldehyde group is slightly more acidic than one α to an ester group.

14.30 (a) ⇌ $CH_3CO_2C_2H_5$ + $CH_3C(=O)\bar{C}HCO_2C_2H_5$ (b) ⇌ $CH_3C(=O)\bar{C}HNO_2$ + H_2O

(c) ⇌ (cyclohexyl anion)$-CO_2C_2H_5$ + $CH_3C(=O)CH_2C(=O)CH_3$

14.31 (a) 2,2-dimethylcyclopentane-1,3-dione (O= ring =O, CH_3 CH_3) (b) $CH_3C(=O)CHCO_2C_2H_5$ with cyclohexyl substituent on the CH

(c) (cyclohexyl)$-CH_2C(=O)CH_3$ from the decarboxylation of $CH_3C(=O)CHCO_2H$ (with cyclohexyl substituent on the CH)

(d) The attacking nucleophile is $^{-}CH(CO_2C_2H_5)$, and the leaving group is the tosylate ion (^{-}OTs). The reaction proceeds by an S_N2 mechanism (back-side displacement), resulting in the trans product:

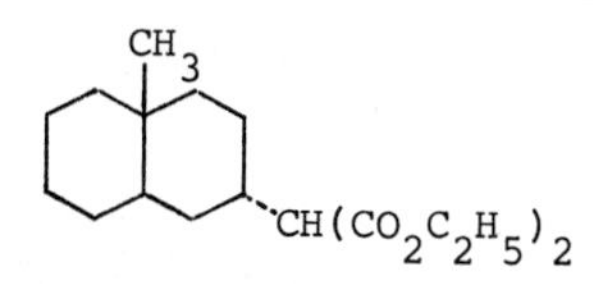

(e) $CH_2[CH(CO_2C_2H_5)_2]_2$

(f) $HO_2CCH_2CH_2CH_2CO_2H$ from the decarboxylation of $(HO_2C)_2CHCH_2CH(CO_2H)_2$

(g) $CH_3\overset{O}{\overset{\|}{C}}CHCO_2C_2H_5$ with $CH_2\underset{\underset{O}{\|}}{C}CH_3$ on the CH   (h) $CH_3\overset{O}{\overset{\|}{C}}CH_2CH_2\overset{O}{\overset{\|}{C}}CH_3$

(i) O_2N–$C_6H_3(NO_2)$–$\underset{\underset{CN}{|}}{C}HCO_2C_2H_5$ by nucleophilic aromatic substitution

14.32 (a) $CH_2(CO_2C_2H_5)_2 \xrightarrow[(2)\ C_5H_9CH_2Br]{(1)\ NaOC_2H_5}$ C_5H_9–$CH_2CH(CO_2C_2H_5)_2 \xrightarrow[heat]{H_2O, H^+}$

(b) $CH_2(CO_2C_2H_5)_2 \xrightarrow[(2)\ CH_3CHBrCO_2C_2H_5]{(1)\ NaOC_2H_5} C_2H_5O_2C\overset{CH_3}{\overset{|}{C}}HCH(CO_2C_2H_5)_2 \xrightarrow[heat]{H_2O, H^+}$

(c) $CH_3\overset{O}{\overset{\|}{C}}CH_2CO_2C_2H_5 \xrightarrow[(2)\ C_6H_5CH_2Br]{(1)\ NaOC_2H_5} CH_3\overset{O}{\overset{\|}{C}}\underset{\underset{CH_2C_6H_5}{|}}{C}HCO_2C_2H_5 \xrightarrow[heat]{H_2O, H^+}$

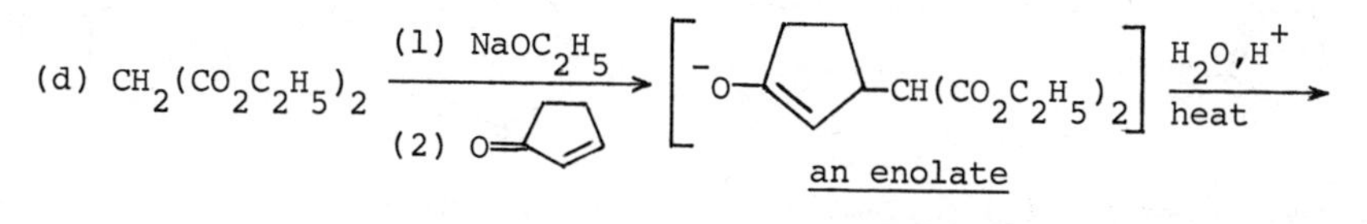

14.33

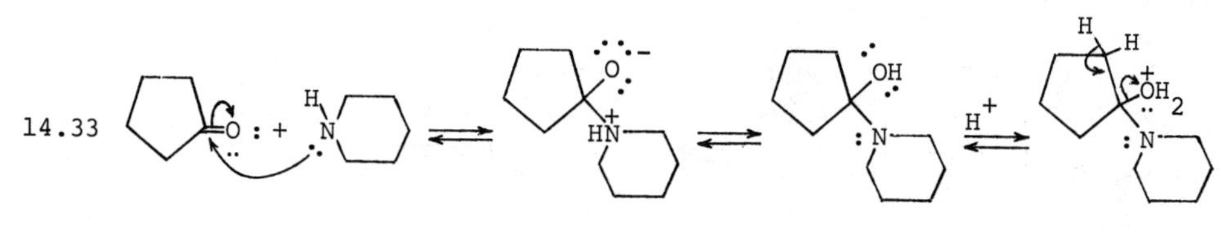

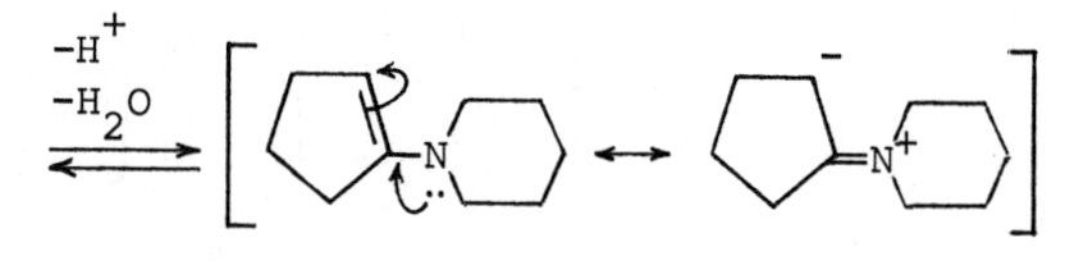

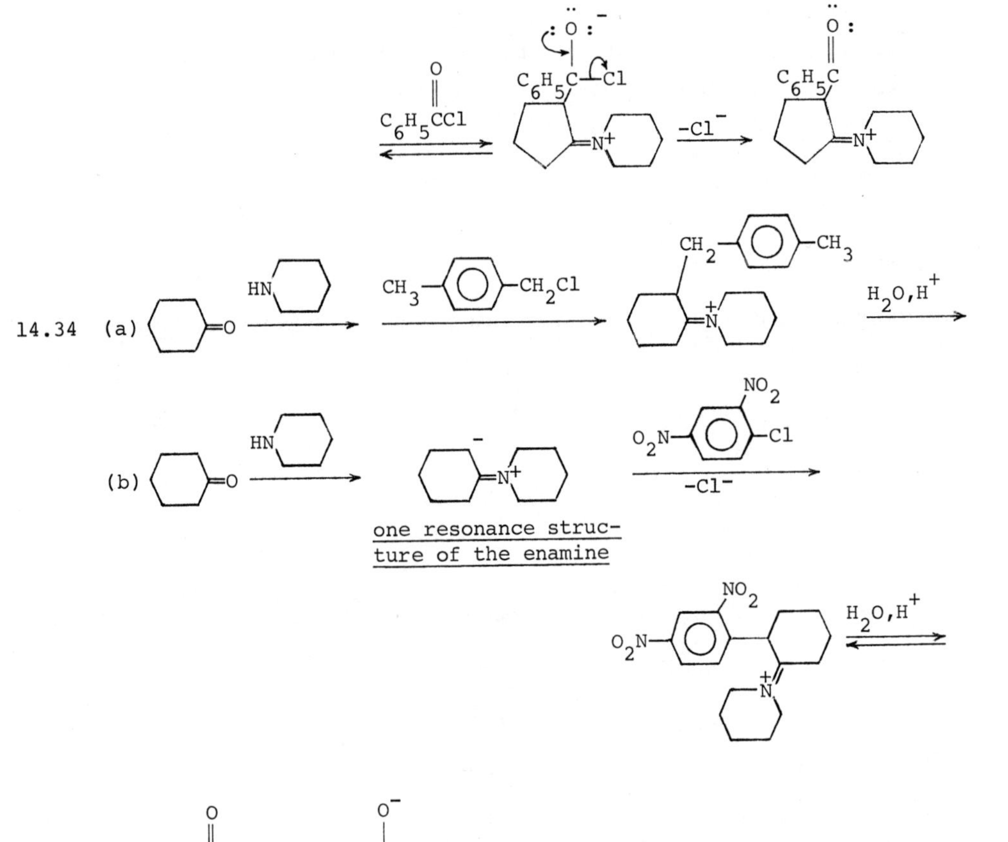

14.35 (a) $^{-}CH_2\overset{O}{\overset{\|}{C}}CH_3 \longleftrightarrow CH_2{=}\overset{O^-}{\overset{|}{C}}CH_3$ Formaldehyde does not contain an acidic hydrogen.

(b) HCHO

(c) $H\overset{O}{\overset{\|}{C}}H + \bar{C}H_2\overset{O}{\overset{\|}{C}}CH_3 \rightleftharpoons CH_3\overset{O}{\overset{\|}{C}}CH_2\overset{O^-}{\overset{|}{C}}H_2 \overset{H_2O}{\rightleftharpoons} CH_3\overset{O}{\overset{\|}{C}}CH_2CH_2OH + OH^-$

14.36

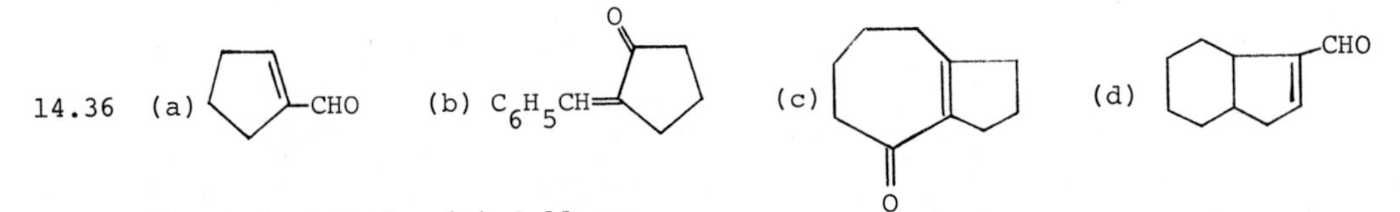

The mechanism for (c) follows:

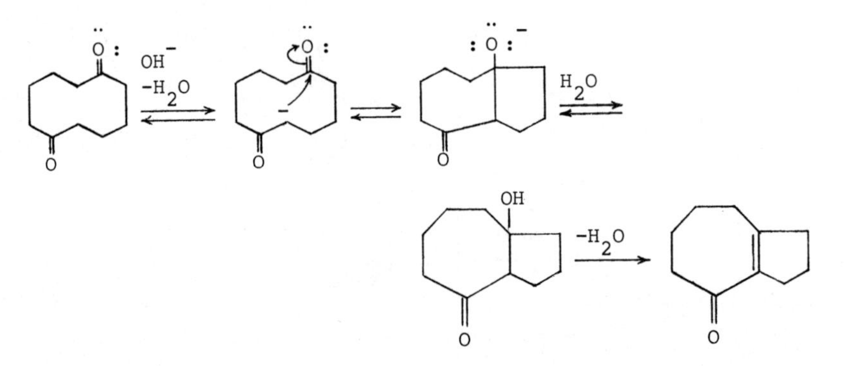

14.37 (a) cyclopentanone + $C_6H_5CH_2CO_2C_2H_5 \xrightarrow{NaOC_2H_5}$ (b) 2-furyl–CHO + $CH_3NO_2 \overset{OH^-}{\rightleftharpoons}$

(c) $C_6H_5CHO + CH_3\overset{O}{\overset{\|}{C}}CH_3 \overset{OH^-}{\rightleftharpoons}$

(d) $CH_3\overset{O}{\overset{\|}{C}}CH_3 \overset{C_6H_5CHO, OH^-}{\rightleftharpoons} C_6H_5CH{=}CH\overset{O}{\overset{\|}{C}}CH_3 \overset{C_6H_5CHO, OH^-}{\rightleftharpoons}$

14.38 (a) 2-naphthyl–CO_2H + 2-naphthyl–CH_2OH (b) $(CH_3)_3CCO_2H + (CH_3)_3CCH_2OH$

In (a) and (b), neither starting aldehyde contains an α hydrogen; therefore, both undergo Cannizzaro reactions.

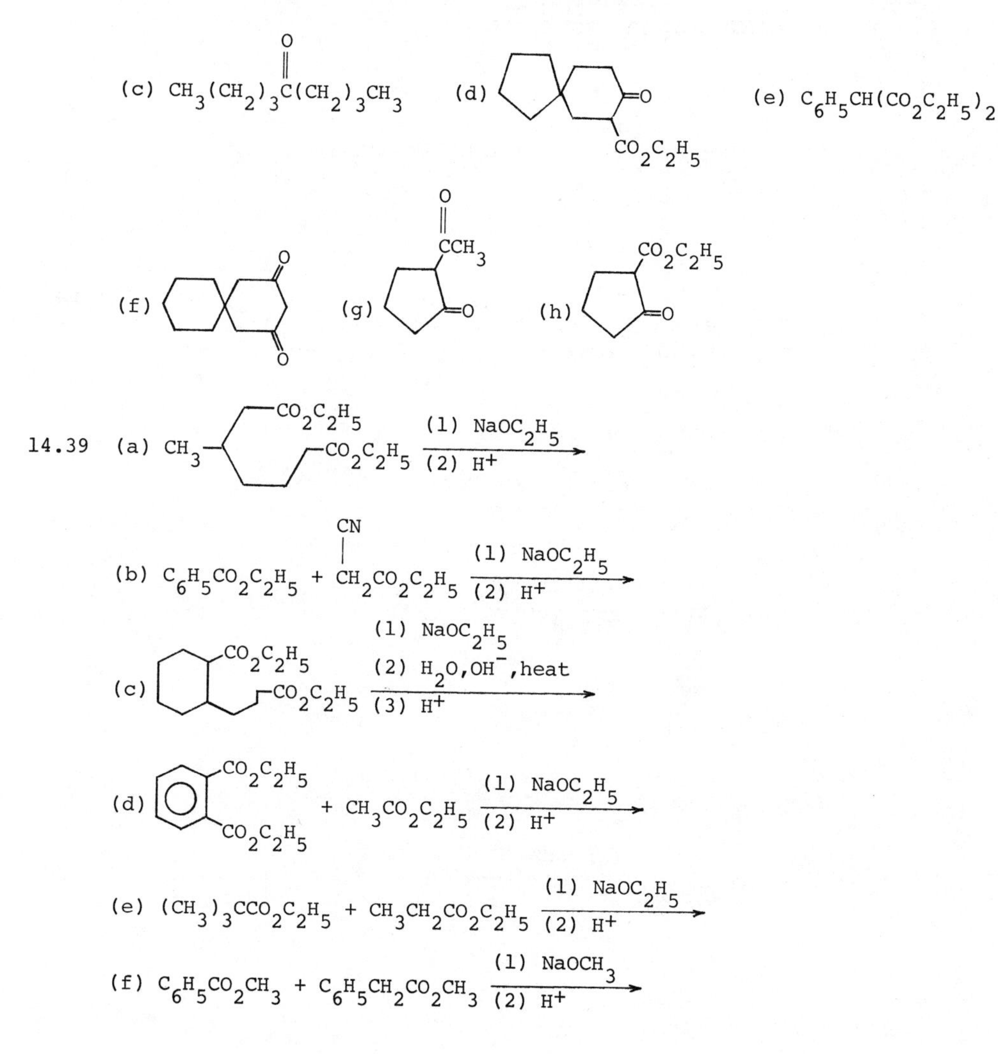

14.40 (a) $CH_2CH_2\overset{O}{\overset{\|}{C}}CH_2CH_3$ with $CH(CO_2C_2H_5)_2$ attached:

$$\underset{\displaystyle |\atop CH(CO_2C_2H_5)_2}{CH_2}CH_2\overset{O}{\overset{\|}{C}}CH_2CH_3$$

(b)

$$\begin{array}{l} CH_2CH_2\overset{O}{\overset{\|}{C}}CH_2CH_3 \\ | \\ C(CO_2C_2H_5)_2 \\ | \\ CH_2CH_2\underset{\underset{O}{\|}}{C}CH_2CH_3 \end{array}$$

(c) $(CH_3\overset{O}{\overset{\|}{C}})_2CH$–(cyclohexan-3-one ring, $=O$)

(d)

$$\begin{array}{l} CH_3CHCH_2CO_2C_2H_5 \\ \quad | \\ C_6H_5CHCO_2C_2H_5 \end{array}$$

14.41 (a) $CH_3\overset{O}{\overset{\|}{C}}CH_2CO_2C_2H_5 + CH_2{=}CHCN \xrightarrow[(2)\ H^+]{(1)\ NaOC_2H_5}$

(b) (cyclohexane-1,3-dione, O, O) $+ CH_2{=}CHCO_2C_2H_5 \xrightarrow[(2)\ H^+]{(1)\ NaOC_2H_5}$

14.42 (a) $C_6H_5CHO + C_6H_5CH_2CO_2C_2H_5 \xrightarrow{NaOC_2H_5} C_6H_5CH{=}\underset{\displaystyle |\atop C_6H_5}{C}CO_2C_2H_5$

(b) (cyclohexanone, $=O$) $\xrightarrow{HN\text{(piperidine)}}$ (1-cyclohexenyl piperidine, –N) $\xrightarrow{CH_3CH{=}CHCH_2Cl}$ (2-($CH_2CH{=}CHCH_3$)cyclohexylidene $=\overset{+}{N}$ piperidinium) $\xrightarrow{H_2O, H^+}$

(c) $CH_2(CO_2CH_3)_2 \xrightarrow[(2)\ (CH_3)_2CHBr]{(1)\ NaOCH_3}$

(d) $C_6H_5CHO + CH_2(CN)_2 \xrightarrow[(2)\ H^+]{(1)\ NaOC_2H_5\ \text{or}\ NaOH}$

(e) $C_6H_5CHO + CH_3\overset{O}{\overset{\|}{C}}CH_2CO_2C_2H_5 \xrightarrow[(2)\ H^+]{(1)\ NaOC_2H_5}$

(f) (cyclohexanone, $=O$) $+ H\overset{O}{\overset{\|}{C}}OC_2H_5 \xrightarrow[(2)\ H^+]{(1)\ NaOC_2H_5}$

(g) $C_2H_5O_2C{-}CO_2C_2H_5 + CH_3CH_2CH_2CO_2C_2H_5 \xrightarrow[(2)\ H^+]{(1)\ NaOC_2H_5}$

(h) (cyclohex-2-enone, $=O$) $+ CH_2(CO_2C_2H_5)_2 \xrightarrow[(2)\ H^+]{(1)\ NaOC_2H_5}$

(i) 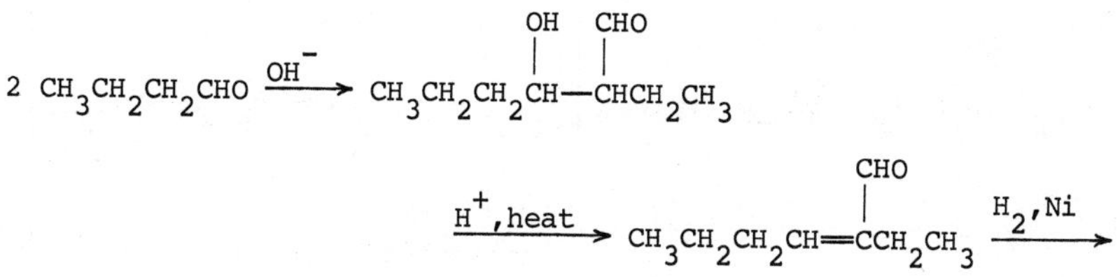 $2\ CH_3CH_2CH_2CHO \xrightarrow{OH^-} CH_3CH_2CH_2CH(OH){-}CH(CHO)CH_2CH_3$

$\xrightarrow{H^+,\text{heat}} CH_3CH_2CH_2CH{=}C(CHO)CH_2CH_3 \xrightarrow{H_2,Ni}$

(j) 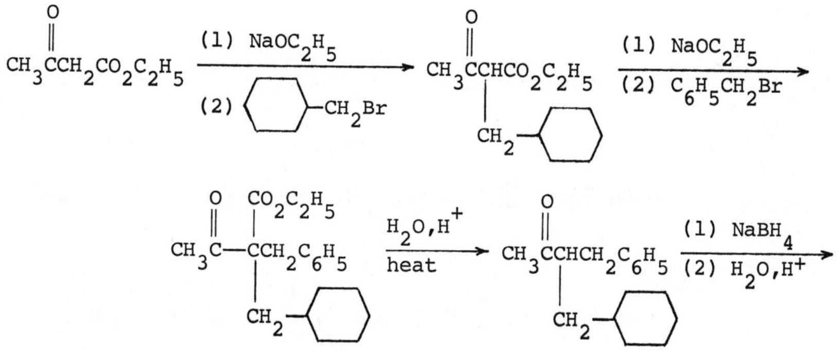 $CH_3C(=O)CH_2CO_2C_2H_5 \xrightarrow[(2)\ C_6H_{11}CH_2Br]{(1)\ NaOC_2H_5} CH_3C(=O)CH(CH_2C_6H_{11})CO_2C_2H_5 \xrightarrow[(2)\ C_6H_5CH_2Br]{(1)\ NaOC_2H_5}$

$CH_3C(=O){-}C(CO_2C_2H_5)(CH_2C_6H_{11})CH_2C_6H_5 \xrightarrow[\text{heat}]{H_2O,H^+} CH_3C(=O)CH(CH_2C_6H_{11})CH_2C_6H_5 \xrightarrow[(2)\ H_2O,H^+]{(1)\ NaBH_4}$

14.43 (a) 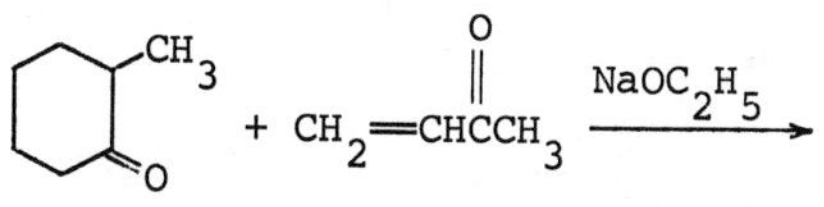 2-methylcyclohexanone $+ CH_2{=}CHC(=O)CH_3 \xrightarrow{NaOC_2H_5}$

(b) 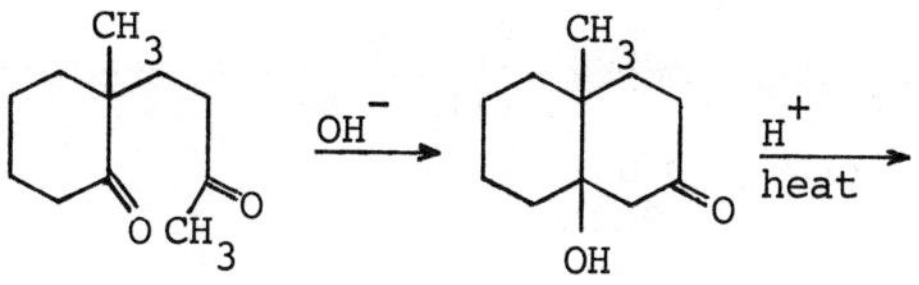 2-methyl-2-(3-oxobutyl)cyclohexanone $\xrightarrow{OH^-}$ bicyclic hydroxy ketone (ring-fusion CH_3 and OH) $\xrightarrow[\text{heat}]{H^+}$

14.44 (a) $CH_3CO_2H + CH_3CH_2OH \underset{}{\overset{\text{dil. HCl,heat}}{\rightleftharpoons}} CH_3CO_2C_2H_5 + H_2O$

(b) 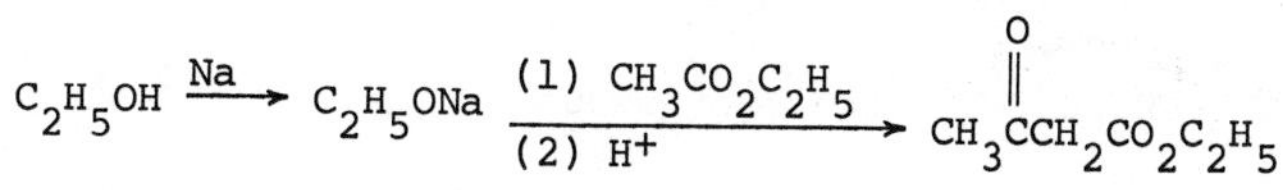 $C_2H_5OH \xrightarrow{Na} C_2H_5ONa \xrightarrow[(2)\ H^+]{(1)\ CH_3CO_2C_2H_5} CH_3C(=O)CH_2CO_2C_2H_5$

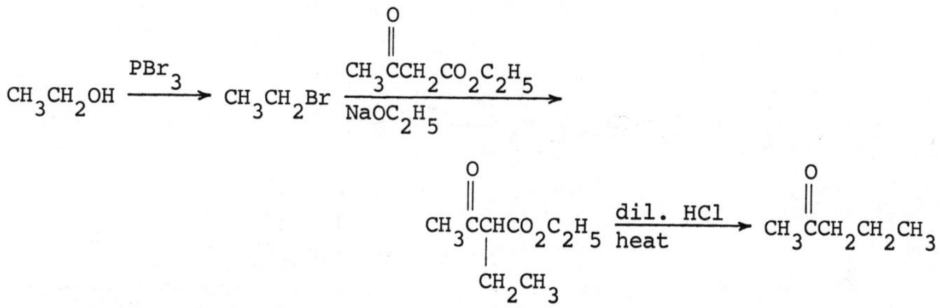

(c) $CH_3CH_2OH \xrightarrow{PBr_3} CH_3CH_2Br \xrightarrow[NaOC_2H_5]{CH_3C(=O)CH_2CO_2C_2H_5}$

$CH_3C(=O)CH(CH_2CH_3)CO_2C_2H_5 \xrightarrow[\text{heat}]{\text{dil. HCl}} CH_3C(=O)CH_2CH_2CH_3$

(d) $CH_2(CO_2C_2H_5)_2 \xrightarrow[(2)\ NaOC_2H_5,C_2H_5Br]{(1)\ NaOC_2H_5,C_2H_5Br} (CH_3CH_2)_2C(CO_2C_2H_5)_2$

$\xrightarrow[\text{heat}]{\text{dil. HCl}} (CH_3CH_2)_2CHCO_2H$

(e) 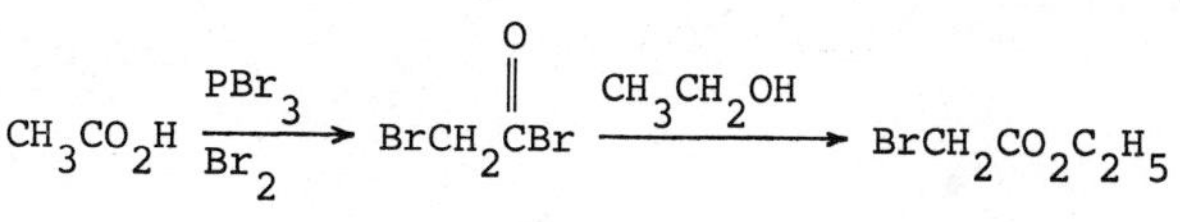 $CH_3CO_2H \xrightarrow[Br_2]{PBr_3} BrCH_2C(=O)Br \xrightarrow{CH_3CH_2OH} BrCH_2CO_2C_2H_5$

14.45

$$(\underline{R})\text{-}C_6H_5\overset{\overset{OH}{|}}{C}HCO_2C_2H_5 \xrightleftharpoons{OH^-} \text{achiral}\left[C_6H_5\overset{\overset{OH}{|}}{\bar{C}}CO_2C_2H_5 \longleftrightarrow C_6H_5\overset{\overset{HO}{|}}{C}{=}\overset{\overset{O^-}{|}}{C}OC_2H_5\right]$$

$$\xrightleftharpoons{H_2O,\,-OH^-} (\underline{R})\text{ and }(\underline{S})\text{-}C_6H_5\overset{\overset{OH}{|}}{C}HCO_2C_2H_5$$

14.46

$$CH_3CHO \xrightarrow{CO_3^{2-}} {}^-CH_2CHO \xrightleftharpoons{HCHO} {}^-OCH_2CH_2CHO \xrightleftharpoons{H_2O} HOCH_2CH_2CHO \xrightleftharpoons{CO_3^{2-}}$$

$$HOCH_2\bar{C}HCHO \xrightleftharpoons{HCHO} HOCH_2\overset{\overset{CH_2O^-}{|}}{C}HCHO \xrightleftharpoons{H_2O} HOCH_2\overset{\overset{CH_2OH}{|}}{C}HCHO \xrightleftharpoons{CO_3^{2-}}$$

$$(HOCH_2)_2\bar{C}CHO \xrightleftharpoons{HCHO} (HOCH_2)_2\overset{\overset{CH_2O^-}{|}}{C}CHO \xrightleftharpoons{H_2O} \text{product}$$

14.47 (a)

$$C_6H_5CH_2\overset{\overset{O}{\|}}{C}CH_3 + CH_2{=}CH\overset{\overset{O}{\|}}{C}CH_3 \xrightarrow{NaOC_2H_5}$$

$$CH_3\overset{\overset{O}{\|}}{C}CH_2CH_2\underset{\underset{C_6H_5}{|}}{C}H\overset{\overset{O}{\|}}{C}CH_3 \xrightarrow[(2)\ H^+,\text{heat}]{(1)\ OH^-} \text{product}$$

Loss of the most acidic proton in the last step would not lead to cyclization because a four-membered ring would result.

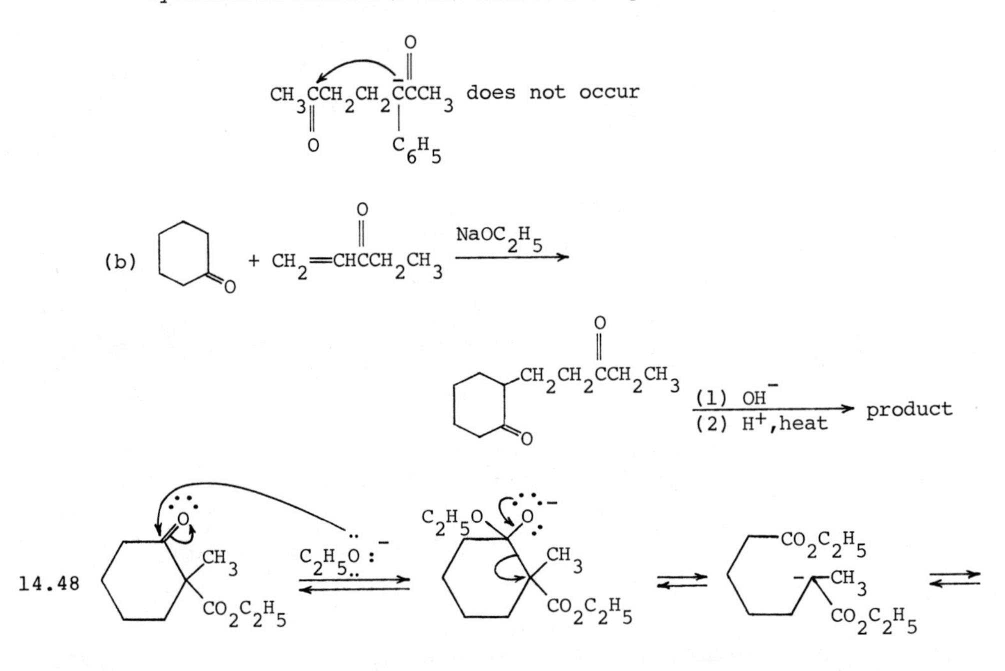

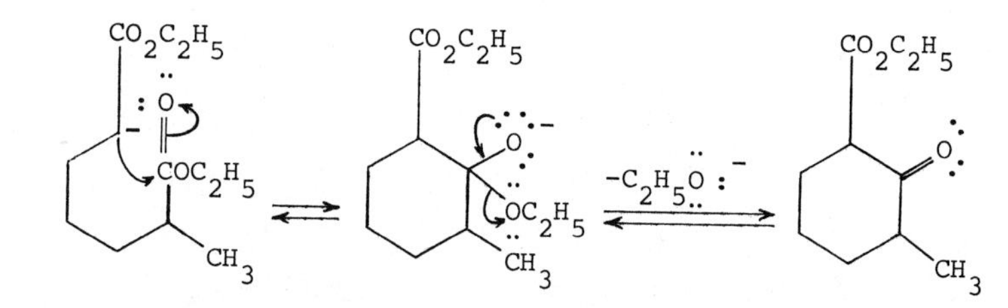

The first steps in this rearrangement are a reverse ester condensation. Then a forward ester condensation yields the rearranged product.

14.49

an enolate

14.50 $CH_2(CO_2C_2H_5)_2$ $\xrightarrow[\text{(2) } Br(CH_2)_{\underline{n}}Br]{\text{(1) } NaOC_2H_5}$ $Br(CH_2)_{\underline{n}}CH(CO_2C_2H_5)_2$ $\xrightarrow{NaOC_2H_5}$

$(CH_2)_{\underline{n}}$ $C(CO_2C_2H_5)_2$ $\xrightarrow[\text{heat}]{H_2O, H^+}$ $(CH_2)_{\underline{n}}$ $CHCO_2H$

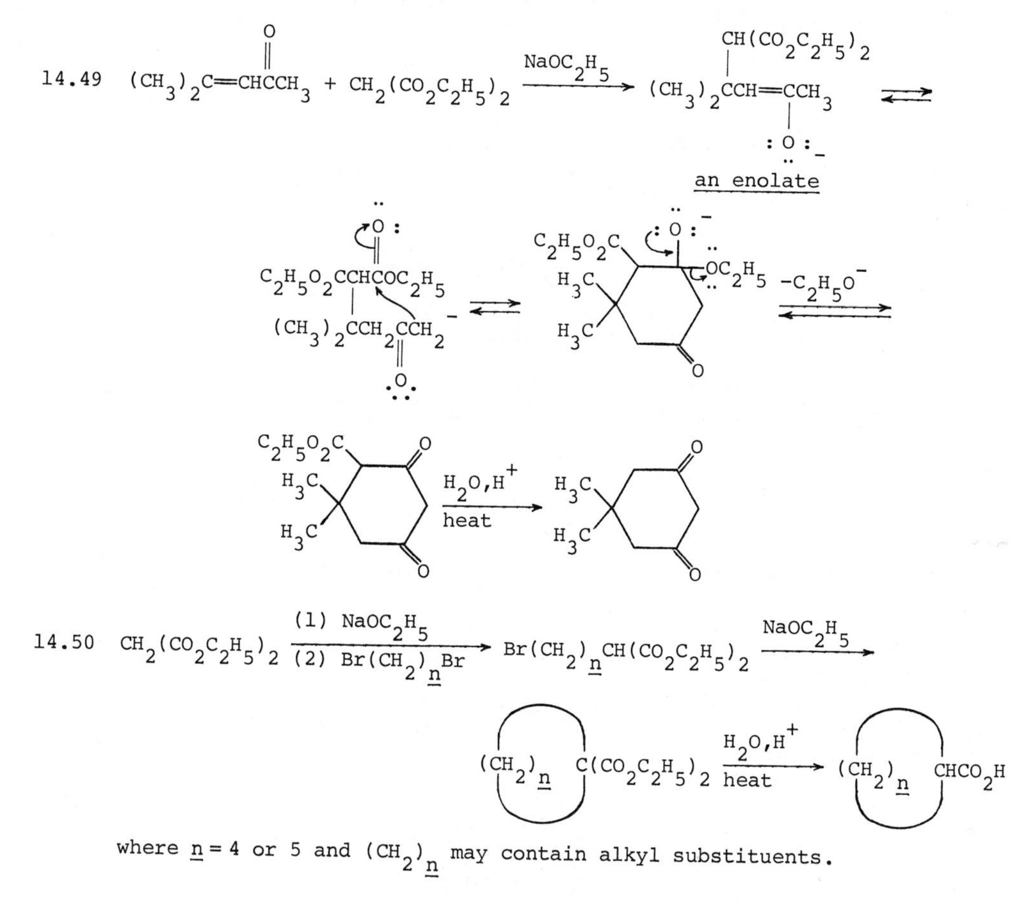

where $\underline{n} = 4$ or 5 and $(CH_2)_{\underline{n}}$ may contain alkyl substituents.

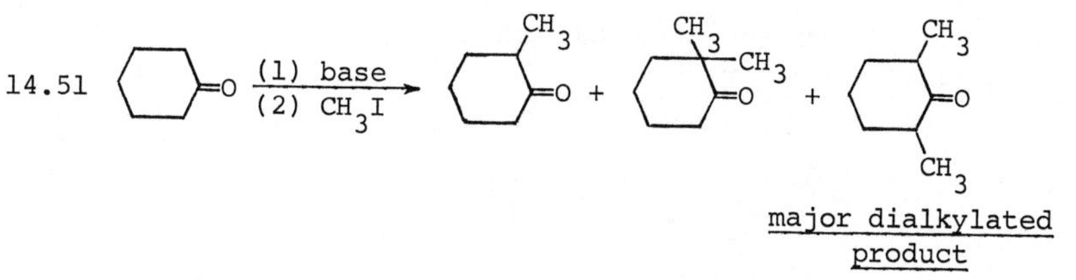

Of the two anions (A and B) leading to dialkylated products, A is in the lowest concentration because of the electron-releasing effect of the methyl group.

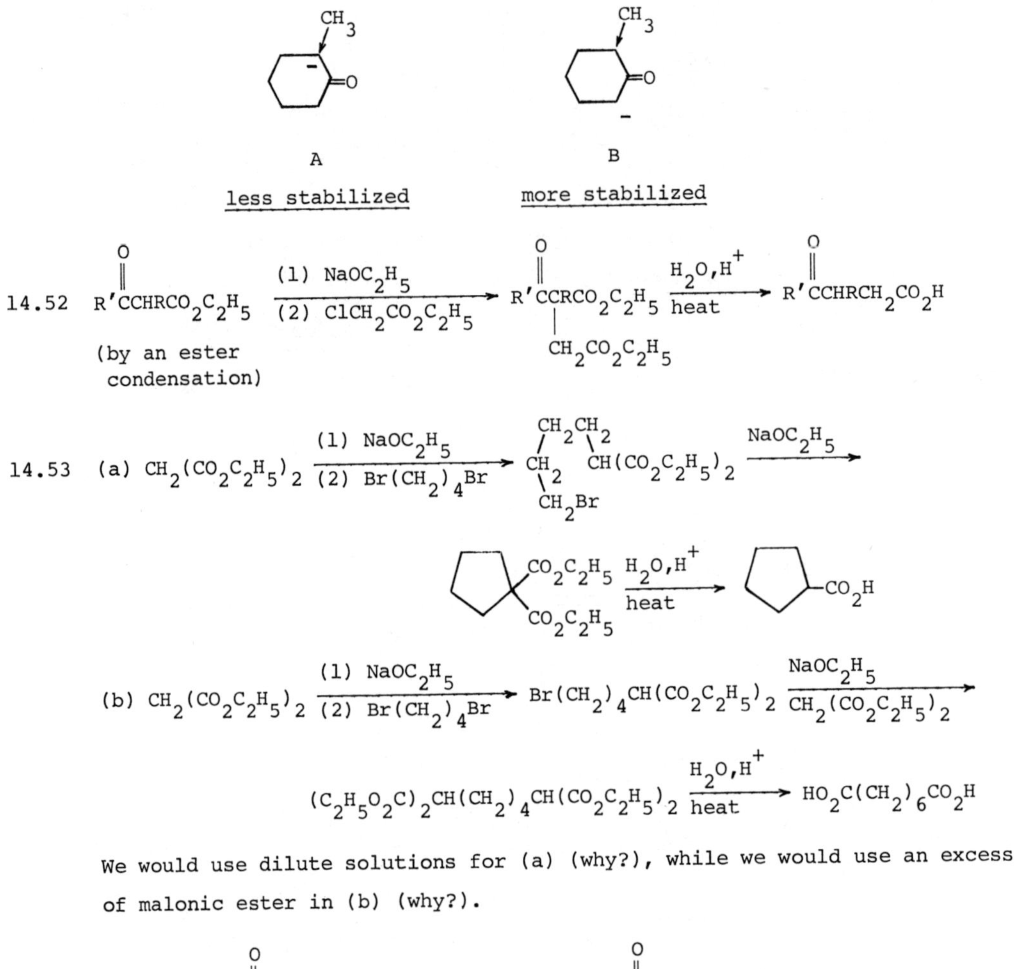

A
less stabilized

B
more stabilized

14.52 $R'CCHRCO_2C_2H_5$ $\xrightarrow[\text{(2) } ClCH_2CO_2C_2H_5]{\text{(1) } NaOC_2H_5}$ $R'CCRCO_2C_2H_5$ $\xrightarrow[\text{heat}]{H_2O,H^+}$ $R'CCHRCH_2CO_2H$

(by an ester
condensation)

$CH_2CO_2C_2H_5$

14.53 (a) $CH_2(CO_2C_2H_5)_2$ $\xrightarrow[\text{(2) } Br(CH_2)_4Br]{\text{(1) } NaOC_2H_5}$

$\xrightarrow{NaOC_2H_5}$

$\xrightarrow[\text{heat}]{H_2O,H^+}$

(b) $CH_2(CO_2C_2H_5)_2$ $\xrightarrow[\text{(2) } Br(CH_2)_4Br]{\text{(1) } NaOC_2H_5}$ $Br(CH_2)_4CH(CO_2C_2H_5)_2$ $\xrightarrow[\text{CH}_2(CO_2C_2H_5)_2]{NaOC_2H_5}$

$(C_2H_5O_2C)_2CH(CH_2)_4CH(CO_2C_2H_5)_2$ $\xrightarrow[\text{heat}]{H_2O,H^+}$ $HO_2C(CH_2)_6CO_2H$

We would use dilute solutions for (a) (why?), while we would use an excess of malonic ester in (b) (why?).

14.54 I: $CH_3CH_2CH(CO_2C_2H_5)CCO_2C_2H_5$ (with C=O) II: $CH_3CH_2CH_2CCO_2H$ (with C=O)

14.55 $C_2H_5O_2CCH_2CH_2CH_2CO_2C_2H_5 \xrightarrow{NaOC_2H_5}$

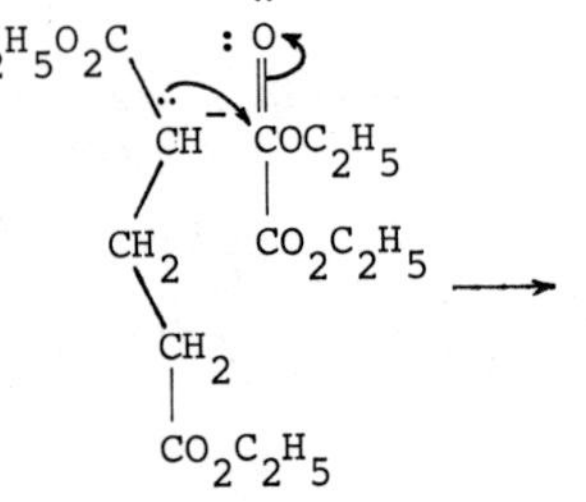

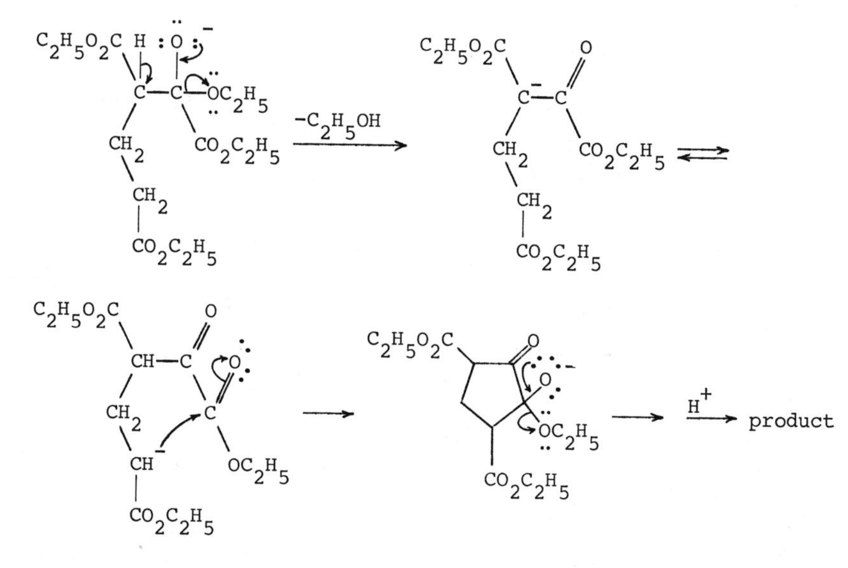

14.56 I, $(C_2H_5O_2C)_2CHCH_2CH_2CH(CO_2C_2H_5)_2$

II, $C_2H_5O_2C$, $C_2H_5O_2C$ [cyclohexane ring] $CO_2C_2H_5$, $CO_2C_2H_5$ III, HO_2C–[cyclohexane ring]–CO_2H

14.57 $CH_2(CO_2C_2H_5)_2 \xrightarrow[(2)\ CH_3CHBrCH_2CH_2CH_3]{(1)\ NaOC_2H_5} CH_3CH_2CH_2CH(CH_3){-}CH(CO_2C_2H_5)_2$

$\xrightarrow[(2)\ CH_2{=}CHCH_2Br]{(1)\ NaOC_2H_5}$ $CH_3CH_2CH_2CH(CH_3)$, $CH_2{=}CHCH_2$ $C(CO_2C_2H_5)_2 \xrightarrow{H_2NC(=O)NH_2}$ product

The last step in this reaction is similar to the reaction of an amine with an ester to yield an amide:

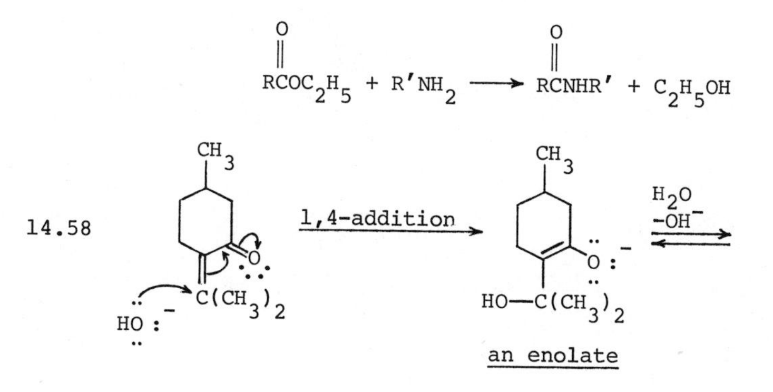

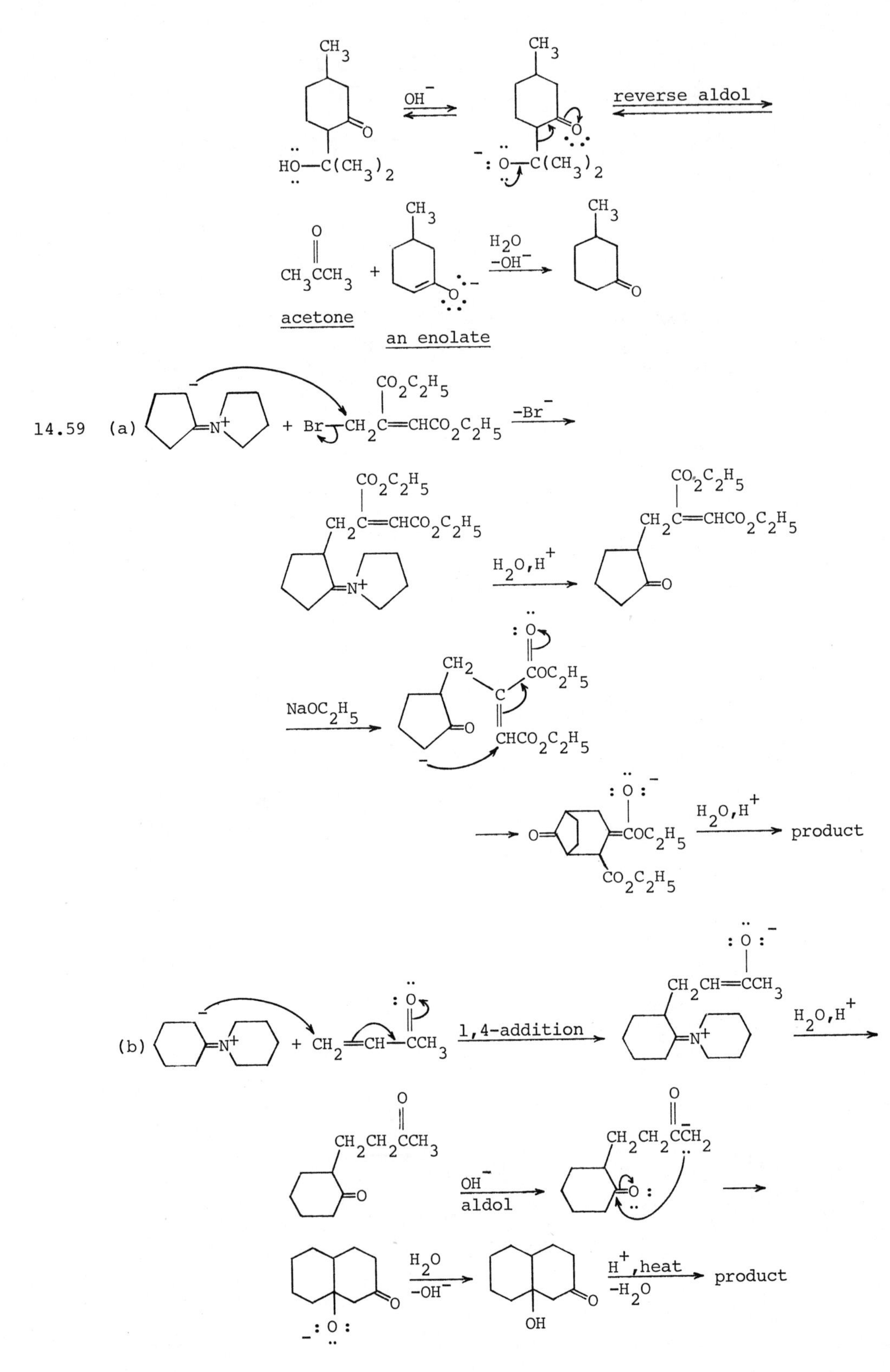

CH3
OH⁻
reverse aldol
HO—C(CH3)2
O—C(CH3)2
CH3CCH3
acetone
an enolate
H2O
-OH⁻
14.59 (a)
CO2C2H5
Br—CH2C═CHCO2C2H5
-Br⁻
H2O,H⁺
NaOC2H5
COC2H5
CHCO2C2H5
product
(b)
CH2═CH—CCH3
1,4-addition
CH2CH═CCH3
CH2CH2CCH3
OH⁻
aldol
CH2CH2CCH2
H2O
-OH⁻
OH
H⁺,heat
-H2O
product

(c) $HCH(=\ddot{O}:) + HN(CH_3)_2 \rightleftharpoons [\,:\ddot{O}:^{-}\text{—}CH_2\overset{+}{N}H(CH_3)_2\,] \rightleftharpoons$

$HOCH_2N(CH_3)_2 \xrightarrow[-H_2O]{H^+} CH_2=\overset{+}{N}(CH_3)_2$

an iminium ion

$C_6H_5CCH_3(=O) \rightleftharpoons C_6H_5C(OH)=CH_2 \xrightarrow{CH_2=\overset{+}{N}(CH_3)_2}$

an enol

$C_6H_5\overset{+}{C}(\ddot{O}H)CH_2CH_2\ddot{N}(CH_3)_2 \xrightarrow{-H^+}$ product

14.60 (a) 2 cyclopentanone $\xrightarrow{OH^-}$ 2-(1-hydroxycyclopentyl)cyclopentanone (HO) $\xrightarrow[\text{heat}]{H^+}$

(b) o-phthalaldehyde (CHO, CHO) + $CH_3CH_2CCH_2CH_3$ (C=O) $\xrightarrow[-H_2O]{OH^-}$ [o-(CHO)C_6H_4—$CH=C(CH_3)—CCH_2CH_3$ (C=O)] $\xrightarrow{OH^-}$

(c) dilute $CH_3CCH_2CO_2C_2H_5$ (C=O) $\xrightarrow[(2)\ Br(CH_2)_4Br]{(1)\ NaOC_2H_5}$ $CH_3CCHCO_2C_2H_5$ (C=O; CH with $CH_2CH_2CH_2CH_2Br$) $\xrightarrow{NaOC_2H_5}$

(d) $C_2H_5O_2CCH_2CH_2CHCHCO_2C_2H_5$ (with $C_2H_5O_2C$ and CH_3 substituents) $\xrightarrow{NaOC_2H_5}$

$C_2H_5O_2C\bar{C}HCH_2CHCHCO_2C_2H_5$ (with $C_2H_5O_2C$ and CH_3 substituents) $\xrightarrow[(2)\ H^+]{(1)\ \text{ester condensation}}$

The starting triester can be prepared by the following sequence:

$C_2H_5O_2CCH_2CHCO_2C_2H_5$ (CH_3) $\xrightarrow{NaOC_2H_5}$ $C_2H_5O_2C\bar{C}HCHCO_2C_2H_5$ (CH_3) $\xrightarrow[(2)\ H^+]{(1)\ CH_2=CHCO_2C_2H_5}$

(e) 2-oxocyclopentanecarboxylic acid (HO_2C) $\xrightarrow[\text{heat}]{CH_3CH_2OH,H^+}$ ethyl 2-oxocyclopentanecarboxylate ($C_2H_5O_2C$) $\xrightarrow[(2)\ CH_2=CHCCH_2CH_3\ (C=O)]{(1)\ NaOC_2H_5}$

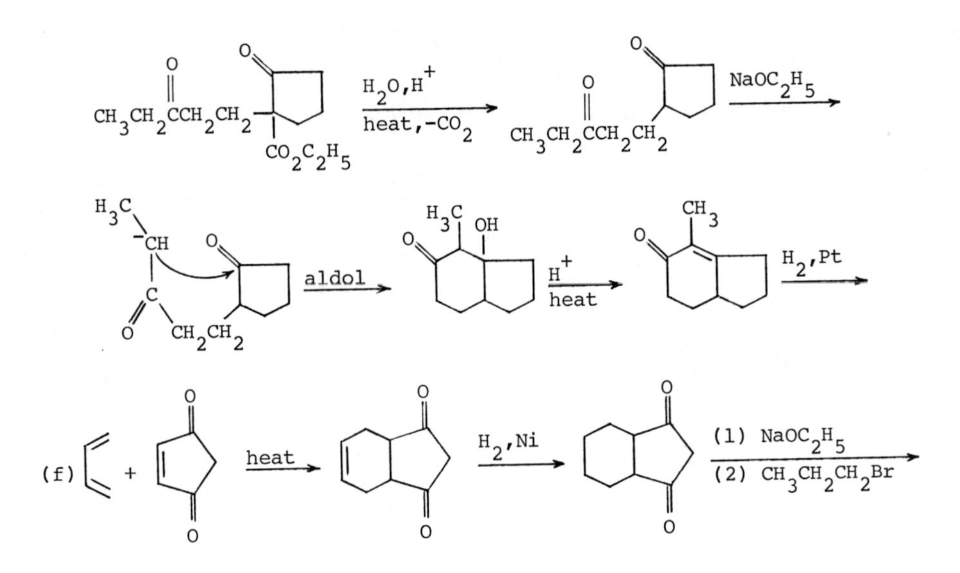
O
CH3CH2CCH2CH2
CO2C2H5
H2O,H+
heat,-CO2
NaOC2H5
H3C
CH
C
CH2CH2
aldol
OH
H+
heat
CH3
H2,Pt
(f)
heat
H2,Ni
(1) NaOC2H5
(2) CH3CH2CH2Br

CHAPTER 15

Amines

Some Important Features

Amines are compounds in which N is bonded to three other groups (or hydrogen). The nitrogen of an amine contains an unshared pair of electrons; therefore, an amine is basic and can act as a nucleophile.

$$R_3N: + H{-}\ddot{\underset{..}{O}}H \rightleftharpoons R_3\overset{+}{N}H + :\ddot{\underset{..}{O}}H^-$$

an amine

$$R_3N: + H{-}Cl \rightleftharpoons R_3\overset{+}{N}H\ Cl^-$$

an amine salt

Although amines can react as nucleophiles with alkyl halides, this reaction often leads to mixtures of products.

$$R_3N: + CH_3{-}I \xrightarrow{S_N2} R_3\overset{+}{N}CH_3\ I^-$$

only product

$$RNH_2 + CH_3I \longrightarrow R\overset{+}{N}H_2CH_3\ I^- + R\overset{+}{N}H(CH_3)_2\ I^- + R\overset{+}{N}(CH_3)_3\ I^-$$

For this reason, other methods for synthesizing complex amines have been developed (see Section 15.5).

Amines are weak bases. Their basicity is determined by the relative stabilization of amine versus amine salt.

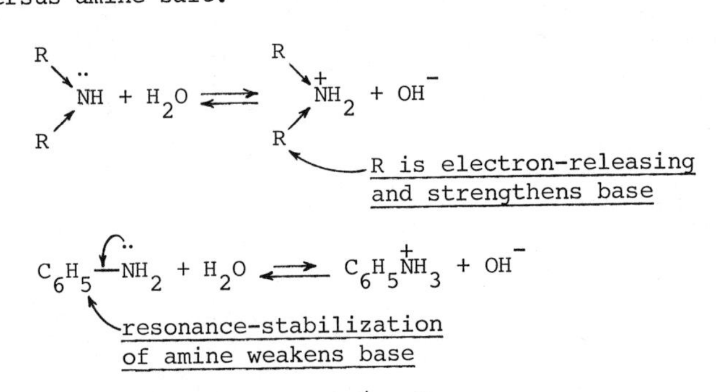

Quaternary ammonium hydroxides ($R_4N^+ \ OH^-$), when heated, usually yield Hofmann elimination products (the least substituted alkenes) because of steric hindrance in the transition state. (You may find it helpful to circle all β hydrogens in the quaternary ammonium hydroxide, then determine the most likely position or positions of elimination.)

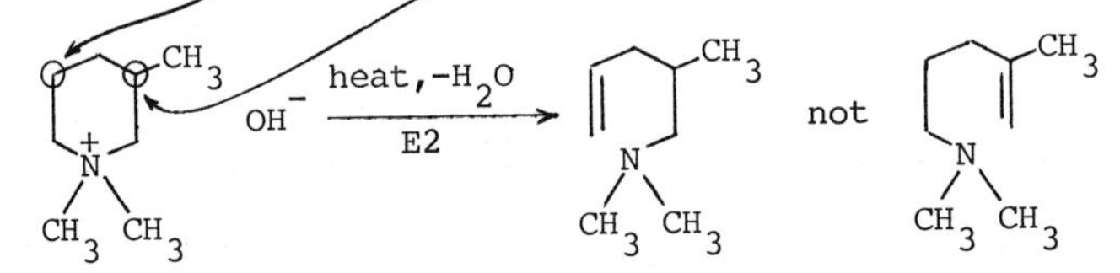

This reaction is the basis of exhaustive methylation, a reaction sequence used in structure determinations (see Section 15.10C).

Some reactions of amines (with acid halides, aldehydes, and ketones) are mentioned in Section 15.8.

Reminders

In determining the relative basicities of amines, always ask yourself about the availability of the unshared electrons for donation to H^+. For example, the electrons of aniline are less available than those of an alkylamine; therefore, aniline is less basic.

Answers to Problems

15.21 (a) 3° amine salt (b) 3° amine (c) quaternary ammonium salt
(d) 1° amine salt

15.22

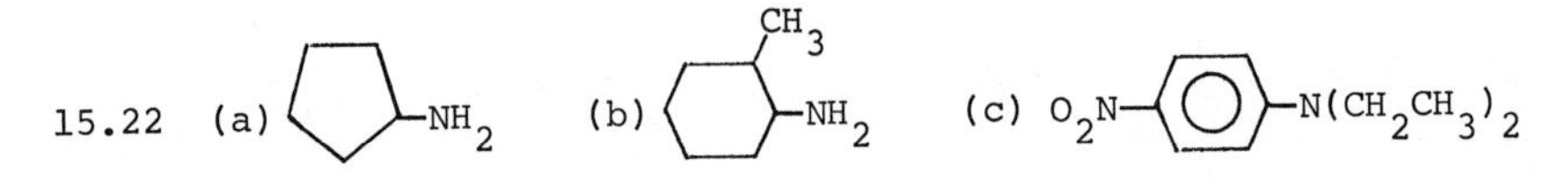

(d) $CH_3(CH_2)_3CHCO_2H$ with $N(CH_3)_2$ on the CH carbon

15.23 (a) N-ethyl-N-methylbenzylamine (b) 1,2-cyclohexanediamine

(c) N,N-dimethylcyclohexylammonium bromide

or N,N-dimethylcyclohexylamine hydrobromide

(d) 4-aminopentanal

15.24 (a) one chiral C: one pair of enantiomers (b) no isolable stereoisomers

(c) achiral (The N does not have four different attached groups.)

(d) one chiral N and one chiral C: four stereoisomers representing two pairs of enantiomers. (The diastereomeric pairs are also geometric isomers.)

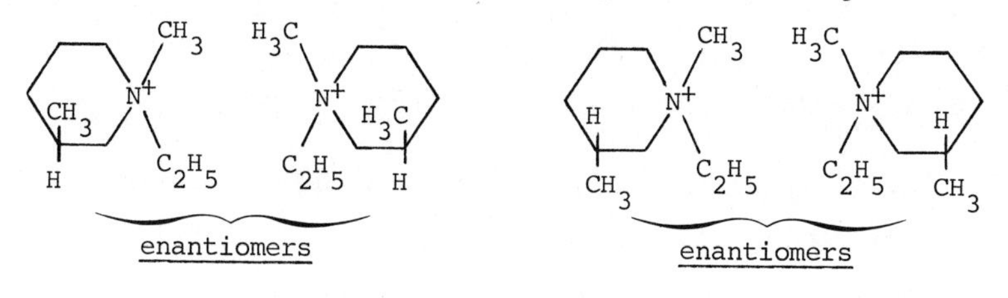

(e) one chiral N: one pair of enantiomers

(f) one double bond: one pair of geometric isomers, or diastereomers:

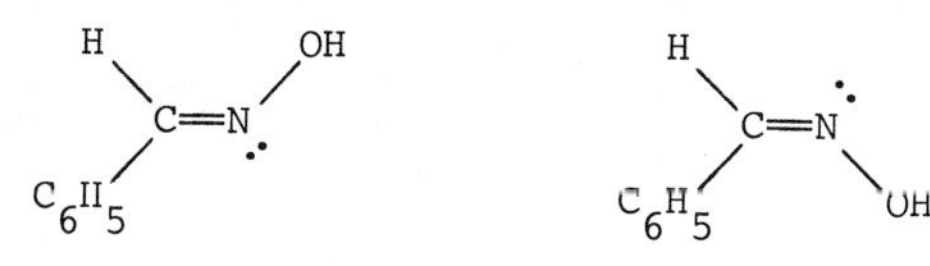

15.25 (a), (b), (c), (d), and (e). Compound (f) has no unshared electrons, which are necessary for an amine to act as a nucleophile.

15.26 (a) Cyclohexylamine forms stronger hydrogen bonds with water than does cyclohexanol. (The partially positive H of H_2O is more strongly held by the more basic N atom.)

(b) Dimethylamine forms hydrogen bonds with itself, while trimethylamine does not.

(c) The branching around the nitrogen of dimethylamine diminishes its ability to hydrogen bond with itself effectively because of steric hindrance.

15.27 (a)

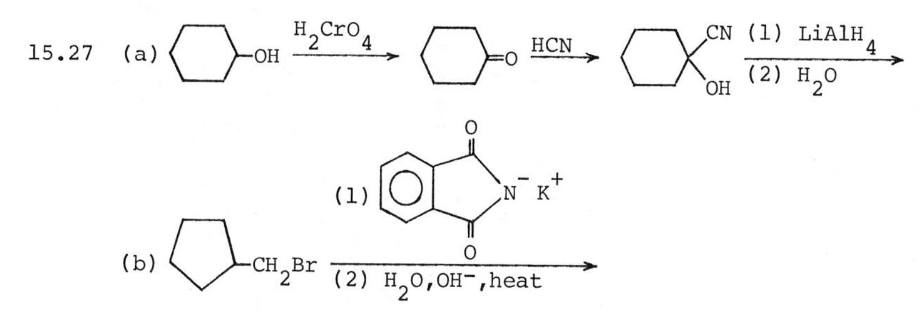

Reaction of the halide with NH_3 would not be as satisfactory as the preceding sequence because of overalkylation. See Section 15.5A.

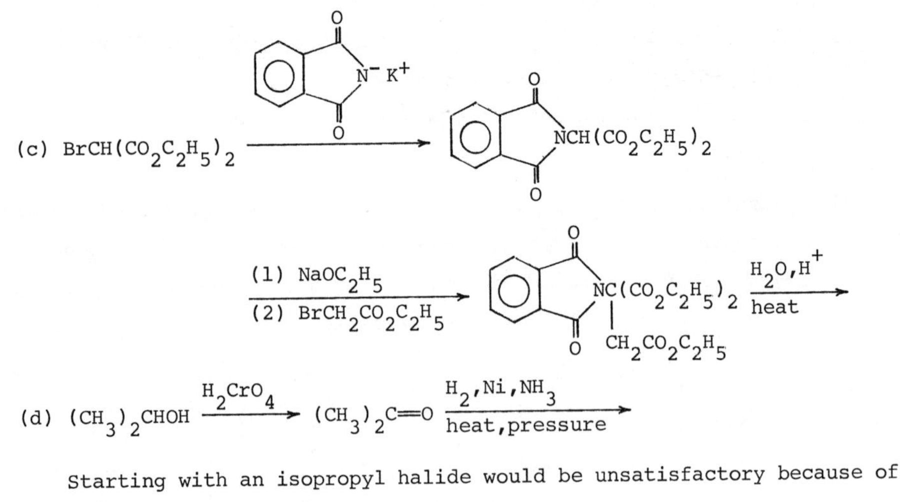

Starting with an isopropyl halide would be unsatisfactory because of elimination side reactions.

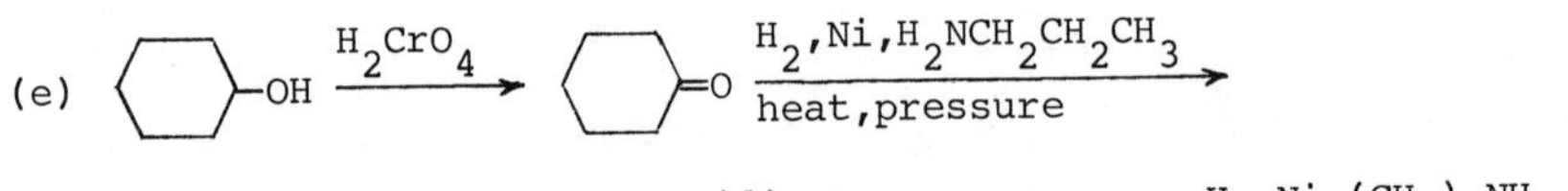

(f) $CH_3CH_2CH_2CH_2OH \xrightarrow{CrO_3 \cdot 2\ \text{pyridine}} CH_3CH_2CH_2CHO \xrightarrow[\text{heat, pressure}]{H_2, Ni, (CH_3)_2NH}$

or $CH_3CH_2CH_2CH_2Br \xrightarrow{\text{potassium phthalimide}} CH_3CH_2CH_2CH_2{-}N(\text{phthalimide})$

$\xrightarrow[\text{heat}]{H_2O, OH^-} CH_3CH_2CH_2CH_2NH_2 \xrightarrow[(2)\ OH^-]{(1)\ 2\ CH_3I}$

15.28 (a) $CH_3(CH_2)_3CH_2OH \xrightarrow{HBr} CH_3(CH_2)_3CH_2Br$

$\xrightarrow[(2)\ H_2O, OH^-, \text{heat}]{(1)\ \text{potassium phthalimide}} CH_3(CH_2)_3CH_2NH_2$

(b) $CH_3(CH_2)_3CH_2Br$ from (a) $\xrightarrow{KCN} CH_3(CH_2)_4CN \xrightarrow[(2)\ H_2O]{(1)\ LiAlH_4} CH_3(CH_2)_4CH_2NH_2$

(c) $CH_3(CH_2)_3CH_2OH \xrightarrow{H_2CrO_4} CH_3(CH_2)_3CO_2H \xrightarrow[(2)\ NH_3]{(1)\ SOCl_2}$

$$CH_3(CH_2)_3\overset{O}{\overset{\|}{C}}NH_2 \xrightarrow{Br_2, OH^-} CH_3(CH_2)_3NH_2$$

15.29 (a) $C_6H_6 \xrightarrow[H_2SO_4]{HNO_3} C_6H_5NO_2 \xrightarrow[HCl]{Fe} C_6H_5\overset{+}{N}H_3\ Cl^- \xrightarrow{OH^-} C_6H_5NH_2$

(b) $C_6H_5\overset{O}{\overset{\|}{C}}NH_2 \xrightarrow{Br_2, OH^-} C_6H_5NH_2$ (c) $C_6H_5NH_2 \xrightarrow{\left(CH_3\overset{O}{\overset{\|}{C}}O\right)_2O} C_6H_5NH\overset{O}{\overset{\|}{C}}CH_3 + CH_3CO_2H$

(d) (cyclic glutaric anhydride) $\xrightarrow{NH_3}$ $\overset{O}{\overset{\|}{C}}NH_2$–$CH_2CH_2CH_2$–$CO_2^-\ NH_4^+$ $\xrightarrow[(2)\ H^+]{(1)\ Br_2,\ ^-OH} H_2NCH_2CH_2CH_2CO_2H$

(e) $(\underline{R})\text{-}CH_3\overset{OH}{\overset{|}{C}}HCH_2CH_3 \xrightarrow{TsCl} (\underline{R})\text{-}CH_3\overset{OTs}{\overset{|}{C}}HCH_2CH_3$

$$\xrightarrow[\text{or phthalimide procedure}]{NH_3} (\underline{S})\text{-}CH_3\overset{NH_2}{\overset{|}{C}}HCH_2CH_3$$

(f) $C_6H_5CH_3 \xrightarrow{KMnO_4} C_6H_5CO_2H \xrightarrow{SOCl_2} C_6H_5\overset{O}{\overset{\|}{C}}Cl \xrightarrow{NH_3}$

$$C_6H_5\overset{O}{\overset{\|}{C}}NH_2 \xrightarrow[(2)\ H_2O]{(1)\ LiAlH_4} C_6H_5CH_2NH_2$$

(g) $CH_3CO_2H \xrightarrow[heat]{CH_3CH_2OH, H^+} CH_3CO_2C_2H_5 \xrightarrow{NH_3} CH_3\overset{O}{\overset{\|}{C}}NH_2$

15.30 (a) Aniline, $C_6H_5NH_2$, is more basic because the bromine of p-bromoaniline is electron-withdrawing and decreases the electron density on the nitrogen atom.

(b) Tetramethylammonium hydroxide, $(CH_3)_4N^+\ OH^-$, is more basic because it is more ionic, comparable to NaOH.

(c) p-Nitroaniline is more basic because it has only one electron-withdrawing —NO_2 group.

(d) Ethylamine is more basic because ethanolamine has an electron-withdrawing —OH group and can also form an internal hydrogen bond, both of which decrease its basicity.

(e) p-Toluidine is more basic because the —CH_3 group is electron-releasing, while the —CCl_3 group is electron-withdrawing.

15.31 In each case, the stronger base holds the proton.

(a) pyridine + piperidinium (NH_2^+) (b) pyridine + H_2O (c) $C_6H_5NH_2$ + $(CH_3)_3NH^+Cl^-$

(d) no appreciable reaction (e) $(CH_3)_4N^+\ {}^-O_2CCH_3 + H_2O$

(f) piperidinium (NH_2^+) $^-O_2CCH_3$ (g) phthalimide anion (N^-) + CH_3OH

15.32 (a) (3), (1), (2) (b) (2), (3), (1) (c) (1), (2), (3) (d) (2), (1)

15.33 (a) Calculate $\underline{K}_b$. $\underline{K}_b = 10^{-3.34} = 10^{0.66} \times 10^{-4} = 4.57 \times 10^{-4}$

Substitute:

$$\frac{[CH_3\overset{+}{N}H_3][OH^-]}{[CH_3NH_2]} = \underline{K}_b$$

$$\frac{[CH_3\overset{+}{N}H_3](0.0100)}{(0.00100)} = 4.57 \times 10^{-4}$$

$$[CH_3\overset{+}{N}H_3] = \frac{(4.57 \times 10^{-4})(10^{-3})}{(10^{-2})} = 4.57 \times 10^{-5}\ \underline{M}$$

(b) $\frac{\underline{x}[OH^-]}{\underline{x}} = 4.57 \times 10^{-4}$

$[OH^-] = 4.57 \times 10^{-4}\ \underline{M}$

$pOH = 3.34$

$pH + pOH = 14$ (definition)

$pH = 14 - 3.34 = 10.7$

15.34 (a) Dissolve in diethyl ether. Wash with dilute acid to remove the amine. Wash with dilute $NaHCO_3$ to remove the acid. The alcohol remains in the ether.

(b) Washing with dilute acid removes the amine.

15.35 The nitrogen of the NCH_3 group is the most basic. The other nitrogens are part of an amide group (not basic) and part of an aromatic ring system (not basic in this case). Even though we have not yet discussed aromatic heterocycles, you should have predicted a very low basicity for this latter N because of delocalization of the unshared electrons.

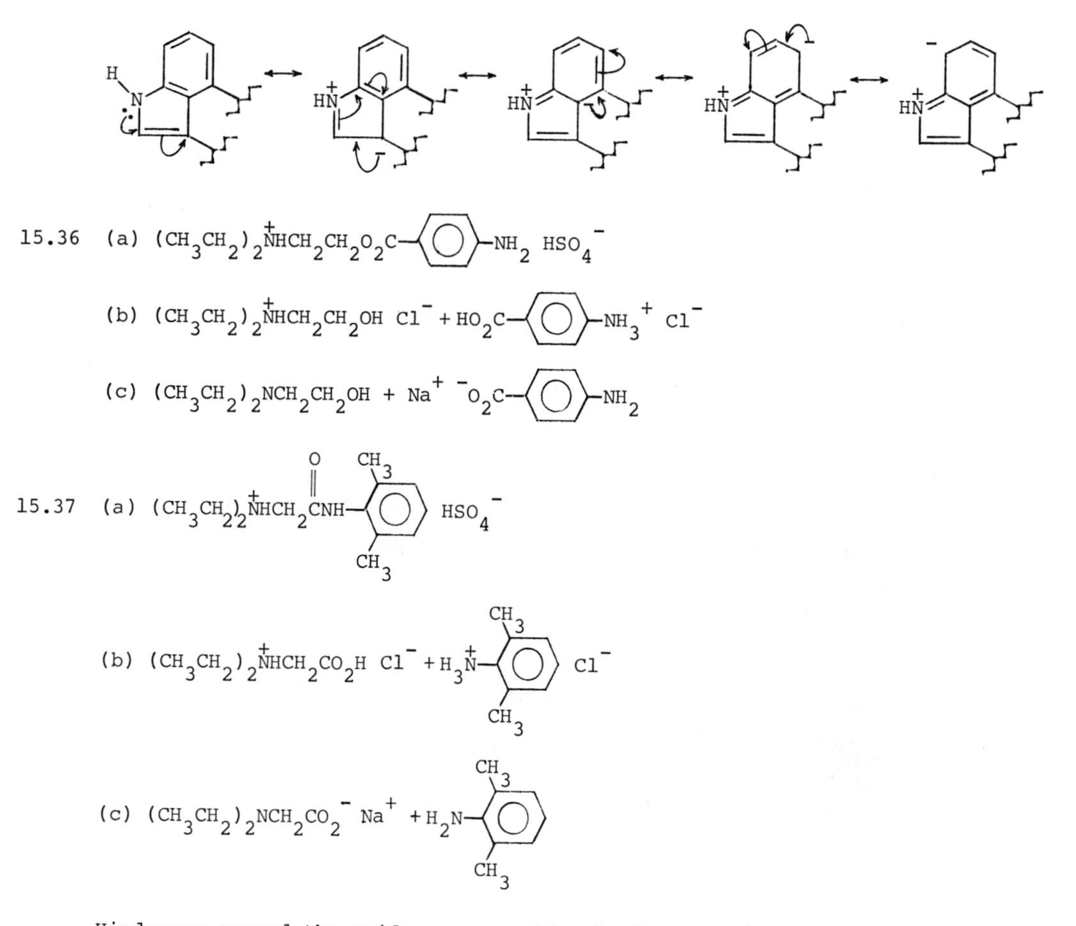

Hindrance around the amide group would make the reactions leading to the products shown in (b) and (c) very slow.

15.38 (a) Treat with one enantiomer of a chiral carboxylic acid, separate the diastereomeric salts, and regenerate the amine with dilute base.

(b) Saponify the ester with dilute NaOH, acidify, treat the resulting racemic carboxylic acid with one enantiomer of a chiral amine, separate the diastereomeric salts, make each salt alkaline, remove the amine by extraction with an organic solvent, and acidify the aqueous solution to regenerate the acid. To regenerate the ester, treat the acid with CH_2N_2 or $CH_3OH + H^+$.

15.39 (a) $CH_3NH_3^+\ I^-$ (b) $(CH_3)_4N^+\ I^-$ (c) $NH_3 + {}^-O_2CCH_2CH_2CO_2^-$

(d) $CH_3NH_2 + {}^-O_2CCH_2CH_2CO_2^-$

15.40 (a) pyrrolidine–$\overset{O}{\overset{\|}{N}C}C_6H_5$ + pyrrolidinium ($\overset{+}{N}H_2$) Cl^- (b) pyrrolidine–$\overset{O}{\overset{\|}{N}C}CH_3 + CH_3CO_2H$ (c) pyrrolidine–$\overset{+}{N}(CH_3)_2\ I^-$

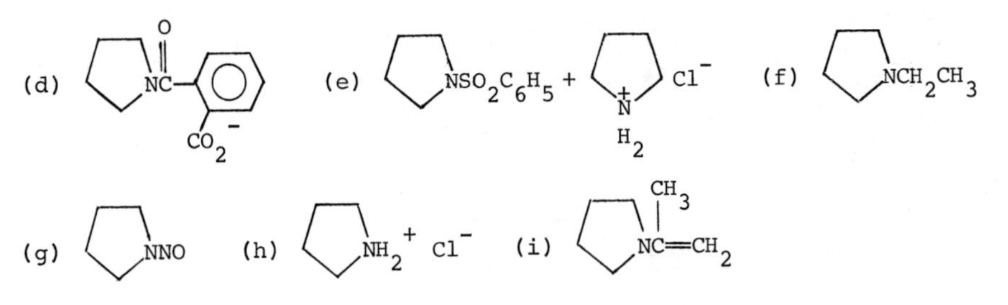

15.41 No reaction for (a), (b), (d), (e), (f), and (i).

(c) $\overset{+}{N}(CH_3)_2$ I^- (h) $\overset{+}{N}HCH_3$ Cl^-

15.42 (a) Treat with a cold aqueous solution of $NaNO_2 + HCl$. Aniline forms a diazonium salt, while n-hexylamine gives off nitrogen gas.

(b) Treat with cold, dilute HCl. n-Octylamine dissolves, while octanamide does not.

(c) Treat with dilute NaOH:

$$(CH_3CH_2)_3NH^+\ Cl^- \xrightarrow{OH^-} (CH_3CH_2)_3N + H_2O + Cl^-$$

amine smell

$$(CH_3CH_2)_4N^+\ Cl^- \xrightarrow{OH^-} \text{no reaction (odorless)}$$

15.43 (a) C_6H_5 N CH_3 (Note the conjugation.) (b) $N(CH_3)_2$ + $N(CH_3)_2$

15.44 + $(CH_3)_3N$

15.45 (a) O NH O $\xrightarrow[(2)\ BrCH(CO_2C_2H_5)_2]{(1)\ OH^-}$ O $NCH(CO_2C_2H_5)_2$ O

$\xrightarrow[(2)\ CH_3CH_2CHBrCH_3]{(1)\ NaOC_2H_5}$ O $CH_3CHCH_2CH_3$ N—$C(CO_2C_2H_5)_2$ O $\xrightarrow[heat]{H_2O,H^+}$

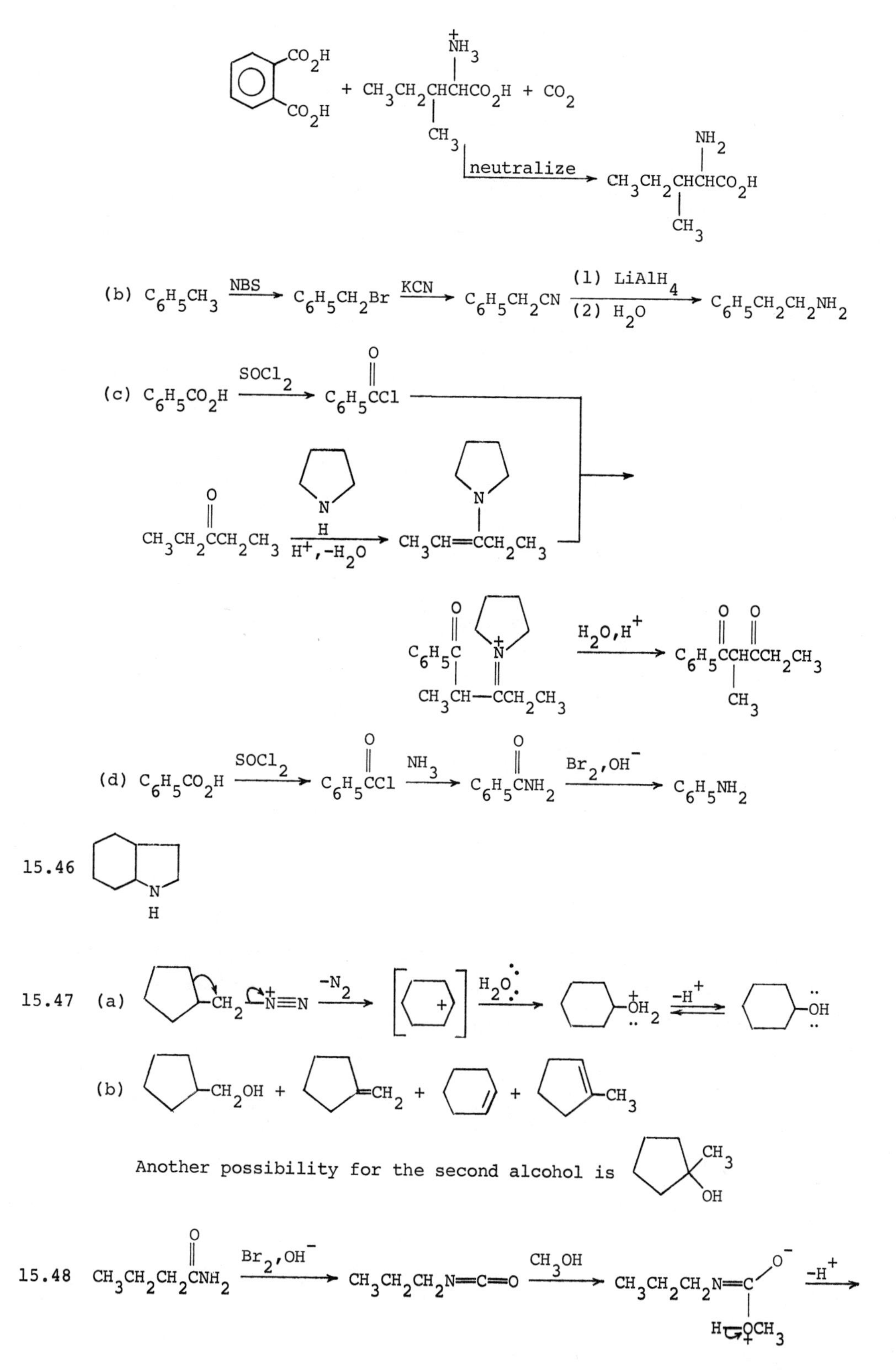
neutralize
(b) NBS KCN (1) LiAlH4 (2) H2O
(c) SOCl2
H+, -H2O
H2O, H+
(d) SOCl2 NH3 Br2, OH−
15.46
15.47 (a) −N2
(b)
Another possibility for the second alcohol is
15.48 Br2, OH− CH3OH −H+

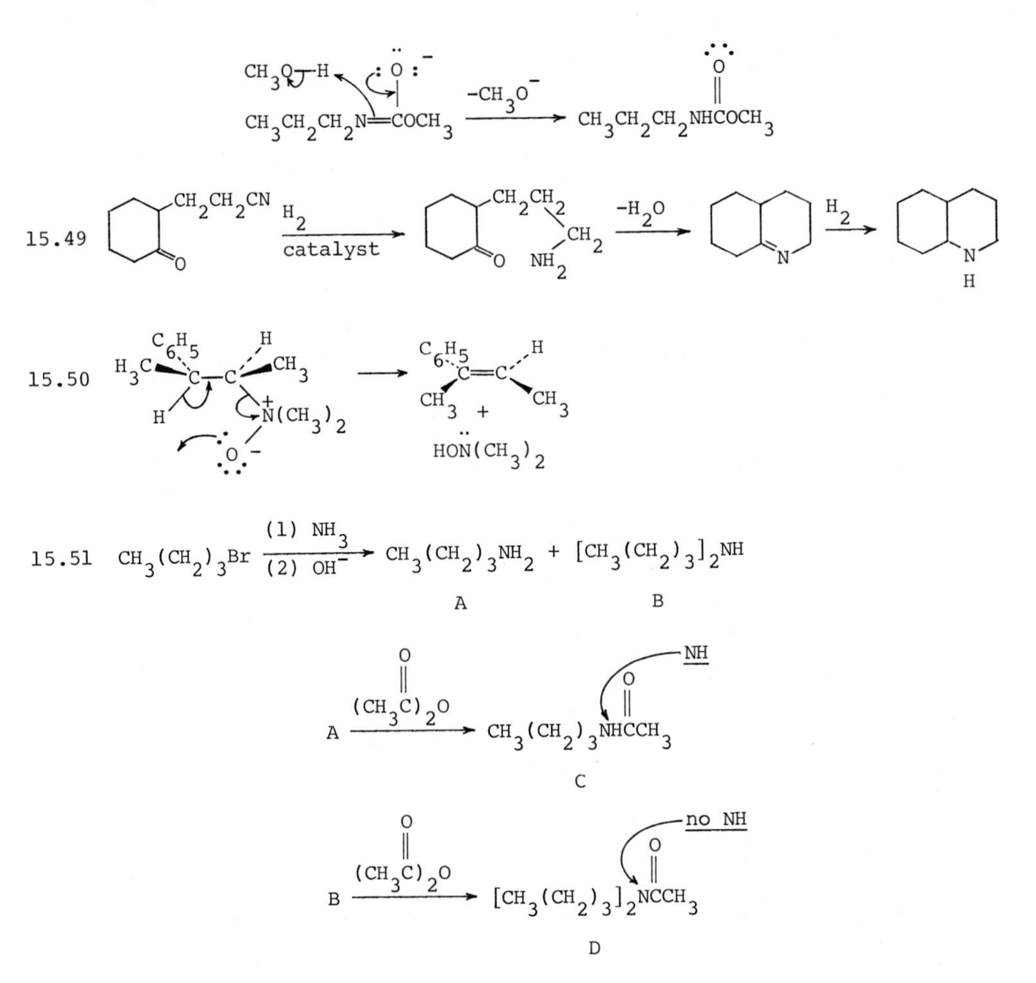

15.52 (a) $C_6H_5CH_2CH_2NH_2$ and (b) $C_6H_5CH(CH_3)NH_2$

Besides nmr absorption from aryl protons, (a) would exhibit two triplets and a singlet (area ratio, 1 : 1 : 1). However, (b) would show a quartet, a doublet, and a singlet (area ratio, 1 : 3 : 2).

15.53 A: [2-methylphenyl-N(CH3)C(O)CH3] B: [2-methylphenyl-NHCH3] C: [2-methylcyclohexyl-NHCH3]

CHAPTER 16

Polycyclic and Heterocyclic Aromatic Compounds

Some Important Features

The polycyclic aromatic compounds, such as naphthalene and anthracene, are not symmetrical as benzene is. Consequently, some carbon-carbon bonds of the polycyclic aromatic compounds have more double-bond character than others.

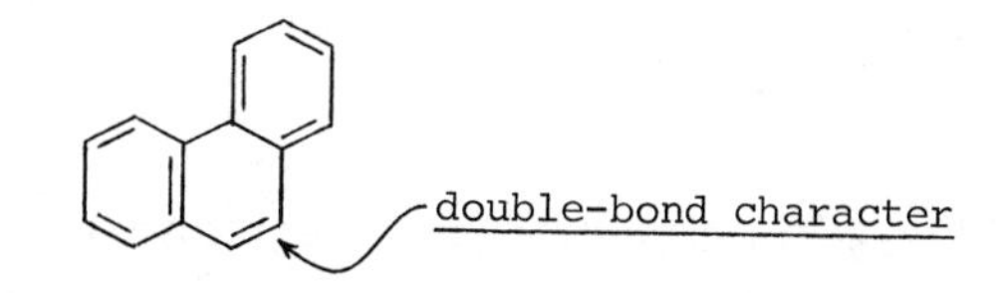

The aromatic polycyclic compounds are more reactive toward electrophiles, oxidizing agents, and reducing agents than is benzene because the intermediates still contain one or more rings with aromatic character.

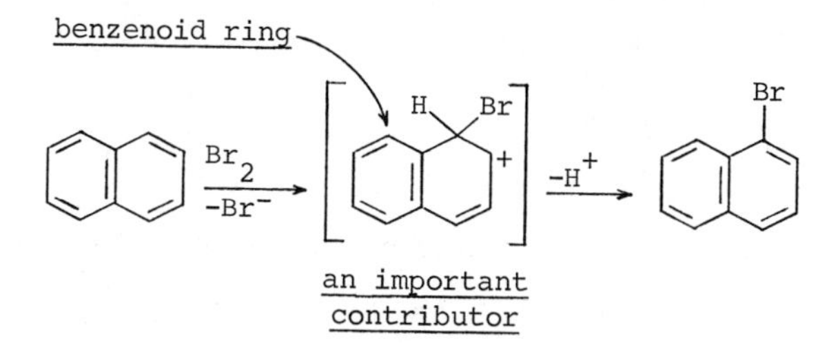

Sulfonation of naphthalene, like the sulfonation of benzene (Section 10.9F), is reversible. Thus, the sulfonation of naphthalene can lead to substitution at the 1- or 2-position, depending on the reaction conditions (see Section 16.5A).

The position of the second substitution on naphthalene is determined by the first substituent.

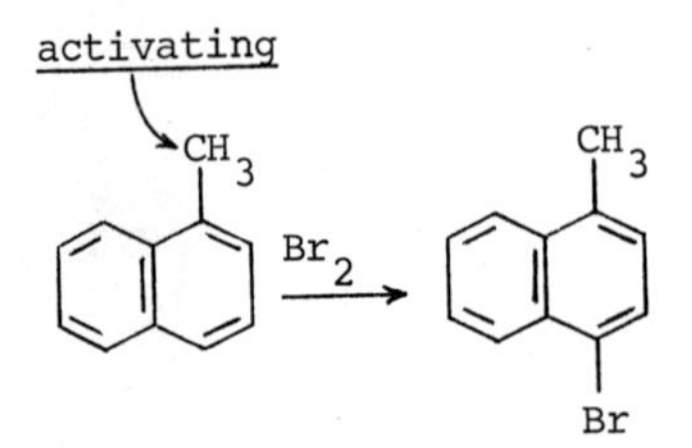

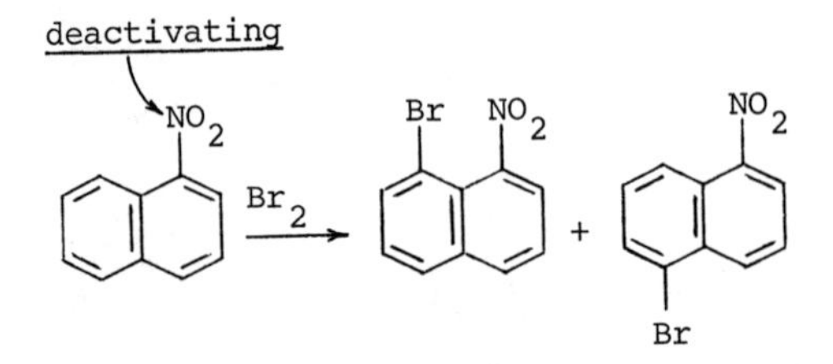

Although pyridine is a weaker base than a tertiary amine (because the N is $\underline{sp}^2$ hybridized), pyridine still reacts with acids or with alkyl halides to yield salts.

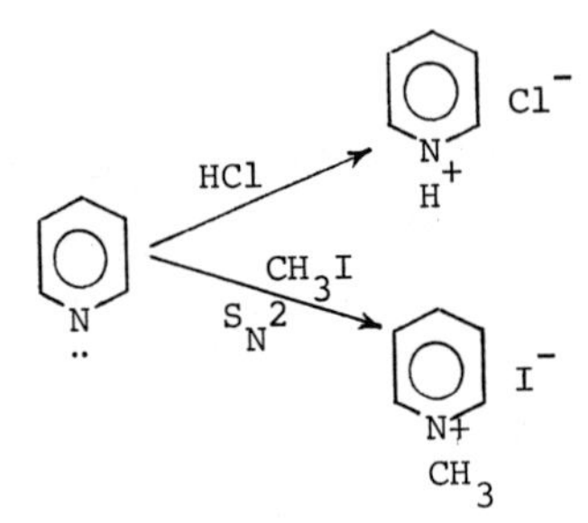

Compared to benzene, pyridine is deactivated to electrophilic substitution and activated to nucleophilic substitution because the nitrogen withdraws electron density from the rest of the ring.

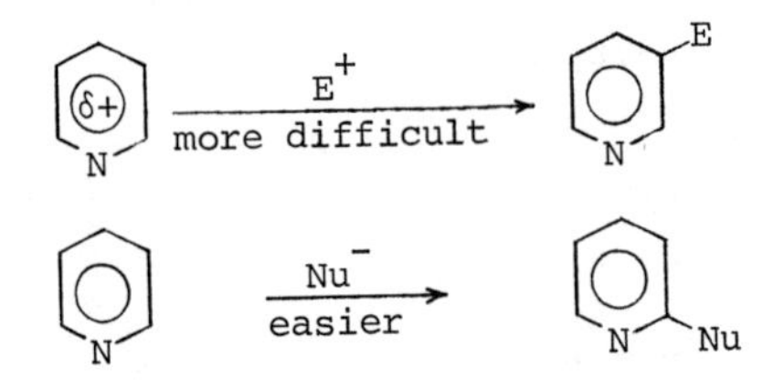

Pyrrole, unlike pyridine, is not basic because it has no unshared electrons; the "extra" pair of electrons on the nitrogen are both contributed toward the aromatic pi cloud.

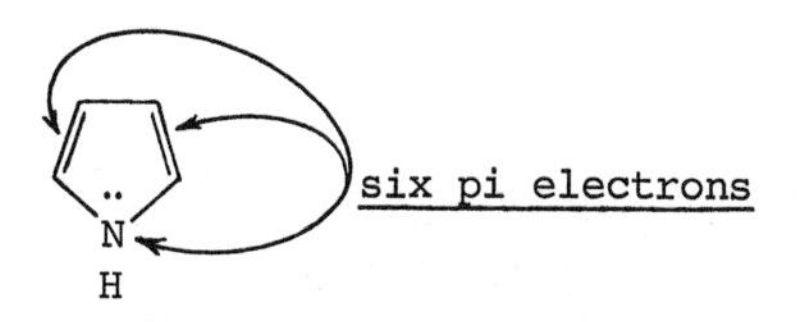

Because the nitrogen is electron-deficient, the rest of the pyrrole ring is electron-rich and undergoes electrophilic substitution more easily than benzene.

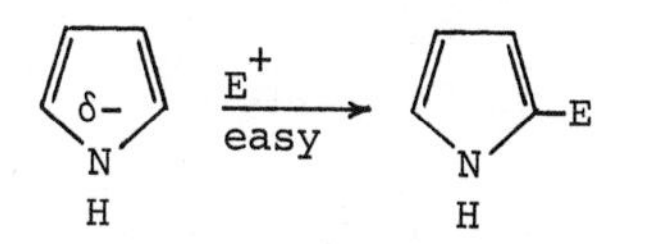

Other important topics in this chapter are quinoline and isoquinoline (Section 16.8), alkaloids (16.10), and nucleic acids (16.11).

Reminders

Review Chapter 10 if aromaticity and aromatic substitution reactions are not clear to you. Review Chapter 15 if the structural features controlling the relative basicities of amines are not clear.

Electrophilic substitution occurs under acidic conditions (or neutral conditions for activated rings). Keep in mind the relative reactivities of aromatic compounds toward electrophilic substitution.

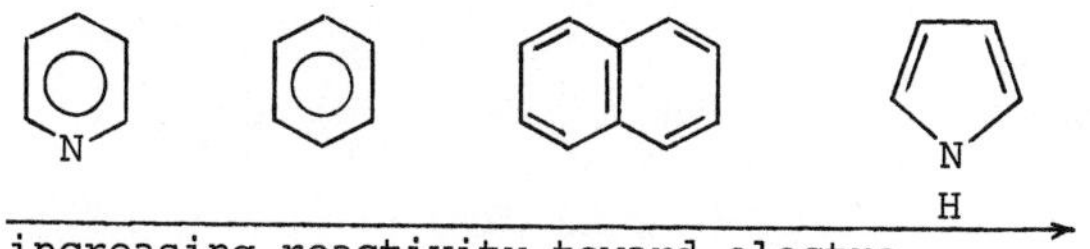

Nucleophilic substitution occurs only with a strong base under any circumstances.

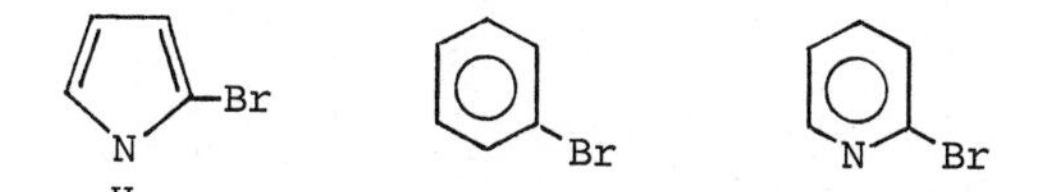

Answers to Problems

16.13 (a) 2-propylnaphthalene (b) 2-propylpyridine (c) 1-phenylnaphthalene
(d) 2,3-dimethylfuran (e) 4-methylquinoline
(f) 6-nitro-1,4-naphthoquinone

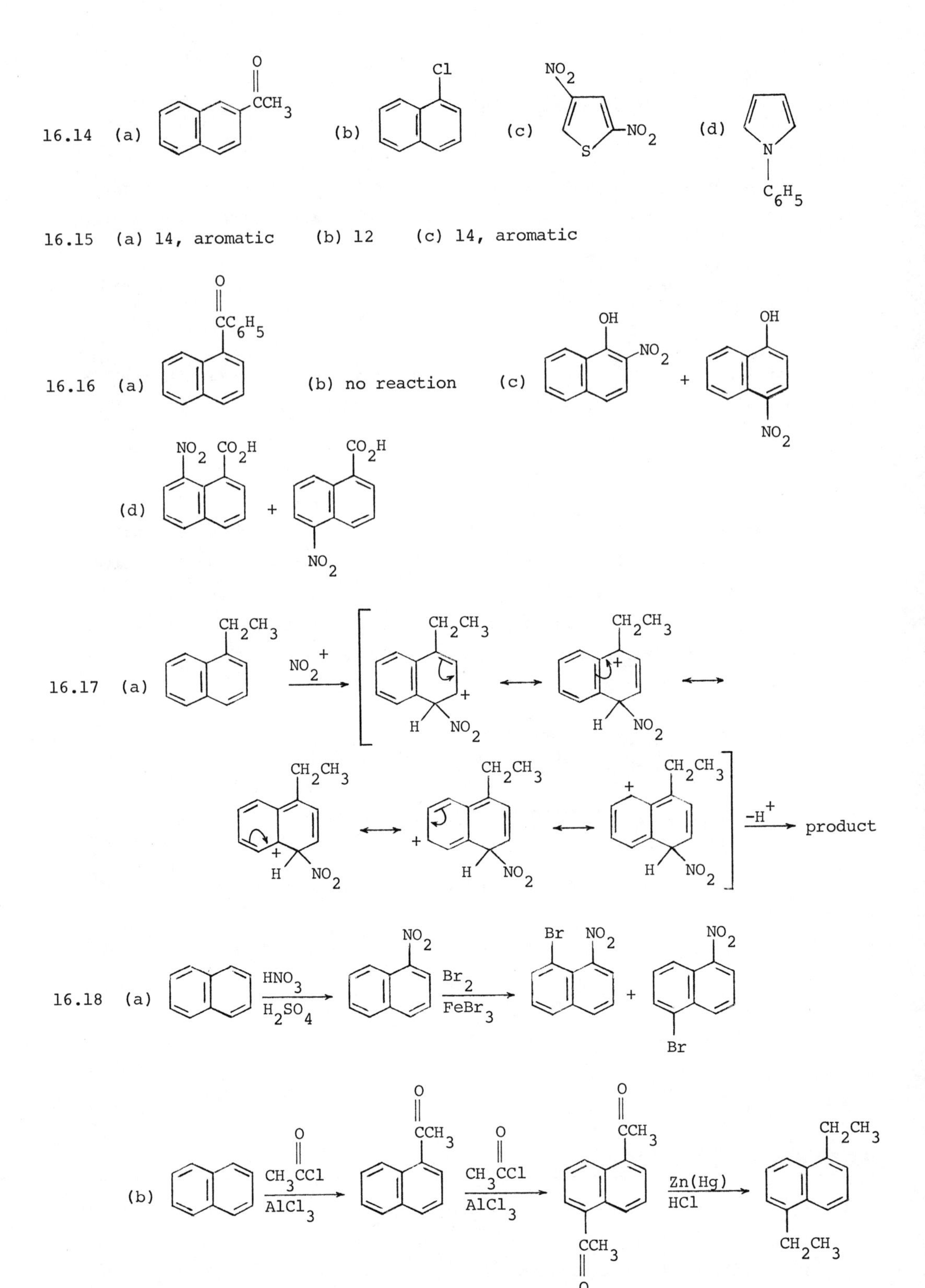
16.14 (a)
O
CCH3
(b)
Cl
(c)
NO2
NO2
S
(d)
N
C6H5
16.15 (a) 14, aromatic (b) 12 (c) 14, aromatic
16.16 (a)
O
CC6H5
(b) no reaction
(c)
OH
NO2
+
OH
NO2
(d)
NO2 CO2H
+
CO2H
NO2
16.17 (a)
CH2CH3
NO2+
H NO2
+
-H+
product
16.18 (a)
HNO3
H2SO4
NO2
Br2
FeBr3
Br NO2
+
NO2
Br
(b)
O
CH3CCl
AlCl3
O
CCH3
O
CH3CCl
AlCl3
O
CCH3
CCH3
O
Zn(Hg)
HCl
CH2CH3
CH2CH3

In (b), we would not expect the 1,8-diacetylnaphthalene to be a major by-product. (Why not?) A direct alkylation using 2 $CH_3CH_2Cl + AlCl_3$ could not be used in (b) because both R groups would substitute on the same ring.

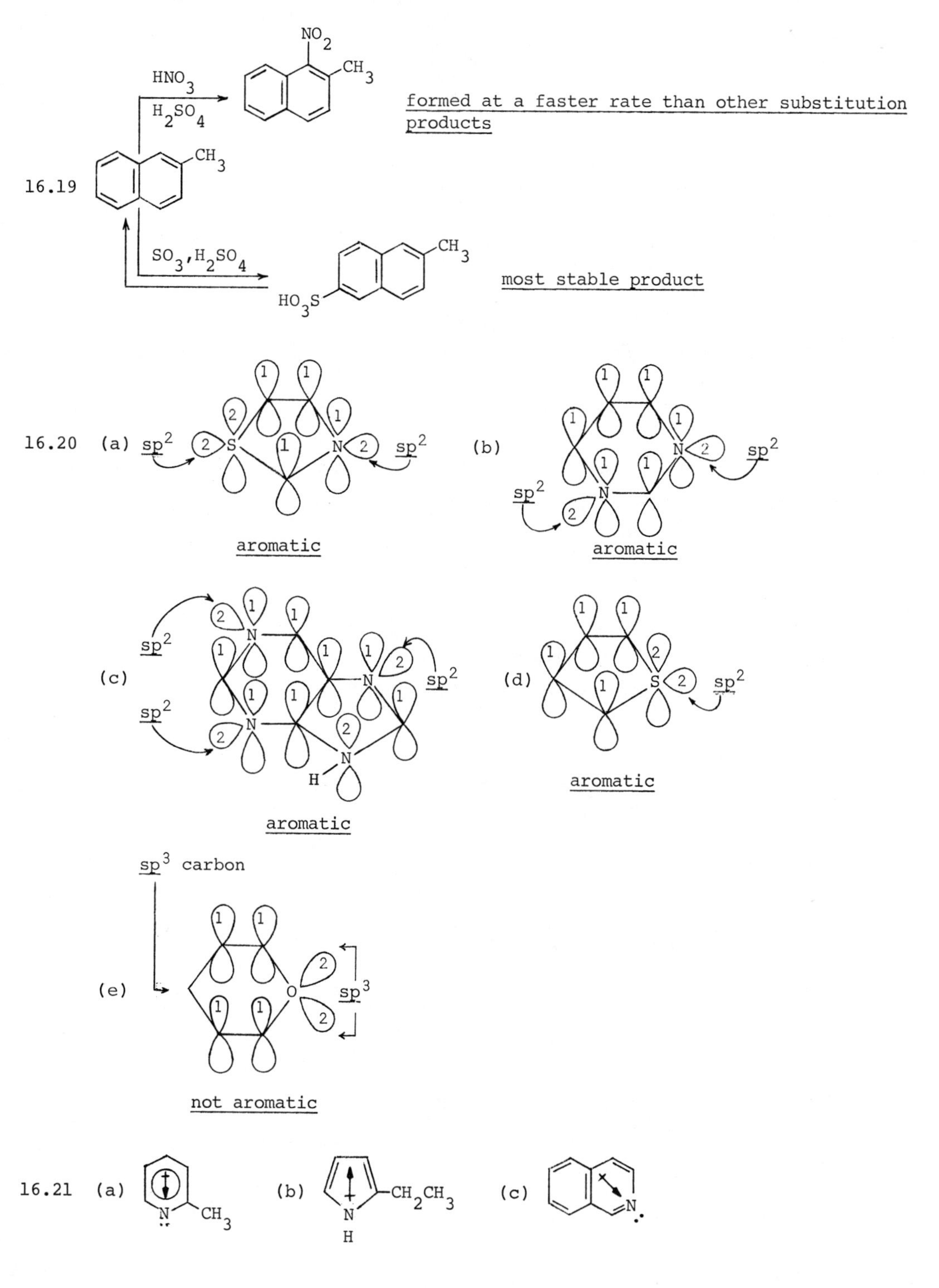

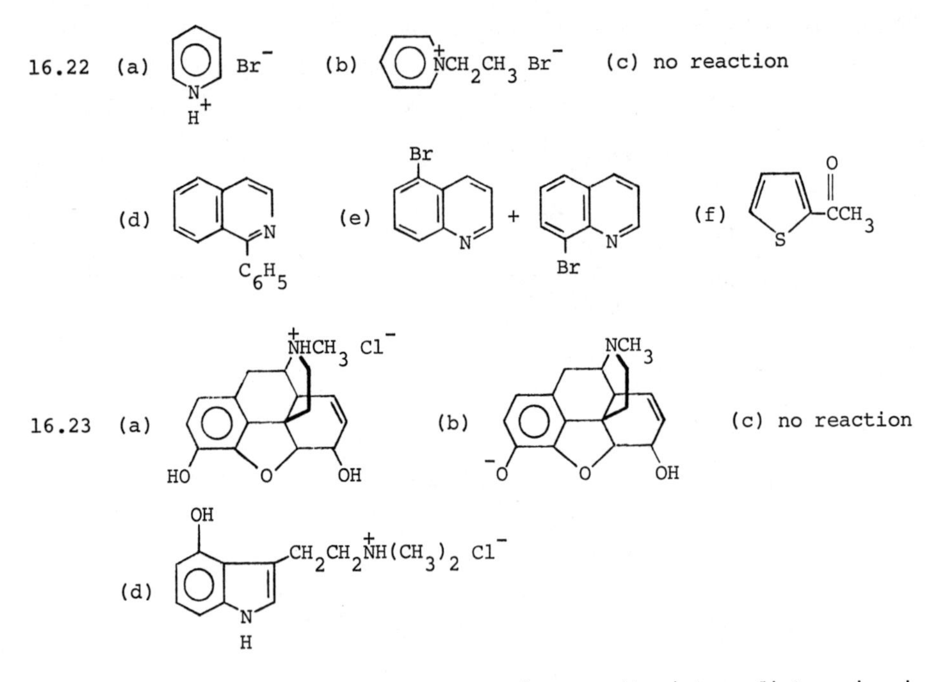

16.24 Reaction (b) has the faster rate because the intermediate anion is stabilized by electron-withdrawal by the chlorine.

16.25 The —CO_2H group is electron-withdrawing and deactivates the ring toward electrophilic substitution.

16.26 The nitrogen in thiazole has a pair of unshared electrons, while the nitrogen in pyrrole does not. In thiazole, the sulfur atom, rather than the nitrogen, provides two electrons for the aromatic pi cloud.

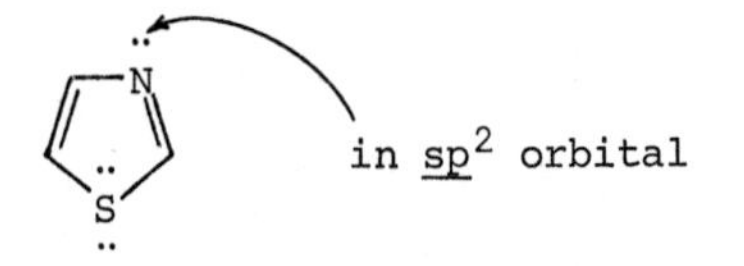

16.27 The negative charge is delocalized:

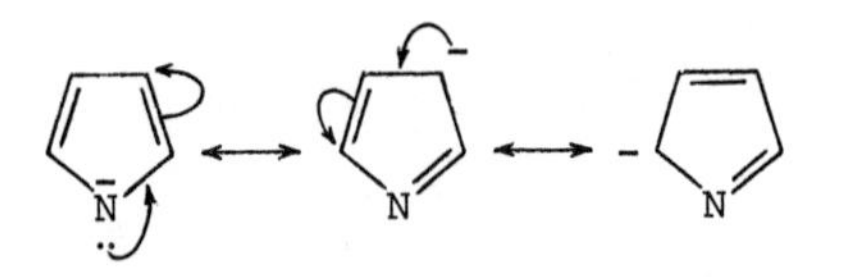

16.28 (b), (c), (a), (d)

(b) is less reactive than (c) because the active positions (2 and 5) are blocked.

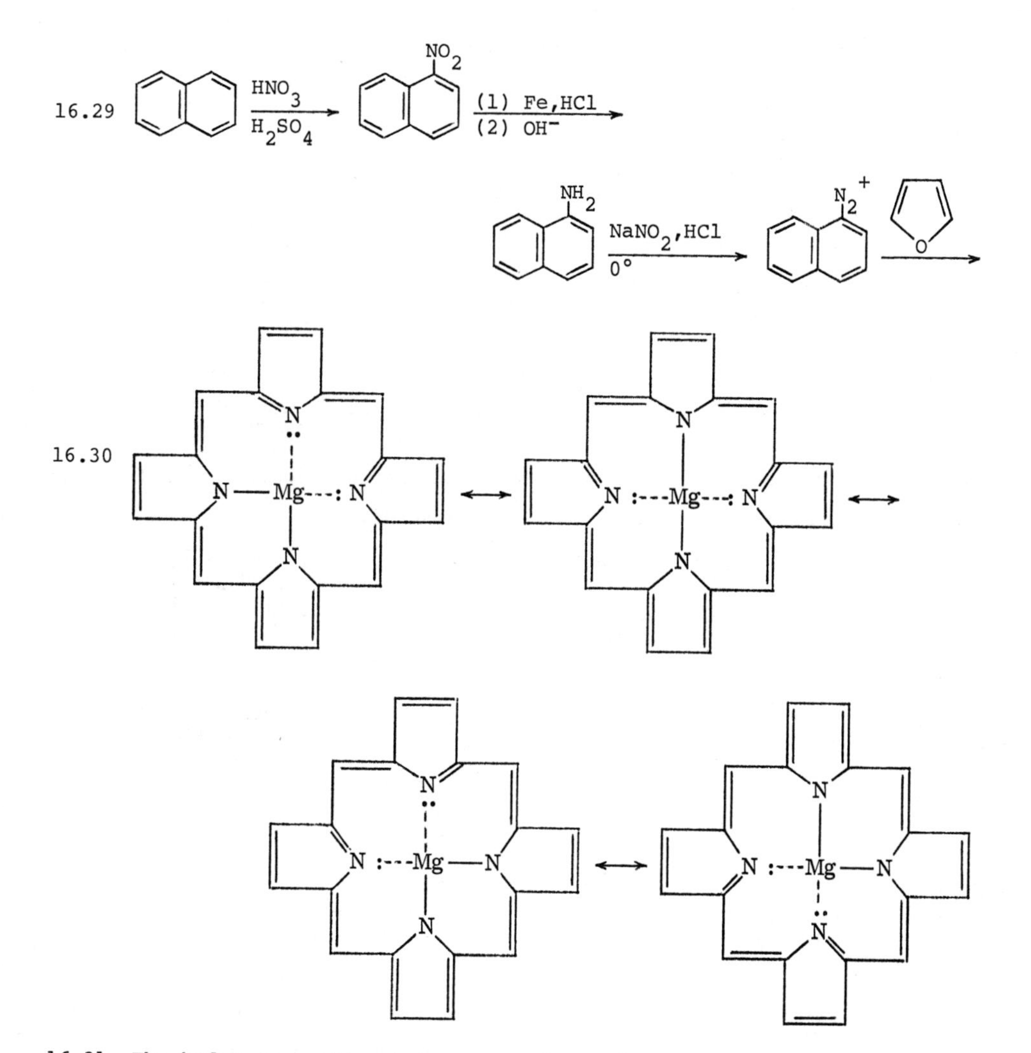

16.31 The hydrogen bonding between guanine and cytosine is shown on page 770 of the text. There are three hydrogen bonds between the two molecules. Cytosine and adenine can form only one hydrogen bond.

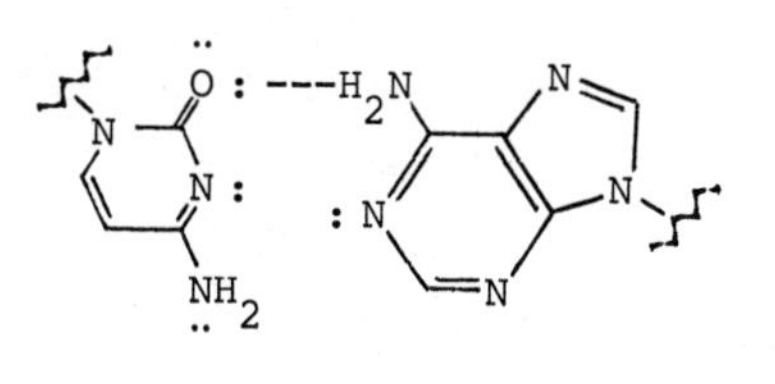

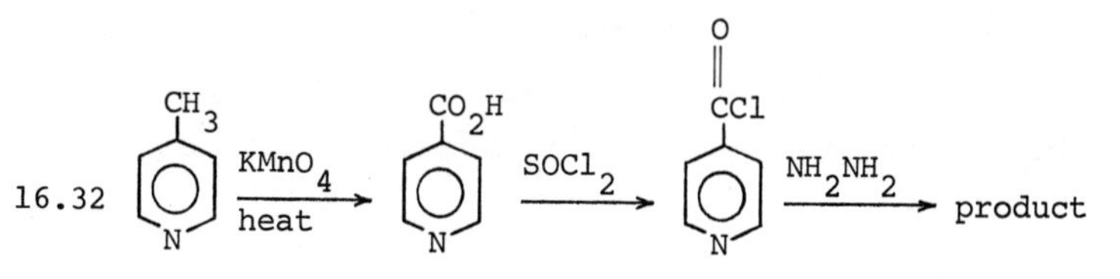

16.33 The 2- and 4-hydroxypyridines undergo tautomerism to lactams that have a high degree of resonance-stabilization. For example,

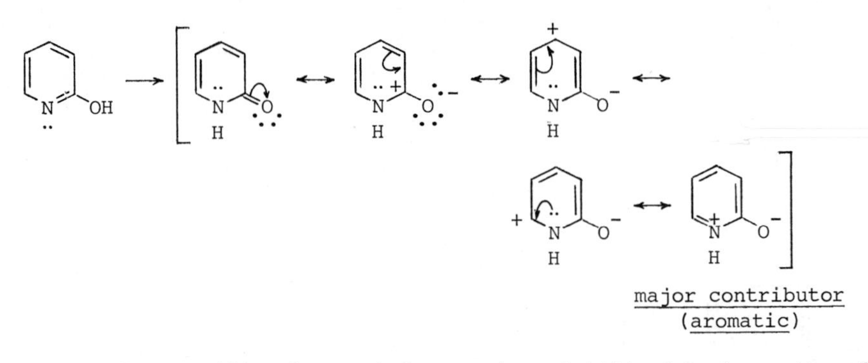

major contributor (aromatic)

The 3-hydroxypyridine does not form such a stabilized lactam; therefore, the enol form (which can act like a phenol) is favored.

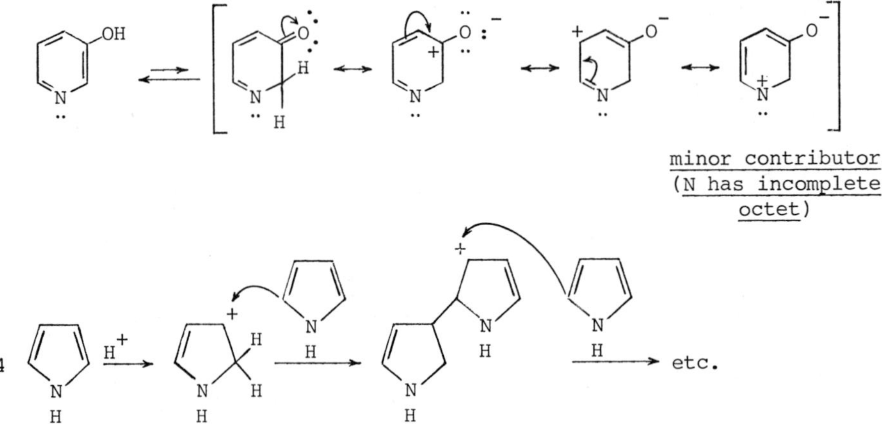

minor contributor (N has incomplete octet)

16.34

etc.

16.35 Substitution in the 3-position is favored because (1) the nitrogen ring is electron-rich, and (2) the intermediate has two resonance structures with the aromatic benzene ring intact (compared to one such structure for 2-substitution).

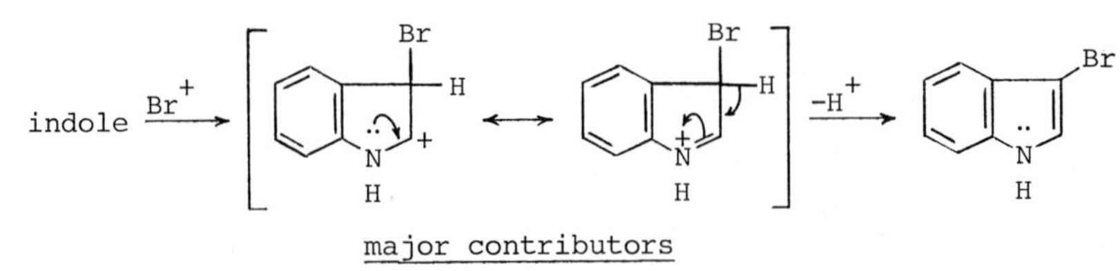

major contributors

16.36 Pyrimidine is less reactive than pyridine toward electrophilic substitution on a ring carbon, but more reactive toward nucleophilic attack because the two nitrogens of pyrimidine decrease the electron density of the ring carbons to a greater extent.

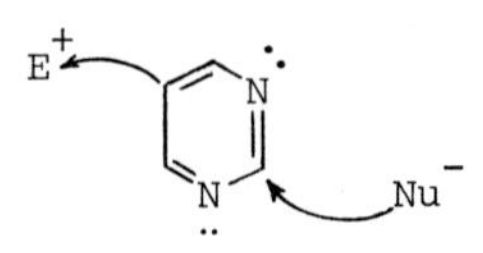

16.37 Imidazole and the imidazolium ion each has six electrons in an aromatic pi cloud, as the p-orbital pictures show. (Note that the H^+ becomes bonded to the N by a pair of sp^2 electrons, not by any of the pi electrons that form the aromatic pi cloud.)

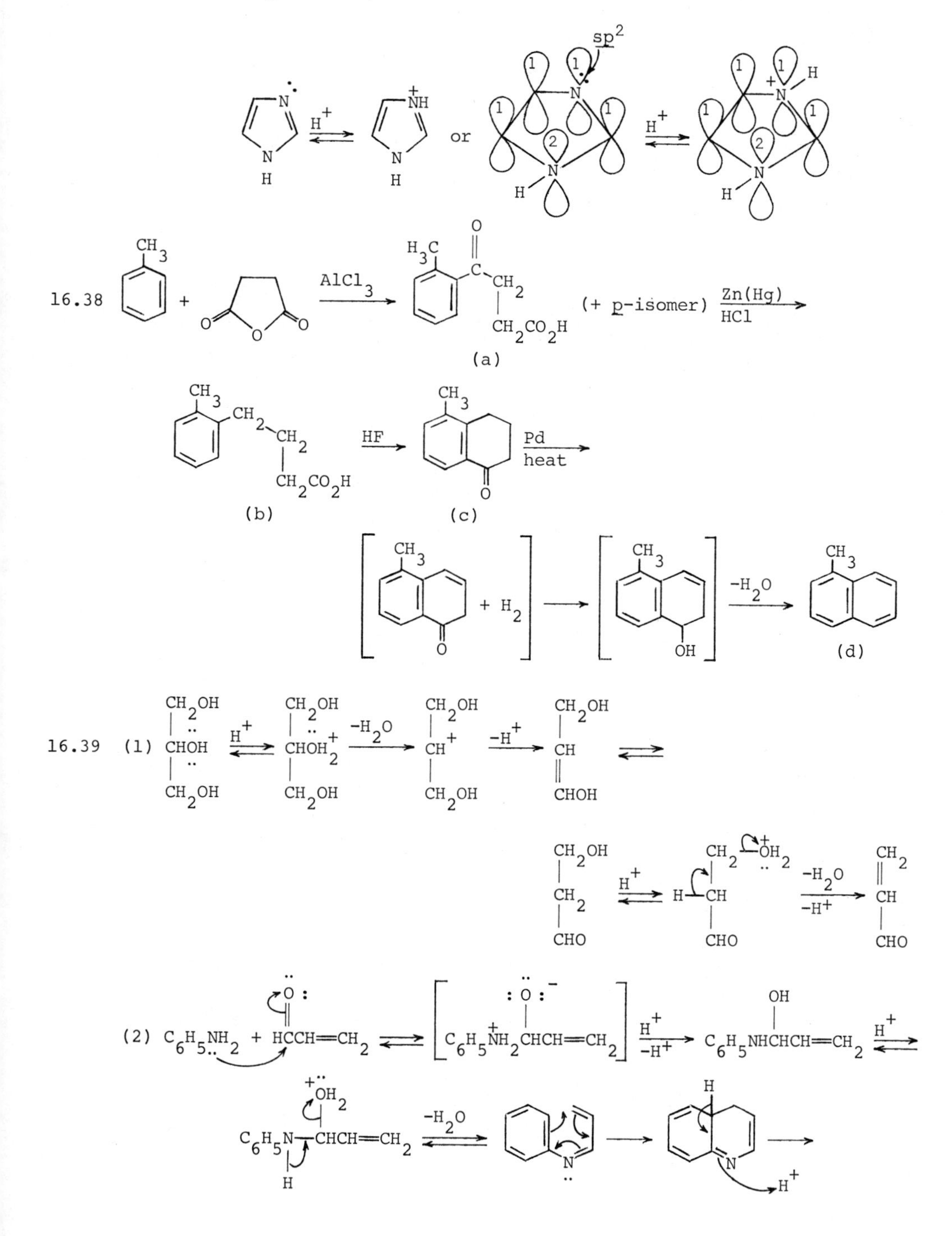

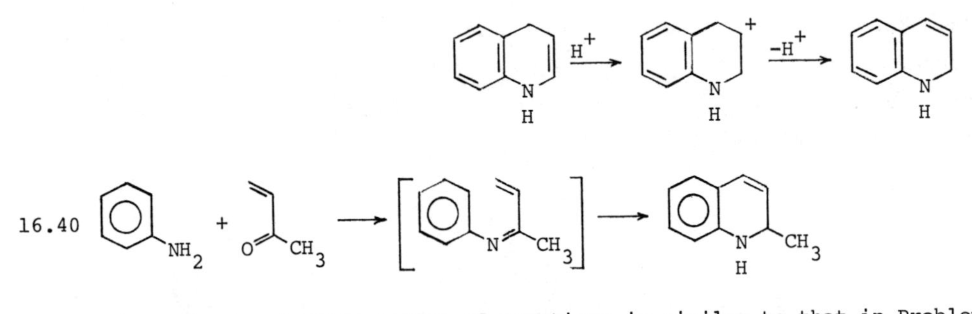

The mechanism in this series of reactions is similar to that in Problem 16.39.

16.41 A, (furan-2-CHO) B, (furan-2-CH_2OH)

The 2-substituted ring is chosen over the 3-substituted ring because only one proton in B shows nmr absorption that is shifted far downfield. (This proton is both aromatic and deshielded by O.)

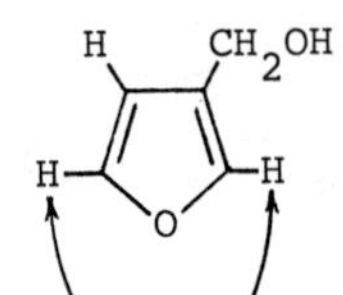

The nmr spectrum of the 3-substituted compound would show downfield aryl absorption for two protons.

CHAPTER 17

Pericyclic Reactions

Some Important Features

Pericyclic reactions are concerted reactions with cyclic transition states involving pi orbitals. The most important pericyclic reactions are cycloaddition reactions, electrocyclic reactions, and sigmatropic rearrangements.

Pericyclic reactions are often stereospecific. Photo-induction and thermal induction of these reactions often yield different products. The frontier orbital approach is one technique used to account for these observations and to predict the course of such reactions. In this technique, the phases of the *p*-orbital components of the HOMO and LUMO are considered. A pericyclic reaction does not proceed readily unless it is symmetry-allowed — that is, the *p*-orbital components of the pertinent molecular orbitals must be of the same phase to undergo overlap and form a new bond.

Cycloaddition reactions are reactions in which the HOMO of one pi system overlaps with the LUMO of another pi system (in the same molecule or in different molecules) to form sigma bonds. The result is a cyclization. A [2 + 2] cycloaddition (two π e^- + two π e^-) is photo-induced and results in a four-membered ring. A [4 + 2] cycloaddition is thermally induced and results in a six-membered ring.

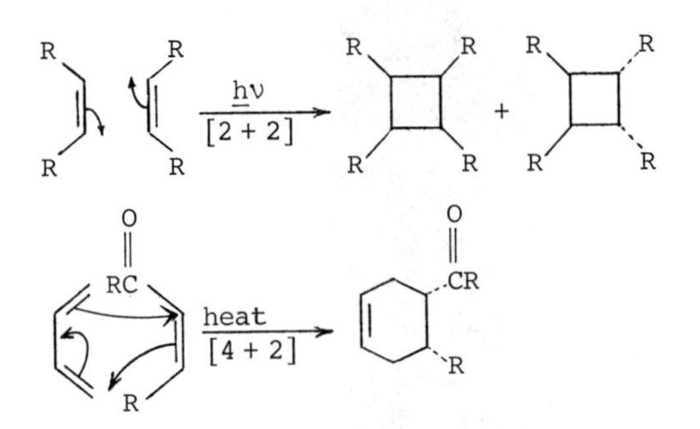

Electrocyclic reactions are reversible cyclizations of conjugated polyenes. These reactions may be either thermally or photo-induced. In properly selected compounds, the stereochemistry of the product is determined by which method of induction is used. To predict the products, we consider the HOMO of the polyene (not of the cyclic product) and determine whether conrotatory or disrotatory motion brings the in-phase p-orbital components together. Table 17.1 in the text summarizes the types of electrocyclic reactions so that you need not consider the molecular orbitals each time you are confronted with an electrocyclic problem.

Cyclization of a (4n + 2) polyene:

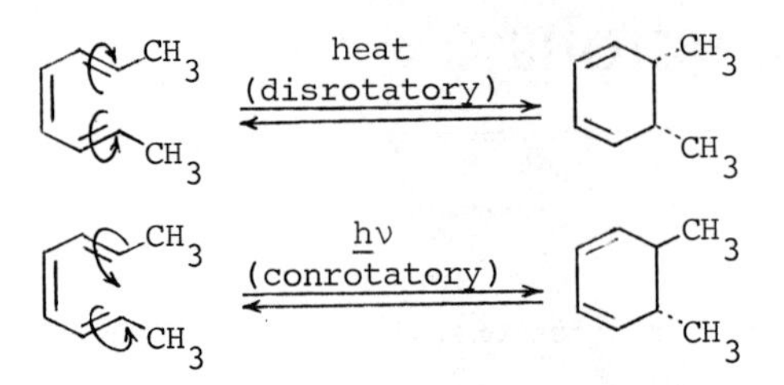

Sigmatropic rearrangements are usually thermally induced and involve migration of groups from one end of a pi system to another portion of the same molecule. To predict whether a sigmatropic reaction is symmetry-allowed, we examine the HOMO of the hypothetical free radical that would be formed if the migrating group were homolytically cleaved. Common symmetry-allowed migrations are those that are [1,5]; [1,7]; and [3,3]. (These classifications are discussed in Section 17.4A.)

Reminders

Keep in mind that electrons are promoted by light, but not by ordinary heating. When a compound is heated, its molecules remain in the ground state.

Models may be helpful in determining the stereochemistry of a pericyclic reaction, especially a Diels-Alder reaction. (You may wish to review the discussion of the stereochemistry of Diels-Alder reactions in Section 9.16.)

Answers to Problems

17.12 (a) cycloaddition (b) sigmatropic (c) electrocyclic
(d) electrocyclic (e) electrocyclic

17.13

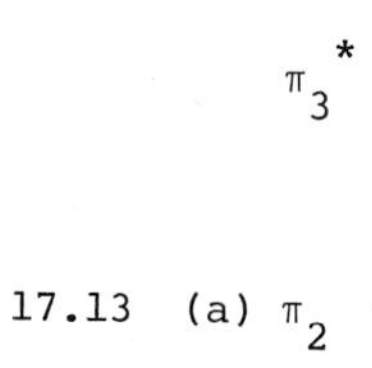

(a)

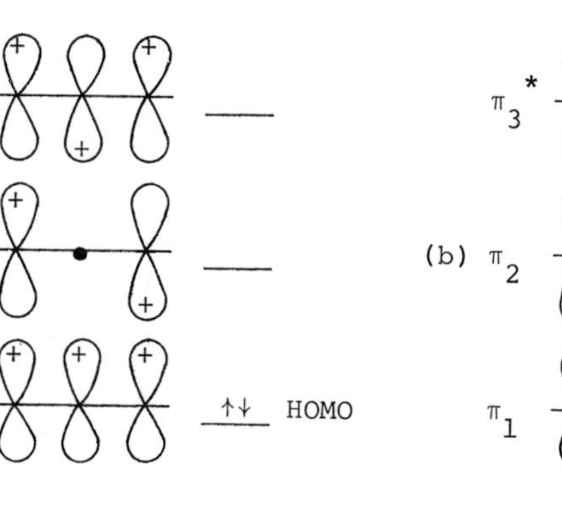

(b)

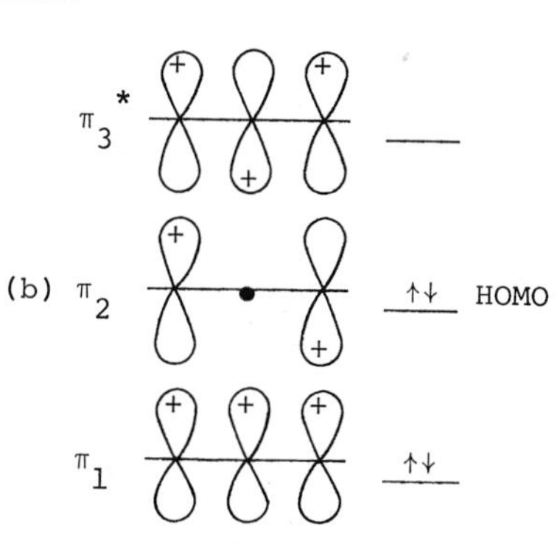

(c)

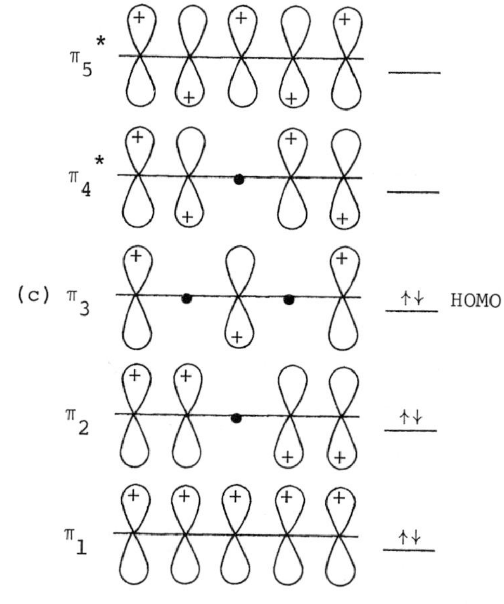

17.14 (a) [2 + 2] (b) [4 + 2] (c) [8 + 2]

17.15 (b) and (c)

17.16 The reaction is a [2 + 2] cycloaddition. Therefore, the skeleton of the product is as shown in the following equation:

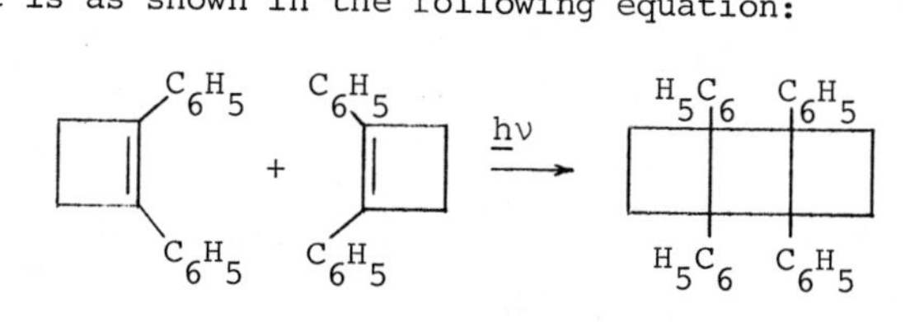

The stereochemistry can be determined by envisioning the end-to-end overlap of the p-orbital components:

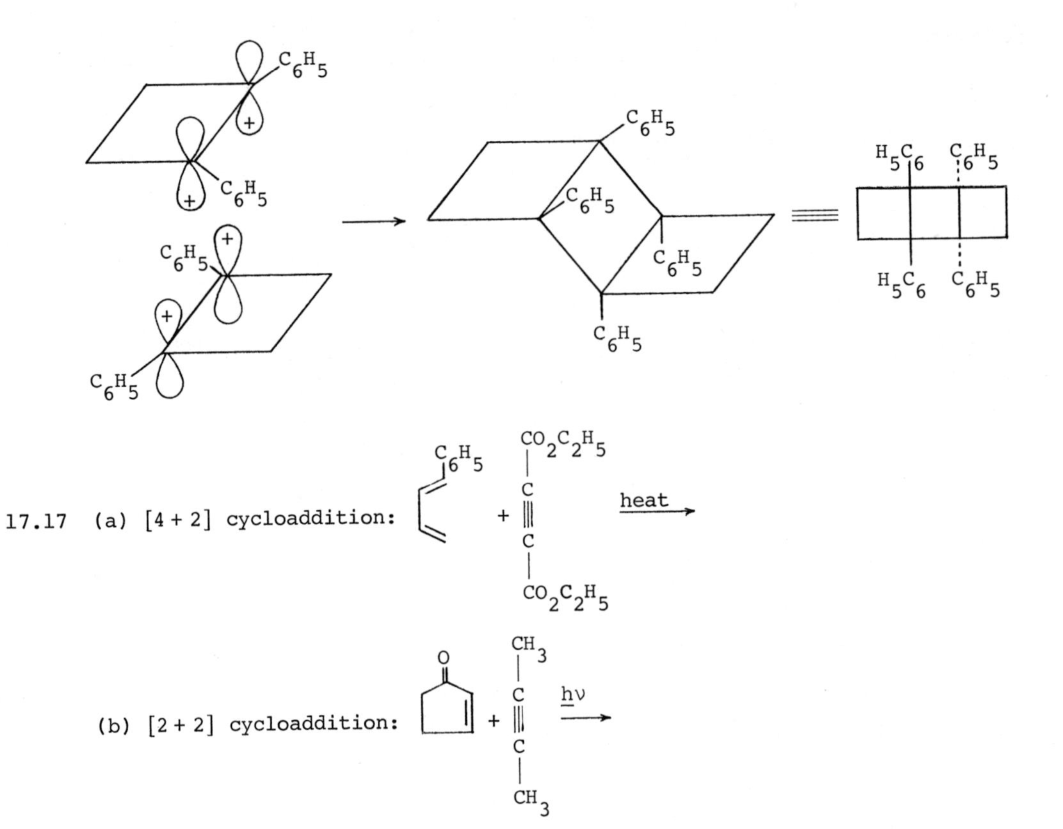

17.18 The mode of rotation can be determined with the aid of Table 17.1 in the text.

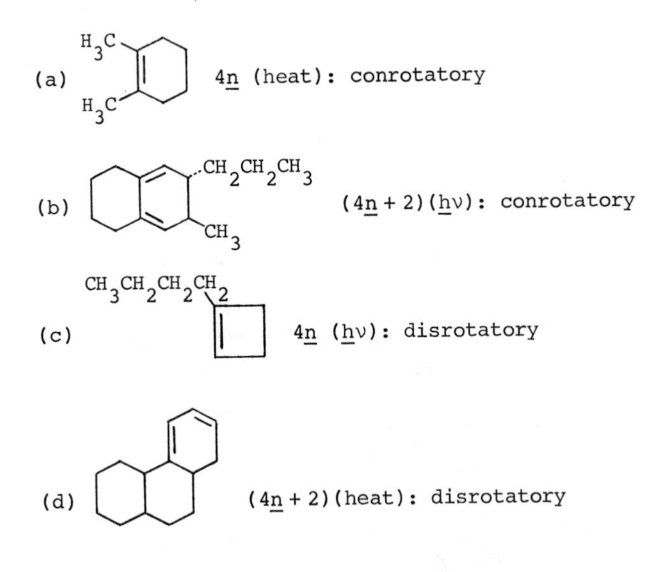

(a) 4\underline{n} (heat): conrotatory

(b) (4\underline{n} + 2)($\underline{h\nu}$): conrotatory

(c) 4\underline{n} ($\underline{h\nu}$): disrotatory

(d) (4\underline{n} + 2)(heat): disrotatory

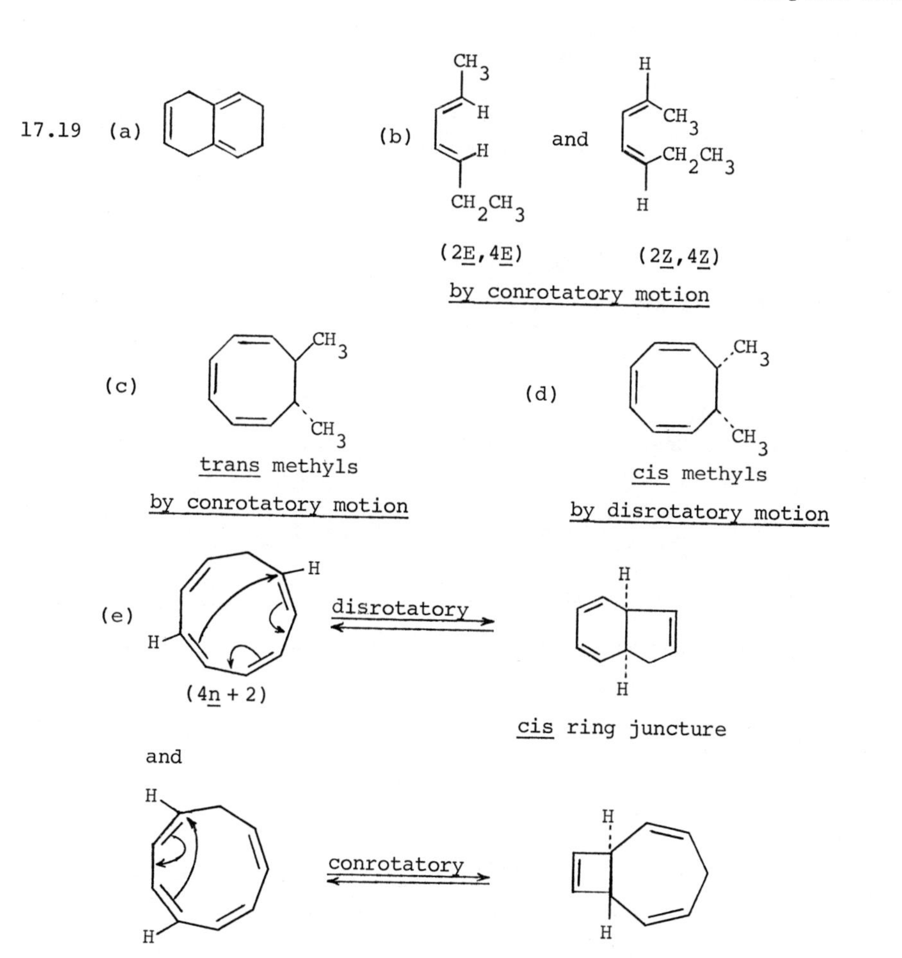

Because electrocyclic reactions are reversible, the first product shown would predominate because it is of lower energy (less-strained ring system).

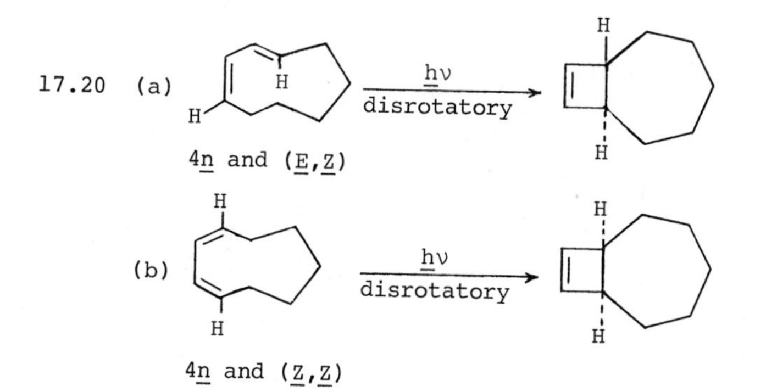

17.21 (a)

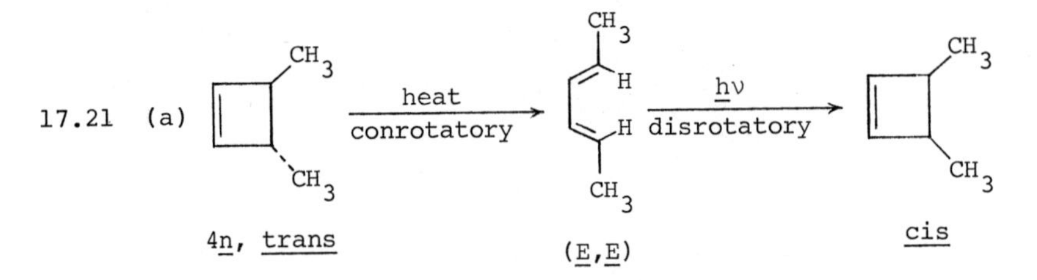

The steps could be reversed, using hν first; the result is the same.

(b)

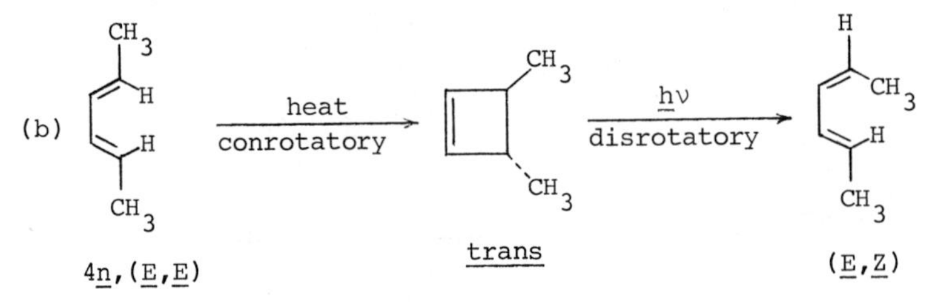

Again, the steps could be reversed.

17.22 In each case, number the migrating group and the alkenyl chain starting at their point of original attachment. Then number the product using the same numbering system.

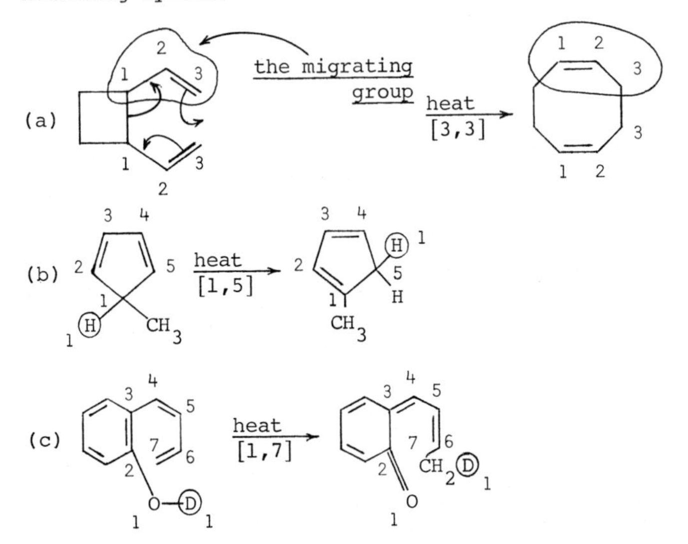

17.23 Inspection of the structure shows that either a [1,3] or a [1,5] sigmatropic rearrangement of deuterium to another position is possible. [1,3]-Sigmatropic rearrangements are rare but [1,5]-sigmatropic rearrangements are common. Therefore, one would predict that the reaction will proceed by a [1,5]-shift.

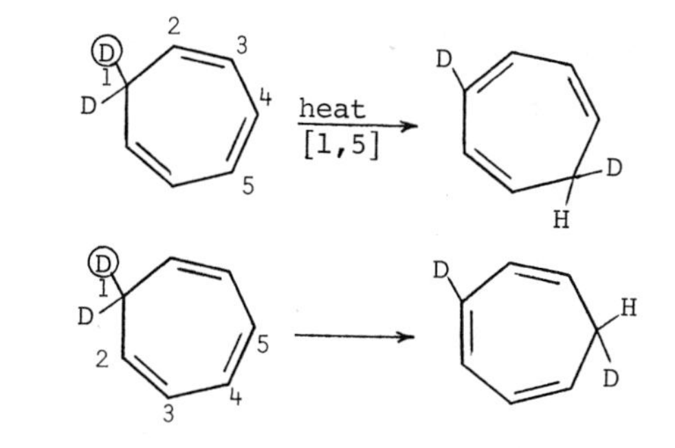

Because of symmetry in the product, either rearrangement shown results in the same compound.

17.24

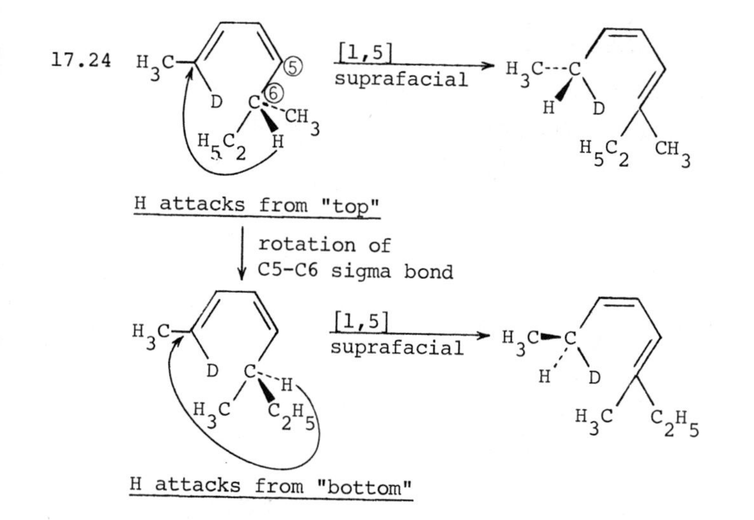

The stereochemistry is easier to see with models.

17.25 (a) 4n electrocyclic reaction, disrotatory, hν.

The reaction must take place with disrotatory motion because conrotatory motion would place one of the H's inside the ring and the other outside.

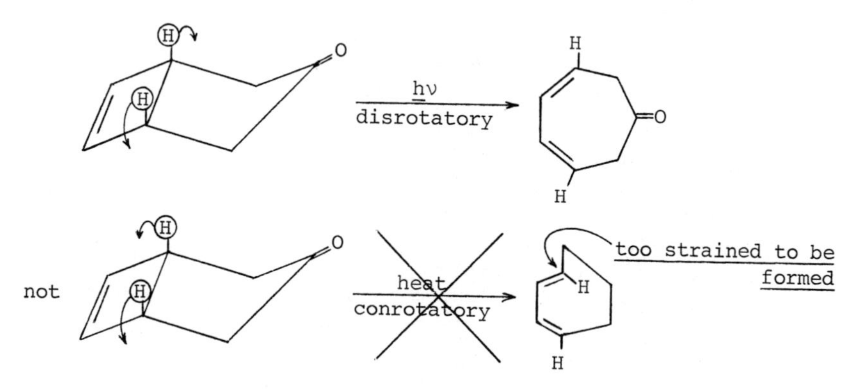

(b) [2 + 2] cycloaddition, hν

(c) [3,3] sigmatropic rearrangement, heat. The rearrangement is easier to see if the structures are redrawn:

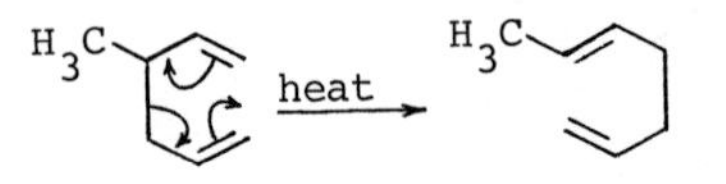

17.26 The thermal reaction is a [4 + 2] cycloaddition, or Diels-Alder reaction.

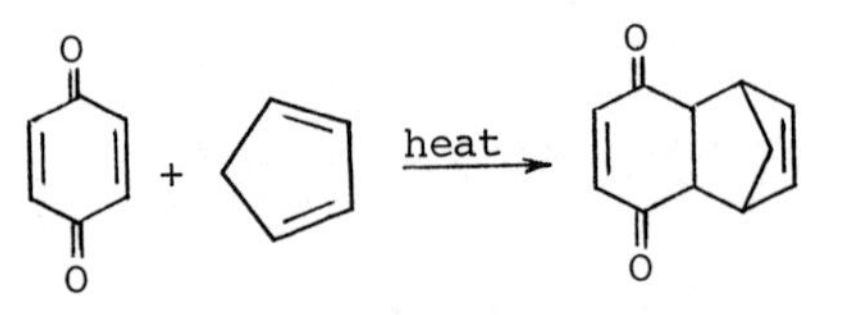

Diels-Alder product

The second reaction is an internal photo-induced [2 + 2] cycloaddition. In order to see how this second reaction takes place, redraw the Diels-Alder product to show the proximity of the double bonds.

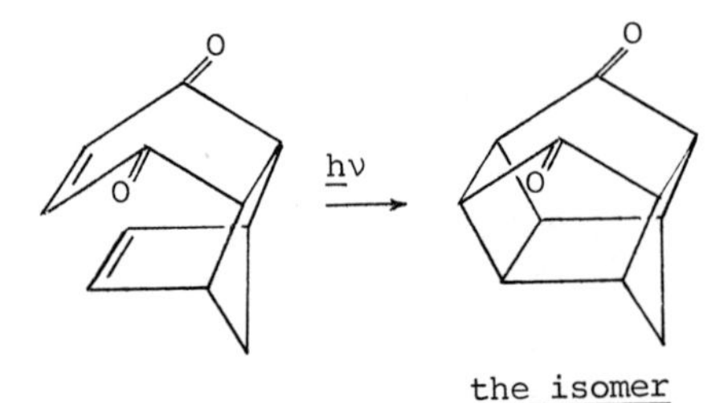

the isomer

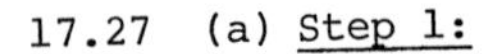

17.27 (a) Step 1:

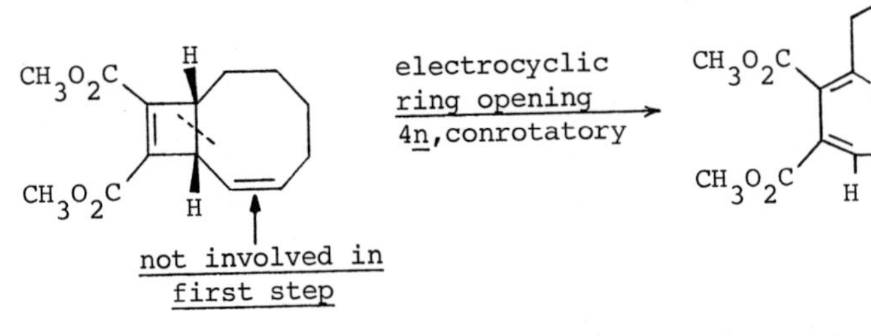

(Note that a ten-membered ring can accommodate a trans double bond with H "inside" the ring.)

Step 2: Redraw the structure.

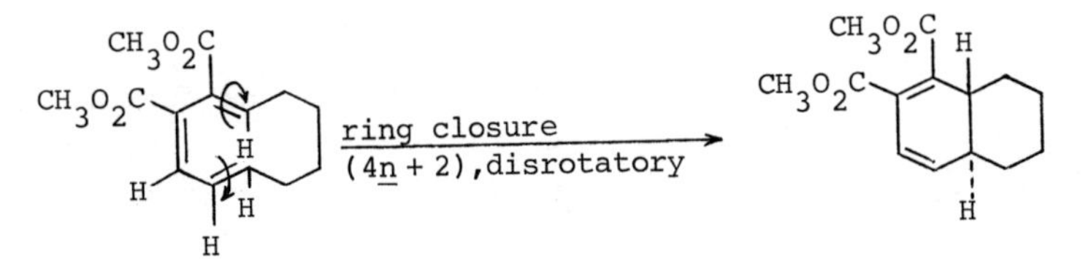

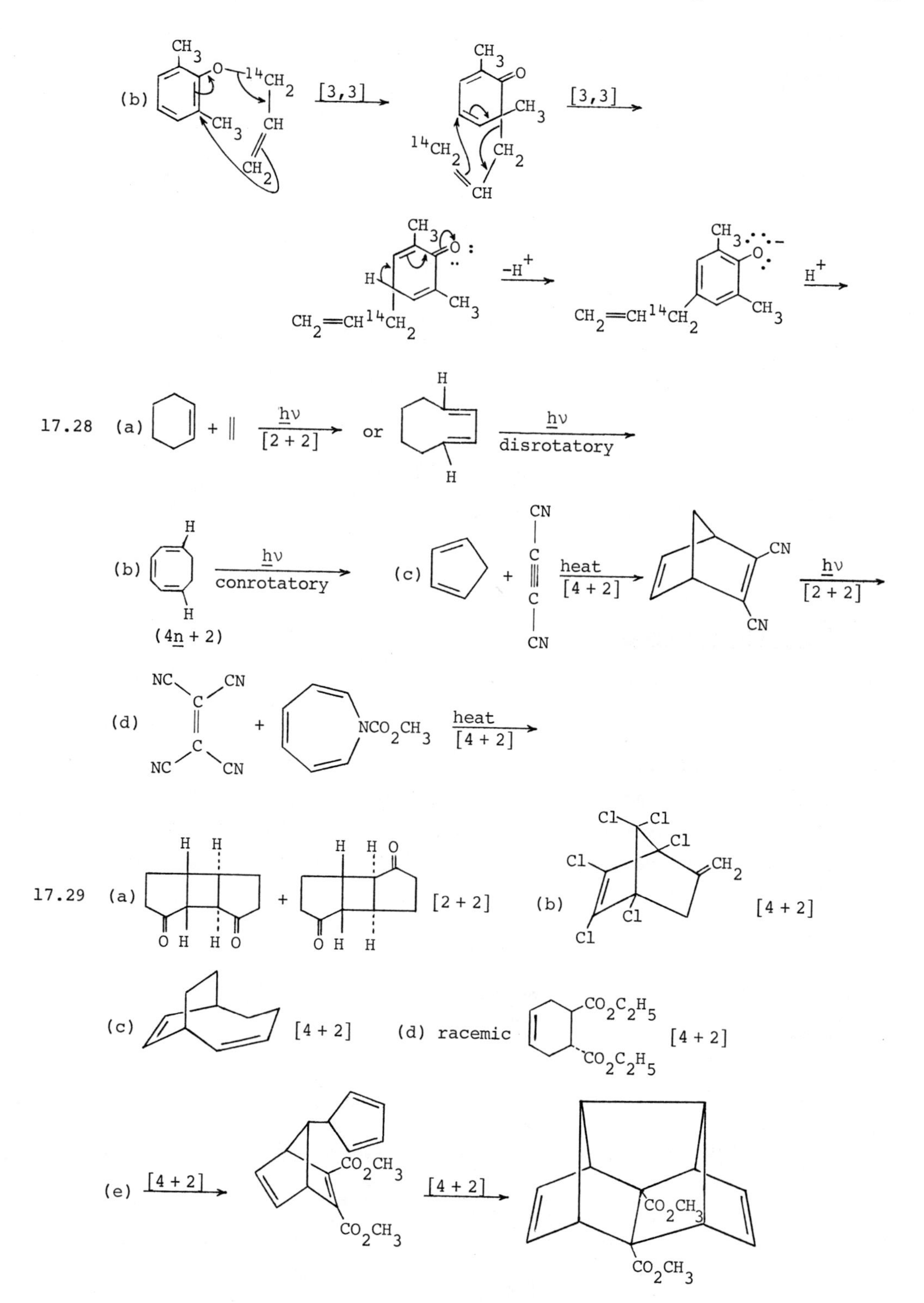

(b)
CH_3
O
$^{14}CH_2$
CH
CH_2
CH_3
[3,3]
CH_3
O
CH_3
$^{14}CH_2$
CH_2
CH
[3,3]
CH_3
O
H
CH_3
CH_2=$CH^{14}CH_2$
$-H^+$
CH_3
O
CH_3
CH_2=$CH^{14}CH_2$
H^+
17.28 (a)
+
hν
[2 + 2]
or
H
H
hν
disrotatory
(b)
H
H
hν
conrotatory
(4n + 2)
(c)
+
CN
C
C
CN
heat
[4 + 2]
CN
CN
hν
[2 + 2]
(d)
NC
CN
C
C
NC
CN
+
NCO_2CH_3
heat
[4 + 2]
17.29 (a)
H H
O H H O
+
H H O
O H H
[2 + 2]
(b)
Cl
Cl
Cl
Cl
CH_2
Cl
Cl
[4 + 2]
(c)
[4 + 2]
(d) racemic
$CO_2C_2H_5$
$CO_2C_2H_5$
[4 + 2]
(e)
[4 + 2]
CO_2CH_3
CO_2CH_3
[4 + 2]
CO_2CH_3
CO_2CH_3

CHAPTER 18

Carbohydrates

Some Important Features

Glucose and the other monosaccharides exist primarily as pairs of cyclic hemi-acetals (anomers), which are in equilibrium with the open-chain carbonyl form in solution. Hemiacetals can undergo aldehyde reactions because of this equilibrium.

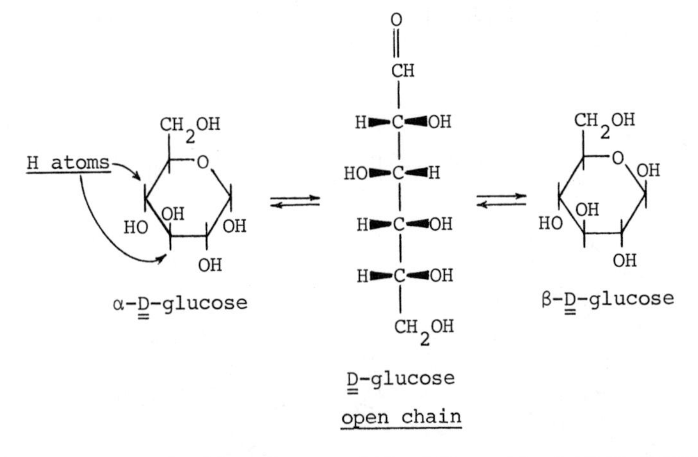

The monosaccharides form acetals (glycosides) when treated with an alcohol. A glycoside is not in equilibrium with the carbonyl form in neutral or alkaline solution; therefore, glycosides do not undergo aldehyde reactions. Glycoside links can be hydrolyzed in acidic solution or with appropriate enzymes.

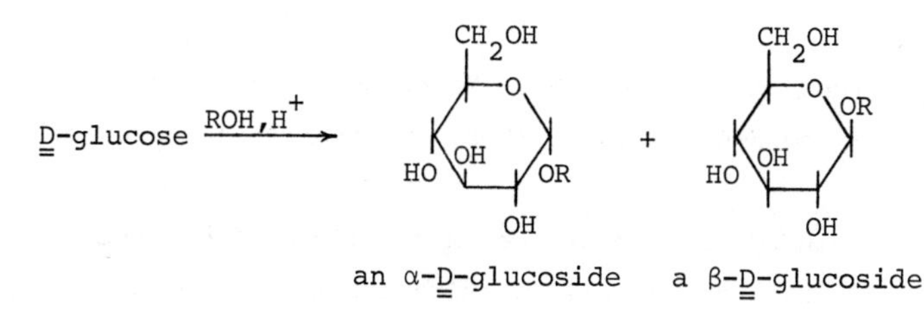

The relative configurations of sugars can be determined by synthesis.

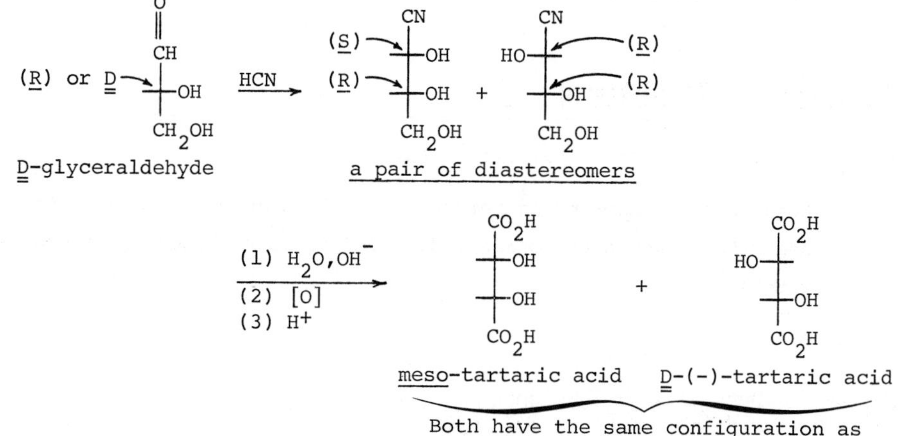

Both have the same configuration as D-glyceraldehyde at carbon 3

The original determination of the structures of some monosaccharides is discussed in Section 18.9.

Some important reactions of monosaccharides are the oxidation to aldonic acids (carbon 1 oxidized), aldaric acids (both carbon 1 and the last carbon oxidized), and uronic acids (only the last carbon oxidized). These oxidations are discussed in Section 18.6. The monosaccharides also can be reduced to alditols (Section 18.7). The hydroxyl groups can be esterified or converted to alkoxyl groups (Section 18.8).

Disaccharides are formed from two monosaccharide units joined by a glycoside link.

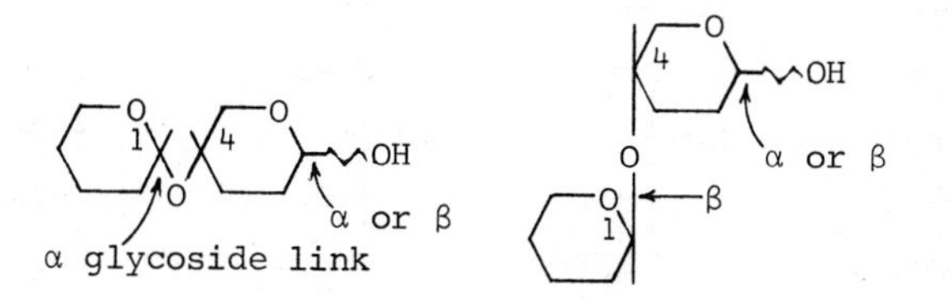

Polysaccharides are composed of many monosaccharide units joined by glycoside links. Partial structures of some important polysaccharides are shown in Section 18.11.

Reminders

Review Fischer projections in Section 4.6C. Memorize the Fischer projections and Haworth formulas for α- and β-D-glucose. Then, interconversions for other sugars are simplified.

If you forget the configurations of the carbon atoms of glucose, draw the favored chair form. β-D-Glucose has all substituents in equatorial positions.

Be sure you know what a meso compound is and how to assign (R) and (S) configurations. (See Chapter 4.)

Answers to Problems

18.21 (a) 4 (b) (3) (c) (1) (d) (2)

Note that an -ose ending refers to an aldehyde or hemiacetal (or ketone or hemiketal), while an -oside ending refers to a glycoside (an acetal or ketal).

18.22 all D

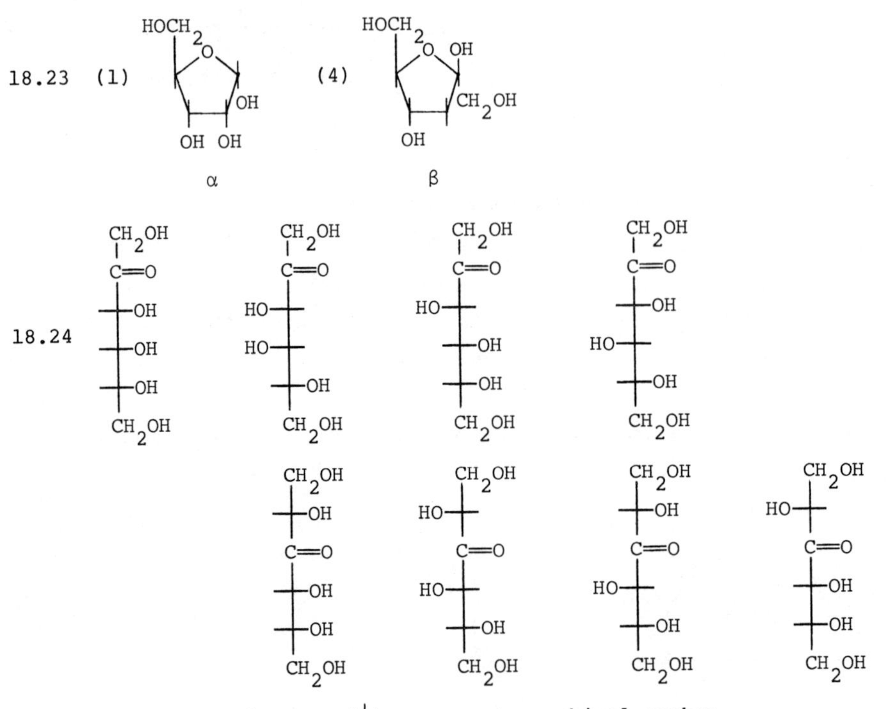

Each cross (+) represents a chiral carbon.

18.25 (a)

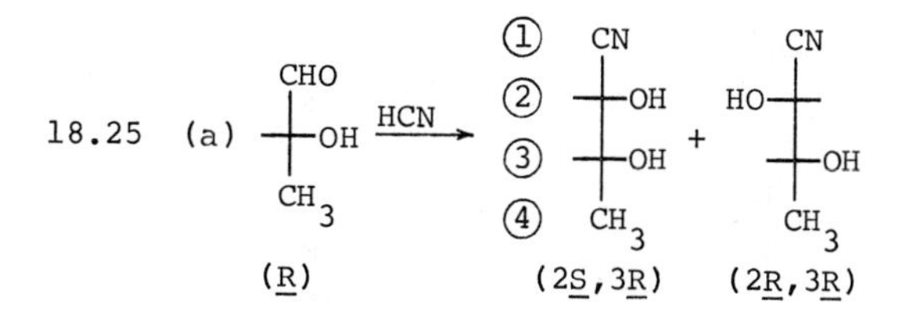

(b) Yes. Although carbon 2 in the product is an equal mixture of (R) and (S), carbon 3 is (R) in each case. The compounds are diastereomers.

18.26

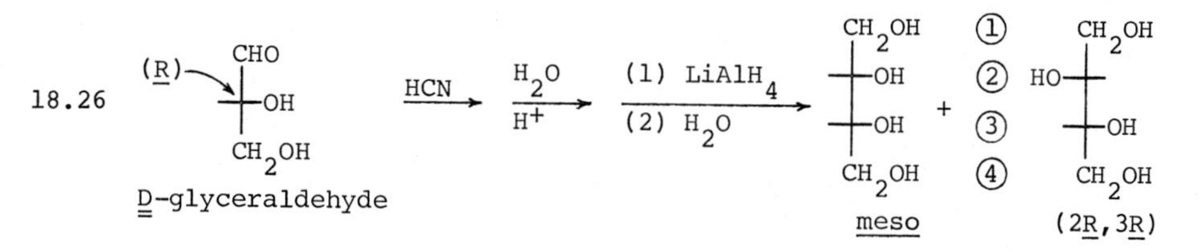

The product mixture is a mixture of diastereomers and thus can be separated by physical means. Each isomer can then be compared with the unknown tetraol. The unknown tetraol is either meso (has no optical rotation); (2R,3R) (same optical rotation as the tetraol synthesized from D-glyceraldehyde); or (2S,3S) (opposite rotation).

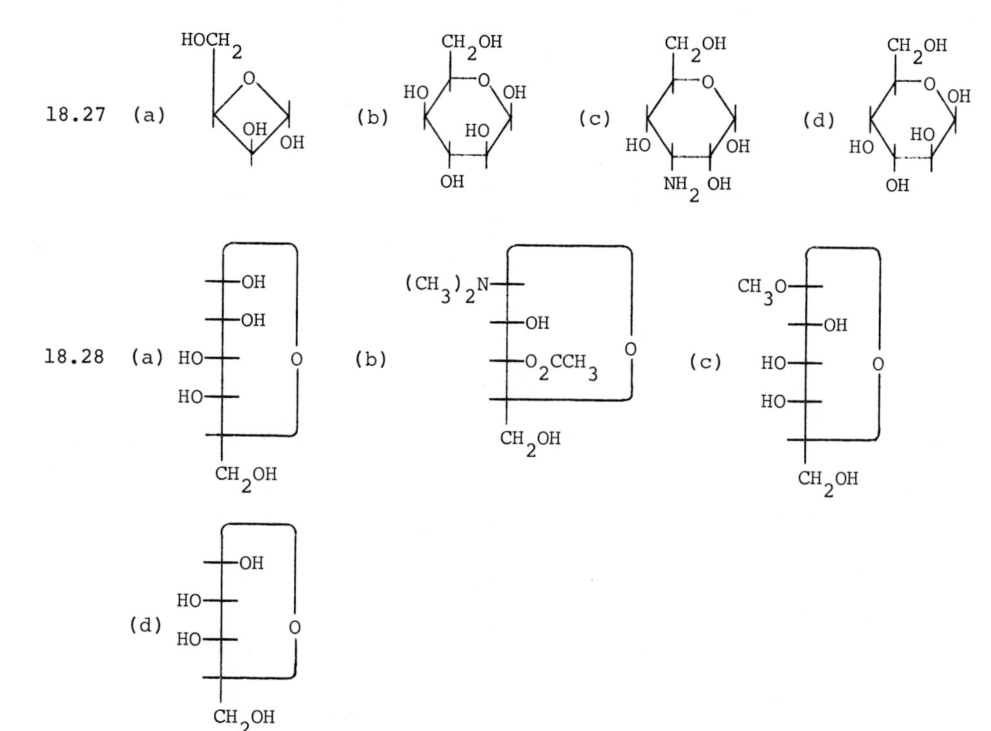

18.29 (a) (b) (c)

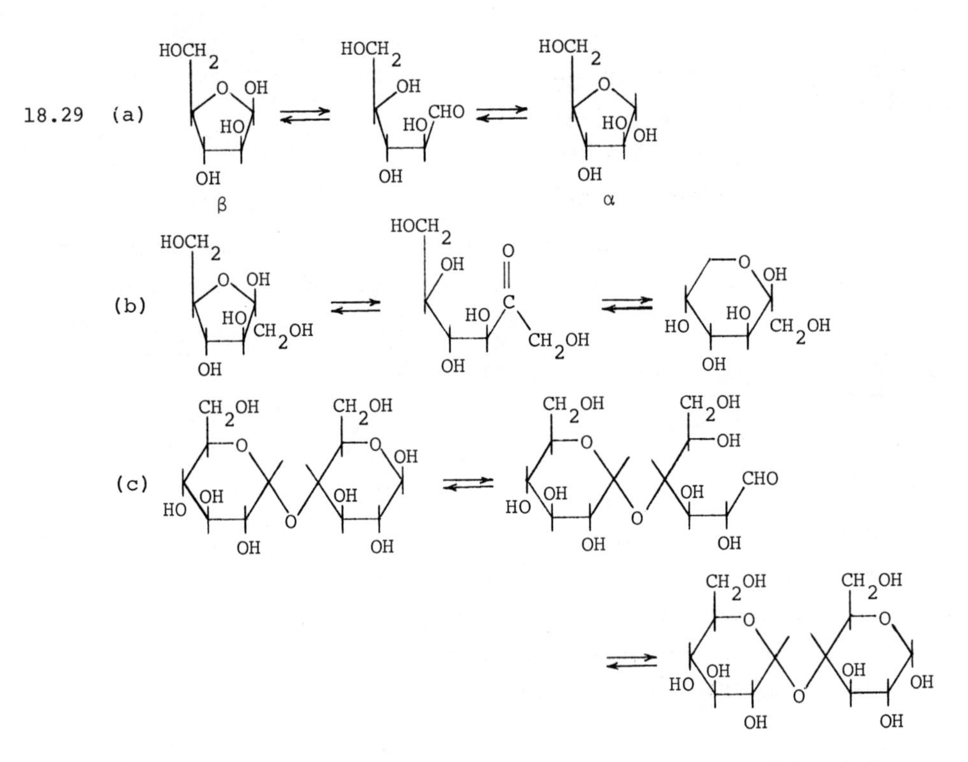

18.30 (a), (d), (e), and (g), because none of these compounds contains a carbonyl or hemiacetal group; all are glycosides, which are not in equilibrium with the carbonyl compounds or the anomers.

18.31 (a), (d), (e), and (g), because a carbonyl or hemiacetal group (not an acetal, or glycoside, group) is required for a sugar to be a reducing sugar.

18.32 (a) HO, CH_2OH, O, HO, OH, OH (b) CH_2OH, O, OH, HO, OH, OH

In each case, the most stable conformation is the one in which the greatest number of groups is equatorial.

18.33 (a) (b) (c)

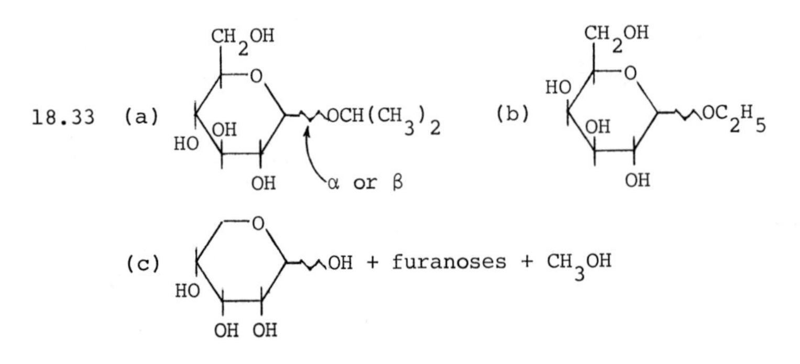

18.34 Tollens reagent contains $Ag(NH_3)_2^+$ and OH^- ions. In base, D-fructose can be converted to D-glucose and D-mannose by way of an enediol intermediate. (These two sugars differ in configuration only at carbon 2.) Oxidation of the aldehyde groups in these sugars therefore yields aldonic acids of both sugars.

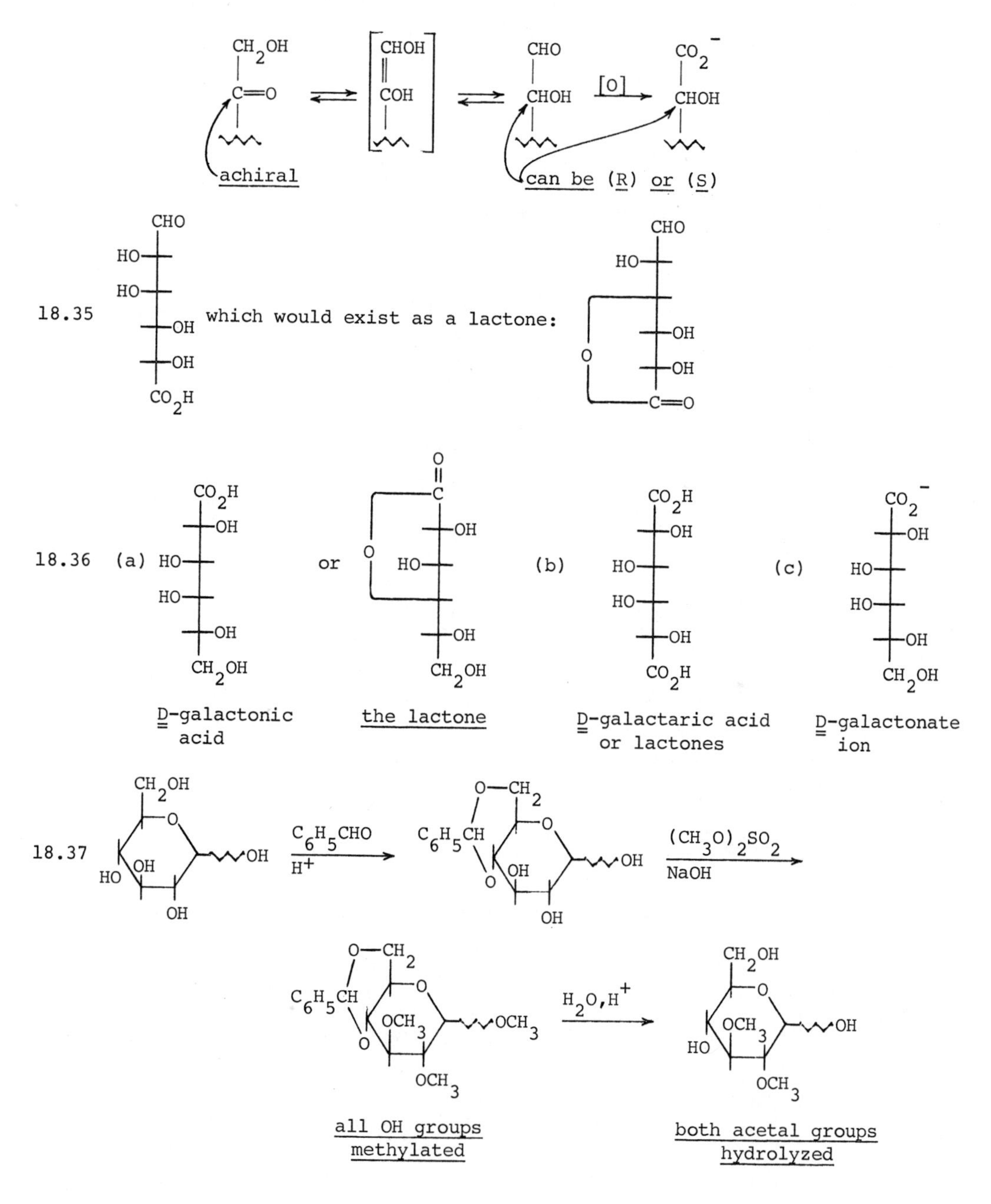

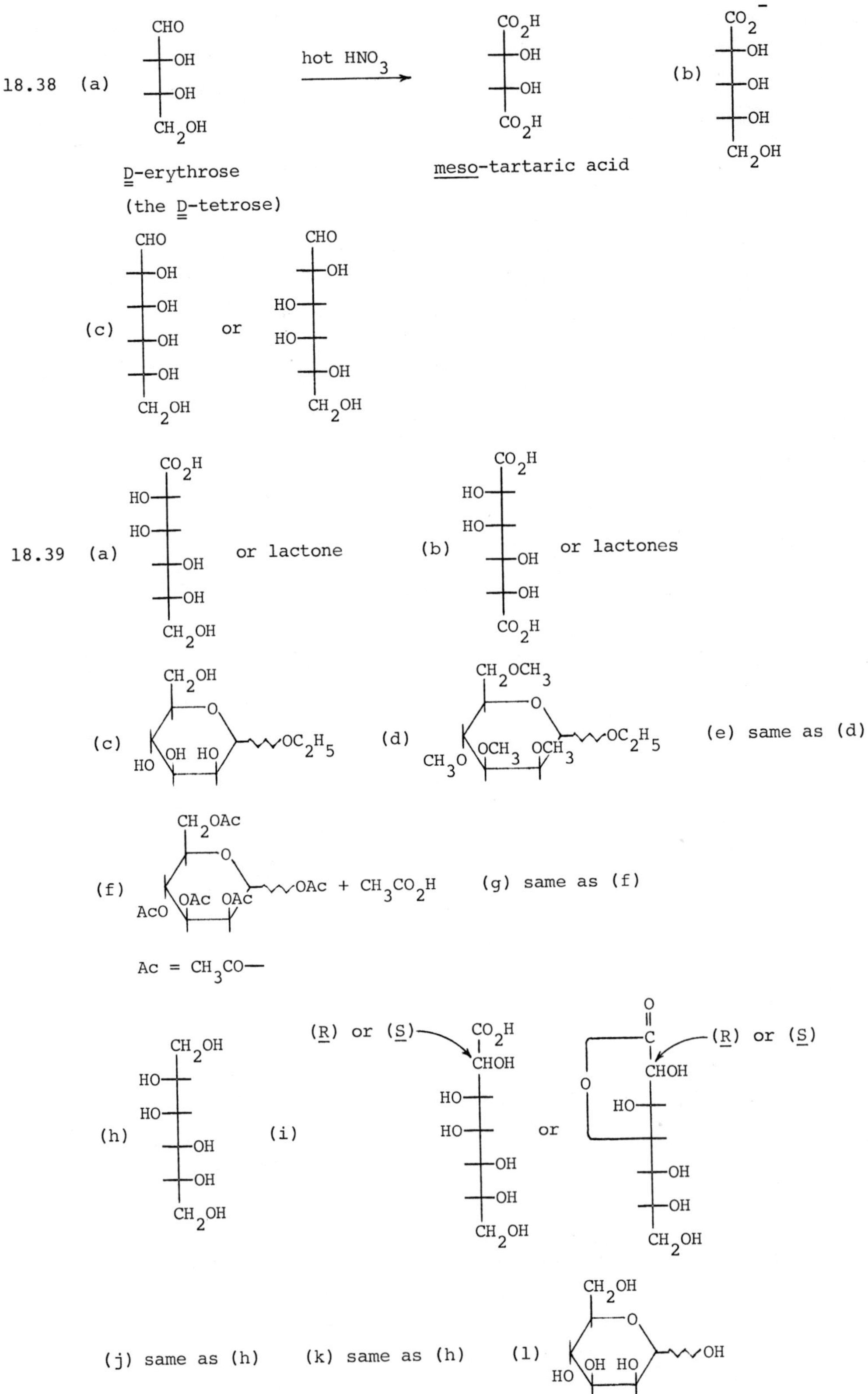

18.38 (a)
CHO
OH
OH
CH_2OH
hot HNO_3
CO_2H
OH
OH
CO_2H
D-erythrose
(the D-tetrose)
meso-tartaric acid
(b)
CO_2^-
OH
OH
OH
CH_2OH
(c)
CHO
OH
OH
OH
OH
CH_2OH
or
CHO
OH
HO
HO
OH
CH_2OH
18.39 (a)
CO_2H
HO
HO
OH
OH
CH_2OH
or lactone
(b)
CO_2H
HO
HO
OH
OH
CO_2H
or lactones
(c)
CH_2OH
O
OC_2H_5
HO
OH HO
(d)
CH_2OCH_3
O
OC_2H_5
CH_3O
OCH_3 OCH_3
(e) same as (d)
(f)
CH_2OAc
O
OAc + CH_3CO_2H
AcO
OAc OAc
(g) same as (f)
Ac = CH_3CO—
(h)
CH_2OH
HO
HO
OH
OH
CH_2OH
(i)
(R) or (S)
CO_2H
CHOH
HO
HO
OH
OH
CH_2OH
or
O
C
(R) or (S)
O
CHOH
HO
OH
OH
CH_2OH
(j) same as (h)
(k) same as (h)
(l)
CH_2OH
O
OH
HO
OH HO

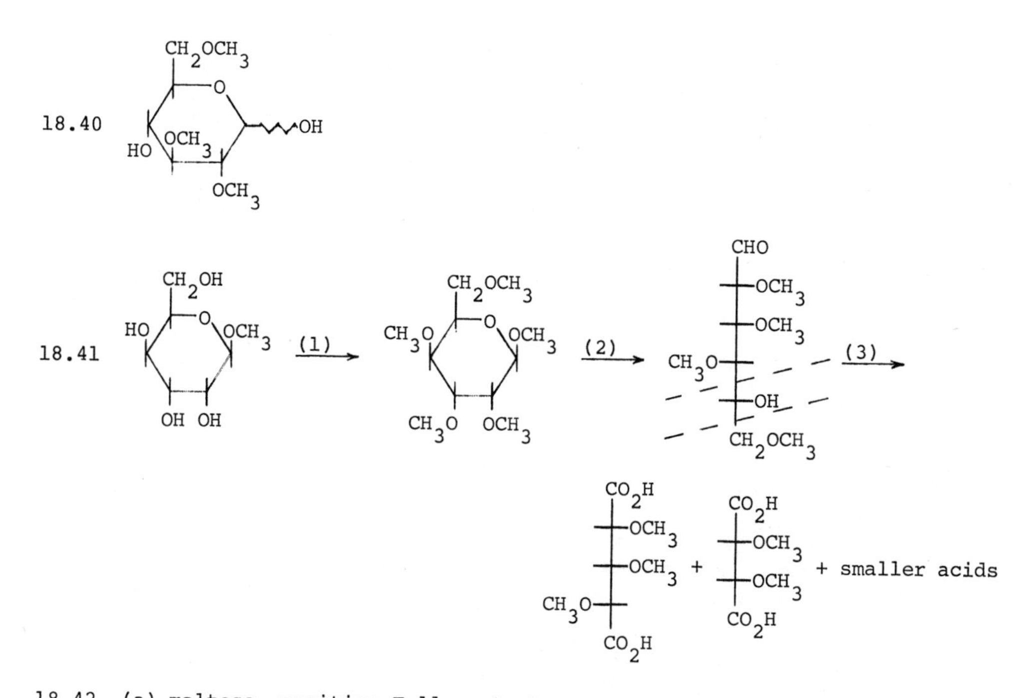

18.42 (a) maltose, positive Tollens test; sucrose, negative.

(b) Treat with HNO_3, heat; D-xylose → *meso*-diacid and D-lyxose → optically active diacid.

18.43 eleven: α,α-1,1′; α,β-1,1′; β,β-1,1′; α-1,2′; β-1,2′; α-1,3′; β-1,3′; α-1,4′; β-1,4′; α-1,6′; β-1,6′

18.44 maltose and

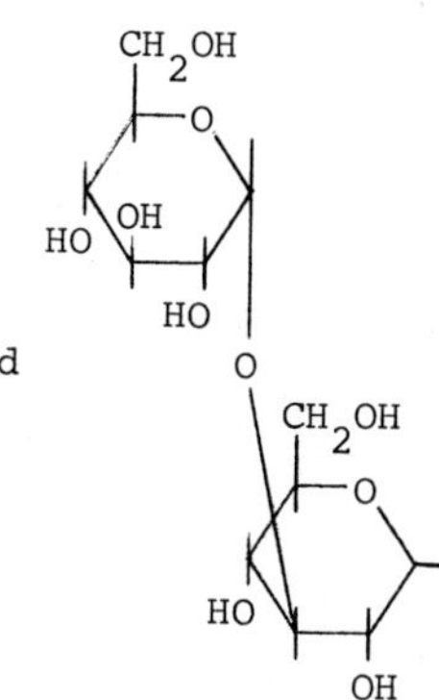

18.45 α,α′-, α,β′-, or β,β′-1,1′-D-glycopyranoside. (Because trehalose is nonreducing, we know that carbon 1 of the first unit is joined to carbon 1 of the second unit.)

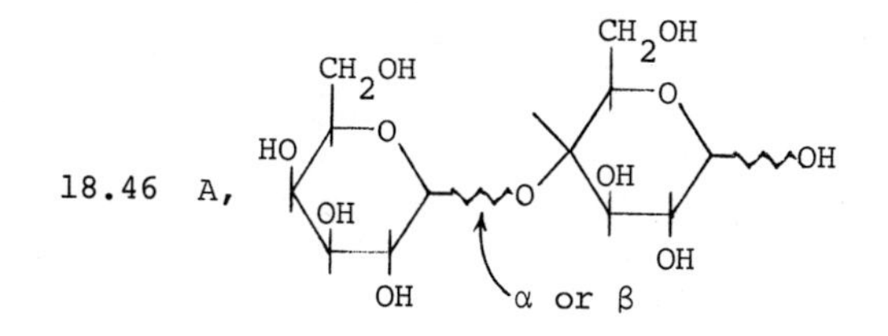

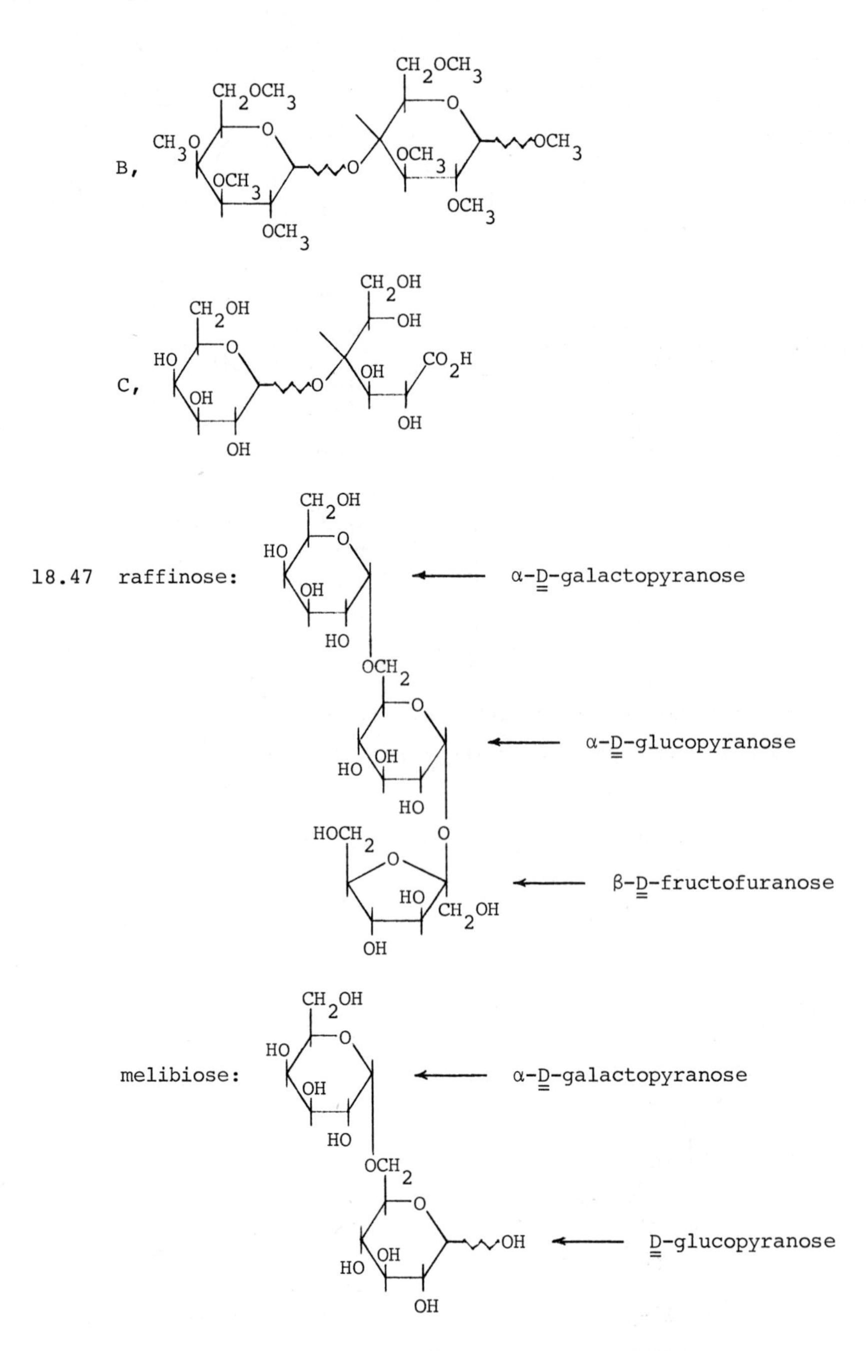
B,
CH2OCH3
CH3O
OCH3
OCH3
O
CH2OCH3
OCH3
OCH3
C,
CH2OH
HO
OH
OH
O
CH2OH
OH
CO2H
18.47 raffinose:
α-D-galactopyranose
OCH2
α-D-glucopyranose
HOCH2
CH2OH
β-D-fructofuranose
melibiose:
α-D-galactopyranose
D-glucopyranose

CHAPTER 19

Amino Acids and Proteins

Some Important Features

Proteins are polyamides formed from L-α-amino acids. These amino acids may be acidic (acidic side chain), basic (basic side chain), or neutral (side chain neither acidic nor basic).

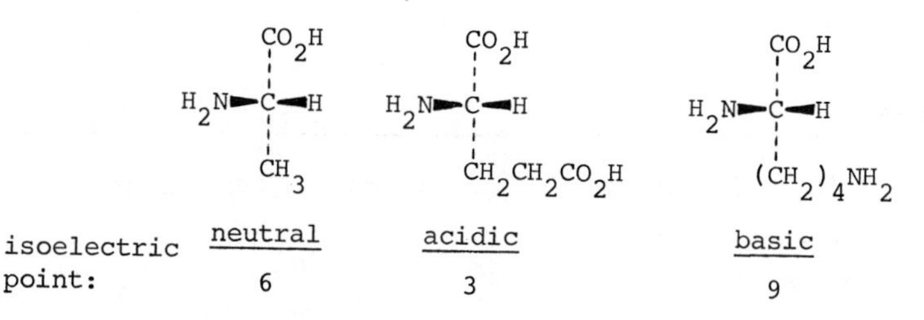

Amino acids exist as dipolar ions:

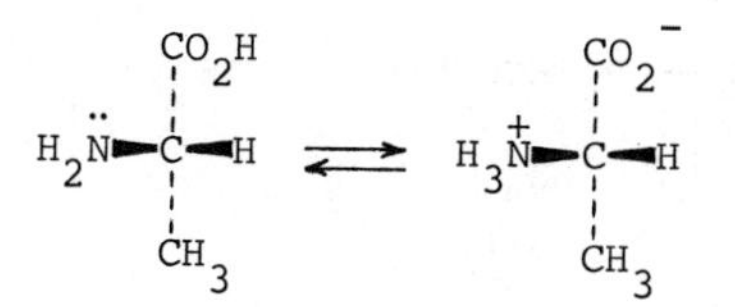

Some techniques for synthesizing amino acids are discussed in Section 19.4.

A small protein molecule is called a peptide. The synthesis of peptides can be accomplished by blocking some reactive groups within the amino acids and then allowing other functional groups to react (Section 19.8). In biological systems, the joining of amino acids into protein chains is accomplished enzymatically by mRNA, ribosomes, and tRNA (Section 19.9).

The order of attachment of amino acids (the primary structure) in peptides and proteins can be determined by partial hydrolysis and terminal residue analysis (Section 19.7).

Hydrogen bonding between NH and C═O groups and side chain interactions allow a protein to assume a secondary structure (the shape of a chain) and possibly a tertiary structure or quaternary structure (interactions between different parts of a chain or between two or more chains). Thus, a chain can form a helix that can fold into a globule or that can interact with other helices. The higher structures can add strength or solubility to a protein. When these higher structures are disrupted by a change in environment (such as a change in pH), the protein is said to be denatured.

Enzymes are proteins that act as biological catalysts. Their specific catalytic action depends upon the unique protein surface presented to substrates and upon the active site, which may be a coenzyme (a nonprotein organic molecule or a metal ion). Many vitamins are coenzymes.

Reminders

In predicting acid-base reactions of amino acids or peptides, look for the most acidic and most basic sites in the molecule.

Remember that any synthesis from an achiral molecule leads to an achiral or racemic product; however, a racemic mixture of enantiomers can be separated by procedures outlined in Chapter 4.

Answers to Problems

19.10 (a) neutral (b) acidic (c) neutral (d) neutral (e) basic (f) neutral

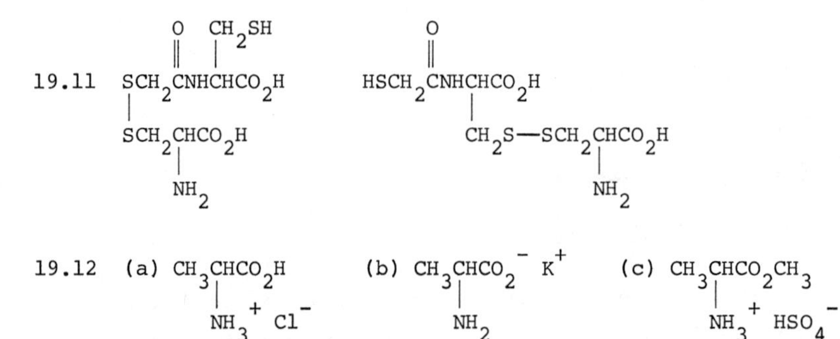

(d) $CH_3CH(NHCOCH_3)CO_2H + CH_3CO_2H$

19.13 $(CH_3)_2CHCH_2CO_2H \xrightarrow[PBr_3]{Br_2} (CH_3)_2CHCHBrCO_2H \xrightarrow{\text{excess } NH_3}$

achiral — racemic

$(CH_3)_2CHCH(NH_2)CO_2^-\ NH_4^+ \xrightarrow{H^+} (CH_3)_2CHCH(NH_2)CO_2H$

racemic — racemic

19.14 ($\underline{R}$)-$CH_3CH(OH)CO_2H \xrightarrow{\text{hot } KMnO_4 \text{ or } H_2CrO_4} CH_3C(=O)CO_2H \xrightarrow[Pd]{H_2, NH_3} CH_3CH(NH_2)CO_2H$

achiral — racemic

19.15 (a) An amino acid contains a carboxylate group ($—CO_2^-$), rather than a carboxyl group ($—CO_2H$).

(b) When the solution is acidified, the carboxyl group is generated.

19.16 (a) $C_6H_5CH_2CHO \xrightarrow[HCN]{NH_3} C_6H_5CH_2CH(NH_2)CN \xrightarrow{H_2O, H^+} C_6H_5CH_2CH(NH_2)CO_2H$

(b) $(CH_3)_2CHCHO \xrightarrow[HCN]{NH_3} (CH_3)_2CHCH(NH_2)CN \xrightarrow{H_2O, H^+} (CH_3)_2CHCH(NH_2)CO_2H$

(c) both racemic

19.17 (a) $CH_3CH(NH_2)CO_2^-\ \overset{+}{N}H_4$ (b) $Cl^-\ \overset{+}{N}H_3CH_2CO_2H$ + $C_6H_4(CO_2H)_2$ (phthalic acid)

(c) $C_6H_5CH_2CH(\overset{+}{N}H_3\ Cl^-)CO_2H$ + $C_6H_4(CO_2H)_2$ (phthalic acid) (d) $(CH_3)_2CHCH_2CH(NH_2)CN$ (e) $C_6H_5CH_2CH(\overset{+}{N}H_3\ Cl^-)CO_2CH_3$

19.18 (a) a neutral amino acid: (3) (b) an acidic amino acid: (4)
(c) a neutral amino acid: (2) (d) a basic amino acid: (1)

Note that cysteine (a) is slightly more acidic than proline (c). The reason is that proline contains a more basic 2° amino group.

19.19 (a) neutral, 6 (b) slightly basic, 8 (See Answer 19.39) (c) acidic, 3
(d) neutral, 6 (e) basic, 10 (See Answer 19.39)

19.20 (a) $H_3\overset{+}{N}CHCO_2^-$ ($CH(CH_3)_2$ side chain) $+ Cl^-$ (b) $H_3\overset{+}{N}CHCO_2^-$ ($(CH_2)_4NH_2$ side chain) $+ Cl^-$

(c) $H_3\overset{+}{N}CHCO_2^-$ ($CH(CH_3)_2$ side chain) $+ Na^+ + H_2O$ (d) $H_3\overset{+}{N}CHCO_2^-$ ($(CH_2)_4\overset{+}{N}H_3$ side chain) $+ Na^+ + H_2O$

In each case, look for the more acidic or more basic group in the reactant molecule.

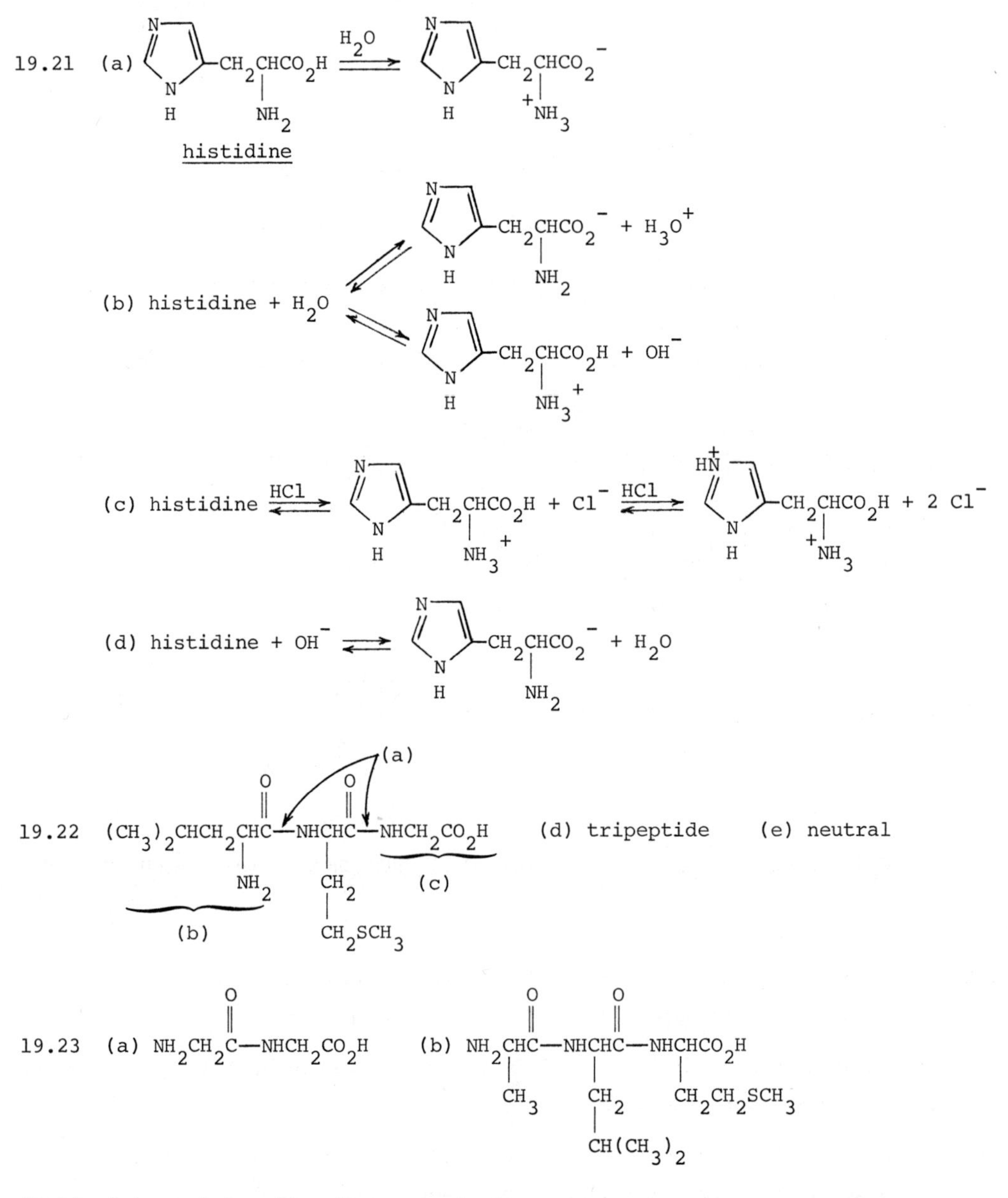

19.24 (a) serylglycylleucine, ser-gly-leu

(b) prolylthreonylmethionine, pro-thr-met

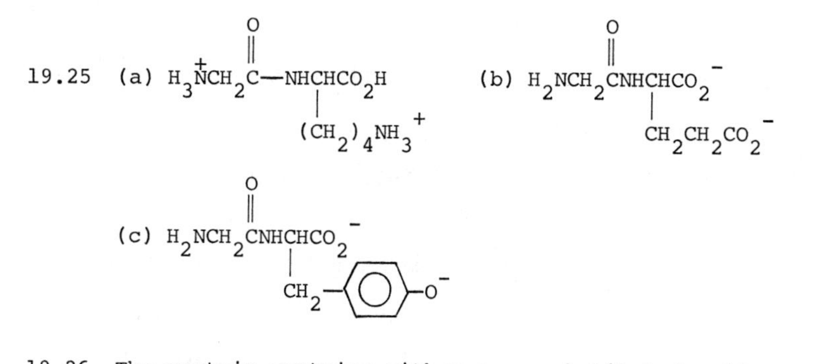

19.26 The protein contains either an unsubstituted amide group, $-\overset{O}{\overset{\|}{C}}NH_2$, or arginine.

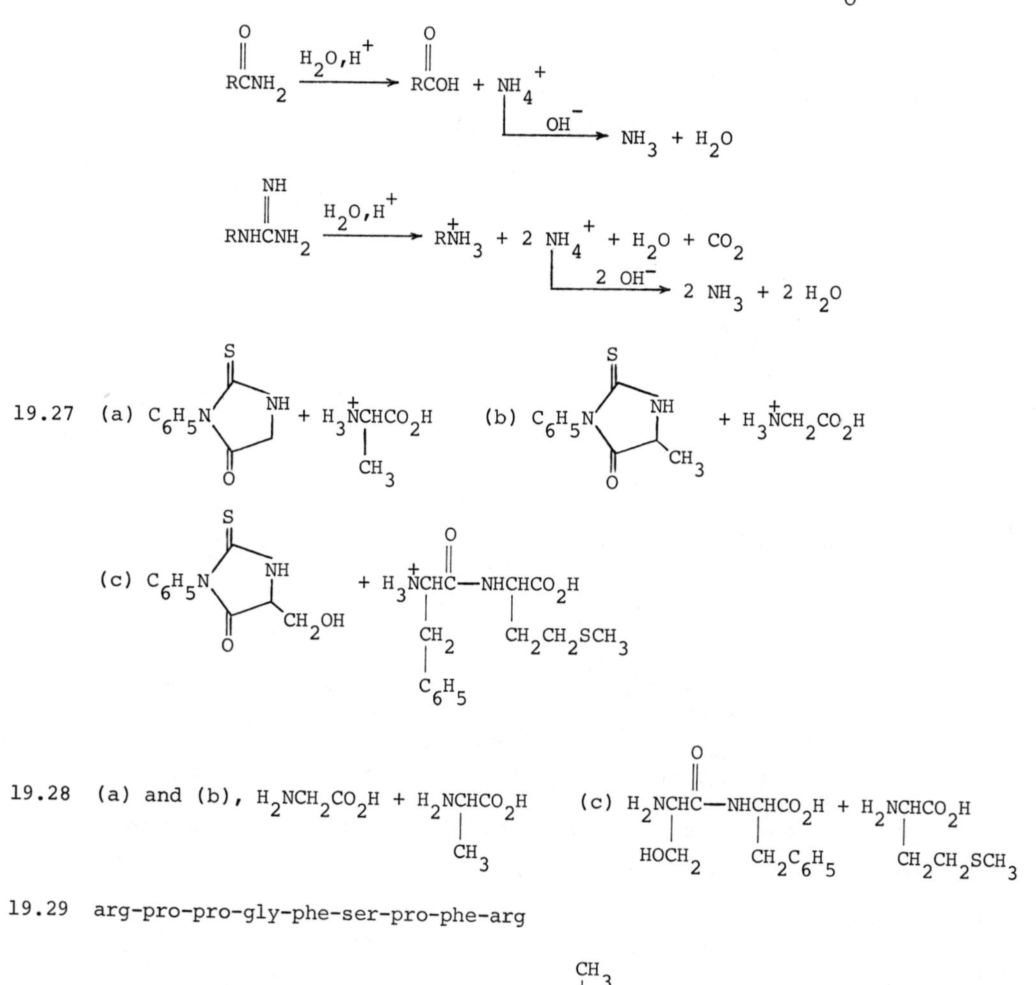

19.29 arg-pro-pro-gly-phe-ser-pro-phe-arg

19.30

$$H_2N\underset{\substack{|\\ CH(CH_3)_2}}{C}HCO_2H \xrightarrow[(a)]{SOCl_2} Cl^- \; H_3\overset{+}{N}\underset{\substack{|\\ CH(CH_3)_2}}{C}HCOCl \xrightarrow[(b)]{H_2N\overset{\substack{CH_3\\ |}}{C}HCO_2H}$$

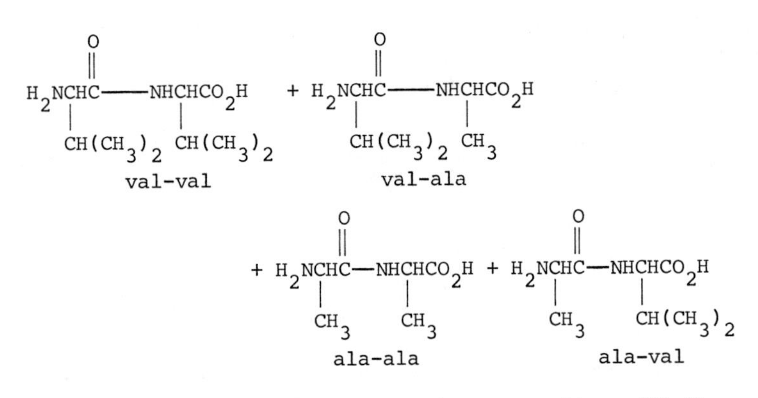

In (b), the acid halide of val can undergo an exchange reaction with the new acid (ala).

$$RCCl + R'COH \rightleftharpoons RCOH + R'CCl$$

(each with C=O)

19.31 (a)

$$\underset{\text{ala}}{CH_3CH(NH_2)CO_2H} \xrightarrow{C_6H_5CH_2O_2CCl} CH_3CH(NHCO_2CH_2C_6H_5)CO_2H \xrightarrow{ClCO_2C_2H_5} CH_3CH(NHCO_2CH_2C_6H_5)CO_2CO_2C_2H_5$$

$$\xrightarrow[\text{or ester}]{H_2NCH_2CO_2H} CH_3CH(NHCO_2CH_2C_6H_5)\overset{O}{\overset{\|}{C}}NHCH_2CO_2H \xrightarrow{H_2,Pd} \underset{\text{ala-gly}}{CH_3CH(NH_2)\overset{O}{\overset{\|}{C}}NHCH_2CO_2H}$$

(b)

$$\underset{\text{phe}}{C_6H_5CH_2CH(NH_2)CO_2H} \xrightarrow[\text{(3) } (CH_3)_2CHCH(NH_2)CO_2H \text{ (4) } H_2,Pd]{\text{(1) } C_6H_5CH_2O_2CCl \text{ (2) } ClCO_2C_2H_5} \underset{\text{phe-val}}{C_6H_5CH_2CH(NH_2)\overset{O}{\overset{\|}{C}}NHCH(CH(CH_3)_2)CO_2H}$$

(c)

$$\text{ala-gly} \xrightarrow[\text{(3) phe-val (4) } H_2,Pd]{\text{(1) } C_6H_5CH_2O_2CCl \text{ (2) } ClCO_2C_2H_5} \text{ala-gly-phe-val}$$

Note that, except for the amino acids, the series of reagents used in (b) and (c) are the same as those in (a).

19.32 (a) TAA, TAG, TAT (b) UAA, UAG, UAU

19.33 (a), (d), (e)

19.34 (a) The acetic acid in vinegar forms hydrogen bonds with protein groups, disrupting the hydrogen bonds that hold the protein molecules together.

(b) Aqueous sucrose can disrupt the hydrogen bonding and tenderize the meat to some extent, but not to the degree that the strongly hydrogen bonding acetic acid can.

19.35 (a) 4 (b) 3, 5, 6 (c) 6 (d) 2 (e) 1 (f) 7

19.36 (b)

19.37

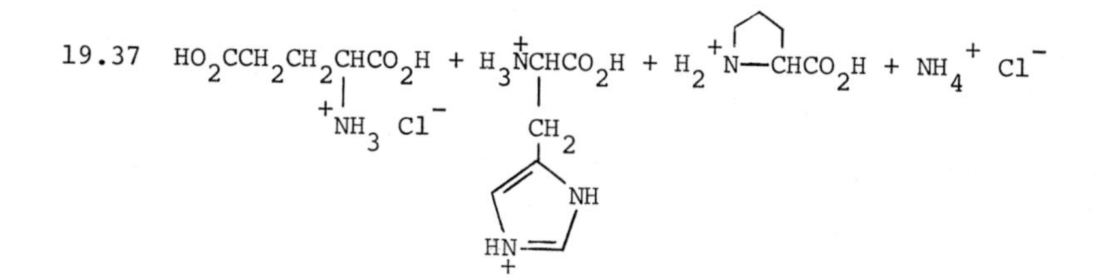

19.38 The peptide is cyclic. For example,

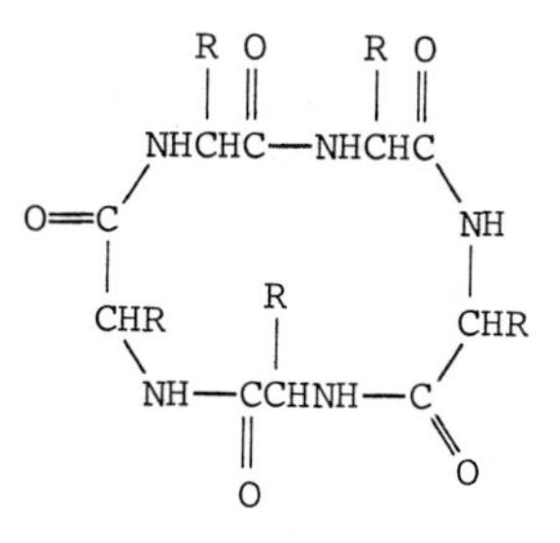

19.39 (a) In addition to the α amino group, lysine has an sp^3 N on its side chain, while histidine has two ring nitrogens. One NH ring nitrogen of histidine is not basic (cf. pyrrole, Section 16.9). The other nitrogen, although basic, is in the sp^2 state and therefore is less basic than the sp^3 nitrogen of lysine.

(b) The side chain of arginine is more basic than that of lysine because of resonance-stabilization of the cation.

lysine: $RCH_2\ddot{N}H_2 + H_2O \rightleftharpoons RCH_2\overset{+}{N}H_3 + OH^-$

arginine: $RNHC(=\ddot{N}H)\ddot{N}H_2 + H_2O \rightleftharpoons$

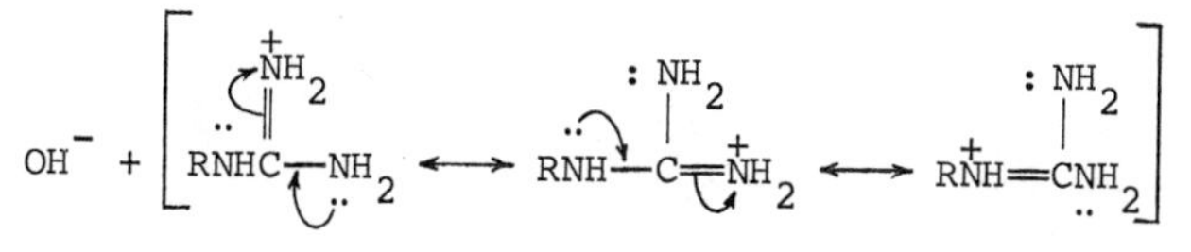

19.40 (a) Because the reaction is slow for acidic and basic D-amino acids, but fast for neutral D-amino acids, we conclude that electronic effects are more important in determining the rate.

(b) For the same reason, we predict that the binding site is nonpolar.

(c) D-Cysteine, a neutral amino acid, undergoes oxidation faster than D-arginine, a basic amino acid.

19.41 To signal valine instead of glutamic acid, the central base in the codon in RNA must be changed from A to U. Therefore, in DNA, the central base corresponding to this codon must be adenine instead of the correct thymine.

19.42 val-[glu-leu]-lys-phe-tyr-asp-ala-gly or val-[leu-glu]-lys-phe-tyr-asp-ala-gly

The two possibilities could be differentiated in a number of ways. One of the simplest is N-terminal analysis (two cycles) to see if leu or gly is the second residue.

19.43 $H_2NCH(CO_2H)CH_2CH_2CH_2NH\text{—}C(=NH)NH_2 \xrightarrow[\text{enzyme}]{H_2O} H_2NC(=O)NH_2 + H_2NCH(CO_2H)CH_2CH_2CH_2NH_2$ (ornithine)

19.44

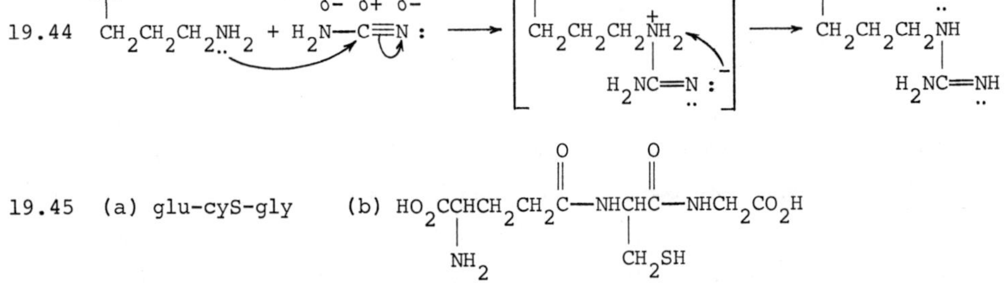

19.45 (a) glu-cyS-gly (b) $HO_2CCH(NH_2)CH_2CH_2C(=O)\text{—}NHCH(CH_2SH)C(=O)\text{—}NHCH_2CO_2H$

CHAPTER 20

Lipids and Related Natural Products

Some Important Features

Some important lipids are the edible fats and oils, terpenes, and steroids. Animal fats are triglycerides of fatty acids containing few carbon-carbon double bonds, while vegetable oils are triglycerides of fatty acids containing a greater number of carbon-carbon double bonds. Saponification of fats or oils yields soaps, RCO_2Na, where R is a long alkyl or alkenyl chain. Twenty-carbon fatty acids are used to biosynthesize prostaglandins (Section 20.4).

Phospholipids are compounds with hydrocarbon chains and a dipolar phosphate-amine group. These compounds form part of cell walls.

Terpenes are compounds with carbon skeletons of head-to-tail isoprene units.

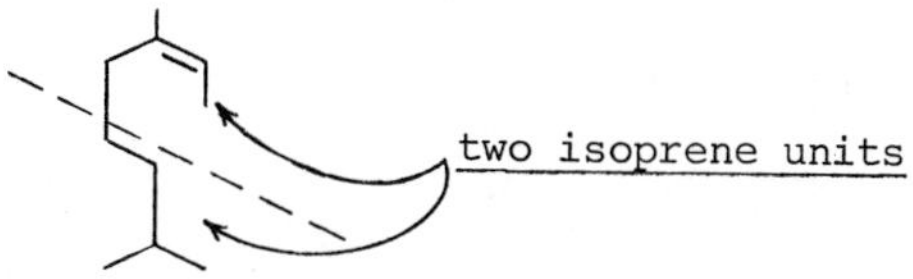

two isoprene units

a monoterpene

Terpenes are found in relatively large amounts in plants. In animals, they are intermediates in the biosynthesis of steroids. Some terpenes act as pheromones (Section 20.6).

Steroids are compounds containing the following ring system:

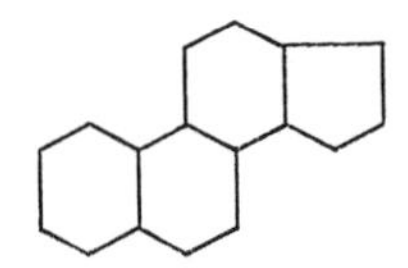

The stereochemistry of steroids is discussed in Section 20.7. Most steroids have trans-ring junctures; the bile acids are a notable exception.

Some steroids act as hormones. A few hormonal steroids, along with cholesterol, are discussed in the text.

Reminders

Although many lipids have complex structures, keep in mind that the reactions of lipids are predictable from their functional groups. For example, triglycerides are esters and may also contain carbon-carbon double bonds. Their reactions are typical of these functional groups. Sphingomyelin (Section 20.3) contains a carbon-carbon double bond, a hydroxyl group, an amide group, an inorganic ester group, and a quaternary nitrogen. Its reactions are typical of these functional groups.

Answers to Problems

20.9 (a) $HOCH_2CH(OH)CH_2OH + 2\ CH_3(CH_2)_{16}CO_2^-\ Na^+ + CH_3(CH_2)_5CH{=}CH(CH_2)_7CO_2^-\ Na^+$

(b) $HOCH_2CH(OH)CH_2OH + 2\ CH_3(CH_2)_{16}CH_2OH + CH_3(CH_2)_{14}CH_2OH$

(c) $CH_2O_2C(CH_2)_7CHBrCHBr(CH_2)_5CH_3$ / $CHO_2C(CH_2)_{16}CH_3$ / $CH_2O_2C(CH_2)_{16}CH_3$ or $CH_2O_2C(CH_2)_{16}CH_3$ / $CHO_2C(CH_2)_7CHBrCHBr(CH_2)_5CH_3$ / $CH_2O_2C(CH_2)_{16}CH_3$

20.10 (a) no reaction (b) $CH_3(CH_2)_5CO_2H + HO_2C(CH_2)_7CO_2H$

(c) $CH_3(CH_2)_4CO_2H + CH_2(CO_2H)_2 + HO_2C(CH_2)_7CO_2H$

(d) $CH_3CH_2CO_2H + 2\ CH_2(CO_2H)_2 + HO_2C(CH_2)_7CO_2H$

20.11 (a) Tripalmitolein decolorizes Br_2/CCl_4 solution; tripalmitin does not.

(b) Beeswax is hydrolyzed to a fatty acid and a water-insoluble monoalcohol; beef fat is hydrolyzed to a fatty acid and glycerol (water-soluble). Also, beef fat will consume approximately twice as much NaOH per gram when saponified.

(c) Paraffin wax (a mixture of alkanes) does not undergo hydrolysis.

(d) Linoleic acid neutralizes dilute aqueous NaOH; linseed oil does not.

(e) Sodium palmitate precipitates with Ca^{2+}; sodium p-decylbenzenesulfonate does not.

(f) A vegetable oil is hydrolyzed to a fatty acid and glycerol; a motor oil (a mixture of hydrocarbons) does not undergo hydrolysis.

20.12 $CH_3(CH_2)_{12}CO_2H$

$$\begin{array}{l} CH_2O_2C(CH_2)_{12}CH_3 \\ | \\ CHO_2C(CH_2)_{12}CH_3 \\ | \\ CH_2O_2C(CH_2)_{12}CH_3 \end{array}$$

20.13 (b) and (c). In each case, the molecule contains an ionic end as well as a hydrocarbon end. Compound (a) does not have a sufficiently large hydrocarbon portion for detergent action.

20.14

$$\begin{array}{ll} ③ & CH_2O_2C(CH_2)_{14}CH_3 \\ & | \\ ② & CHO_2C(CH_2)_{14}CH_3 \\ & | \\ ① & CH_2O_2C(CH_2)_7CH{=}CH(CH_2)_7CH_3 \end{array}$$

Oleic acid must be attached to carbon 1 of glycerol; otherwise, carbon 2 would not be chiral.

20.15 tristearin $\xrightarrow[(2)\ H^+]{(1)\ NaOH, H_2O, heat}$ $CH_3(CH_2)_{16}CO_2H$

tristearin $\xrightarrow[heat, pressure]{H_2, catalyst}$ $CH_3(CH_2)_{17}OH$

$\xrightarrow[heat]{H^+}$ $CH_3(CH_2)_{16}CO_2(CH_2)_{17}CH_3$

20.16 (a) $CH_3(CH_2)_7CH{=}CH(CH_2)_7CO_2^-\ Na^+ + CH_3(CH_2)_{16}CO_2^-\ Na^+$

$+ Na_2HPO_4 + HOCH_2CH_2\overset{+}{N}(CH_3)_3\ Cl^- + HOCH_2CH(OH)CH_2OH$

(b) $CH_3(CH_2)_7CH{=}CH(CH_2)_7CO_2^-\ Na^+ + CH_3(CH_2)_{16}CO_2^-\ Na^+$

$+ Na_2HPO_4 + HOCH_2CH_2NH_2 + HOCH_2CH(OH)CH_2OH$

The only difference between (a) and (b) is the amino alcohol; (a) yields a quaternary amine salt while (b) yields a 1° amine.

20.17 $CH_3(CH_2)_{22}CO_2H$ +

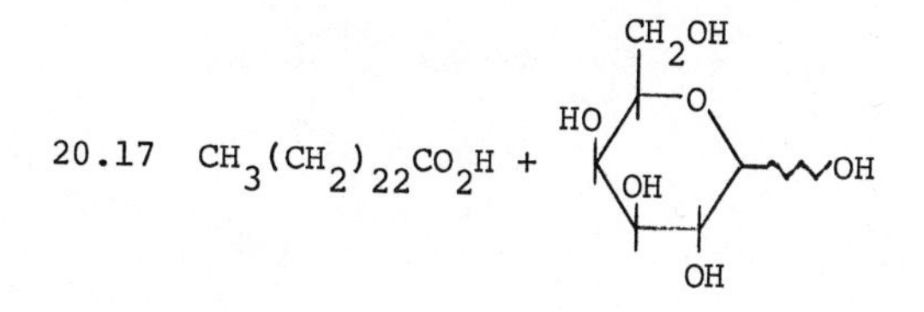

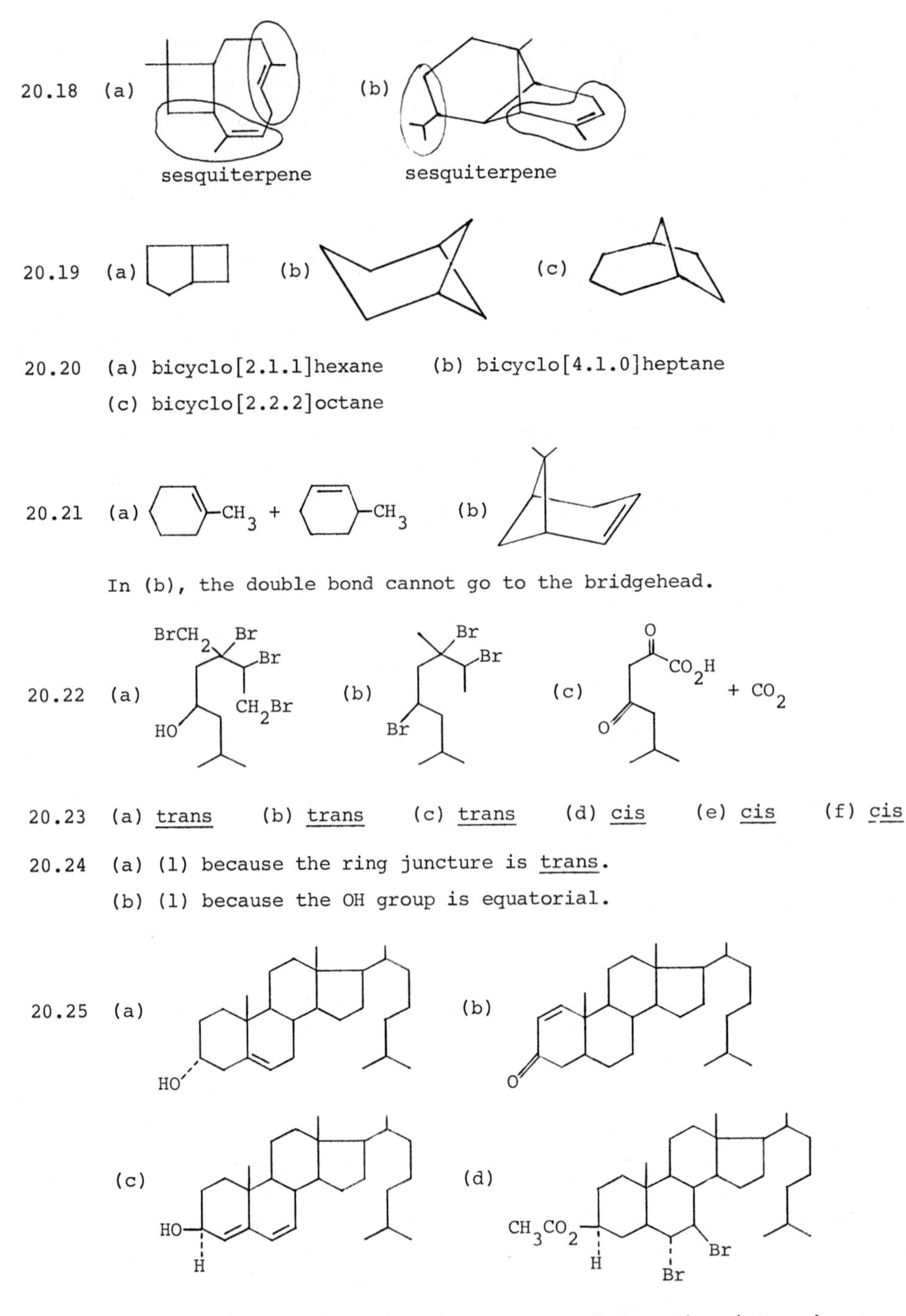

20.20 (a) bicyclo[2.1.1]hexane (b) bicyclo[4.1.0]heptane
(c) bicyclo[2.2.2]octane

In (b), the double bond cannot go to the bridgehead.

20.23 (a) trans (b) trans (c) trans (d) cis (e) cis (f) cis

20.24 (a) (1) because the ring juncture is trans.
(b) (1) because the OH group is equatorial.

20.26 Estradiol is a phenol and can be extracted from the mixture by aqueous NaOH.

20.27

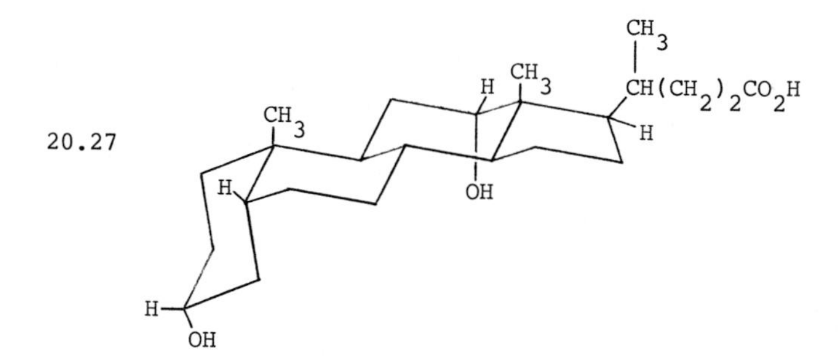

20.28 The OH at position 3 of cholic acid is equatorial, while the other two OH groups are axial. The esterification takes place at the 3-OH group because it is the least sterically hindered.

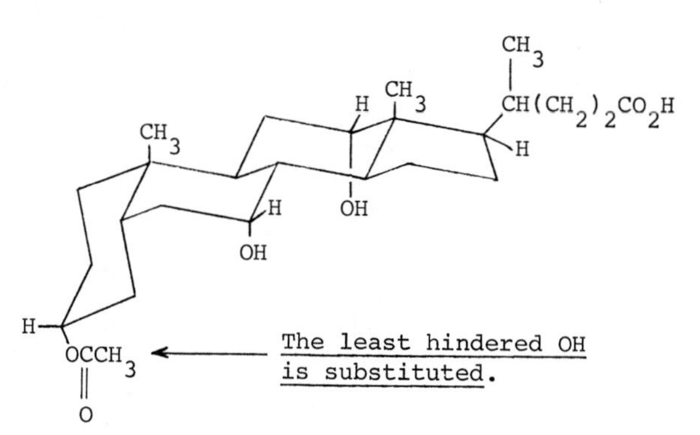

20.29 (a)

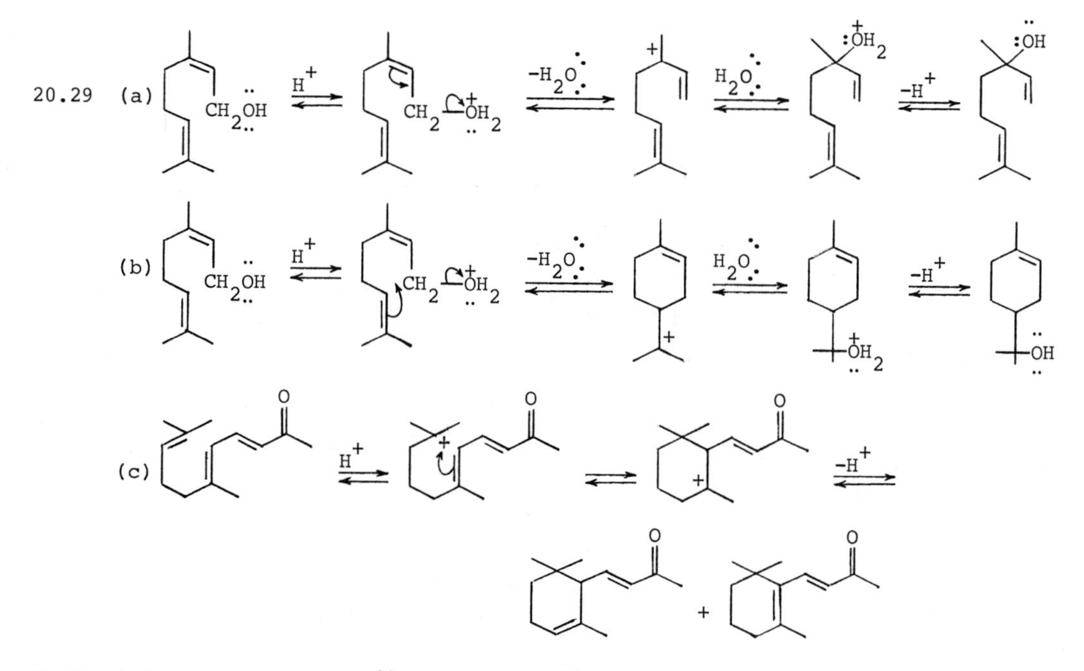

20.30 (a) none because the ^{14}C is lost as $^{14}CO_2$

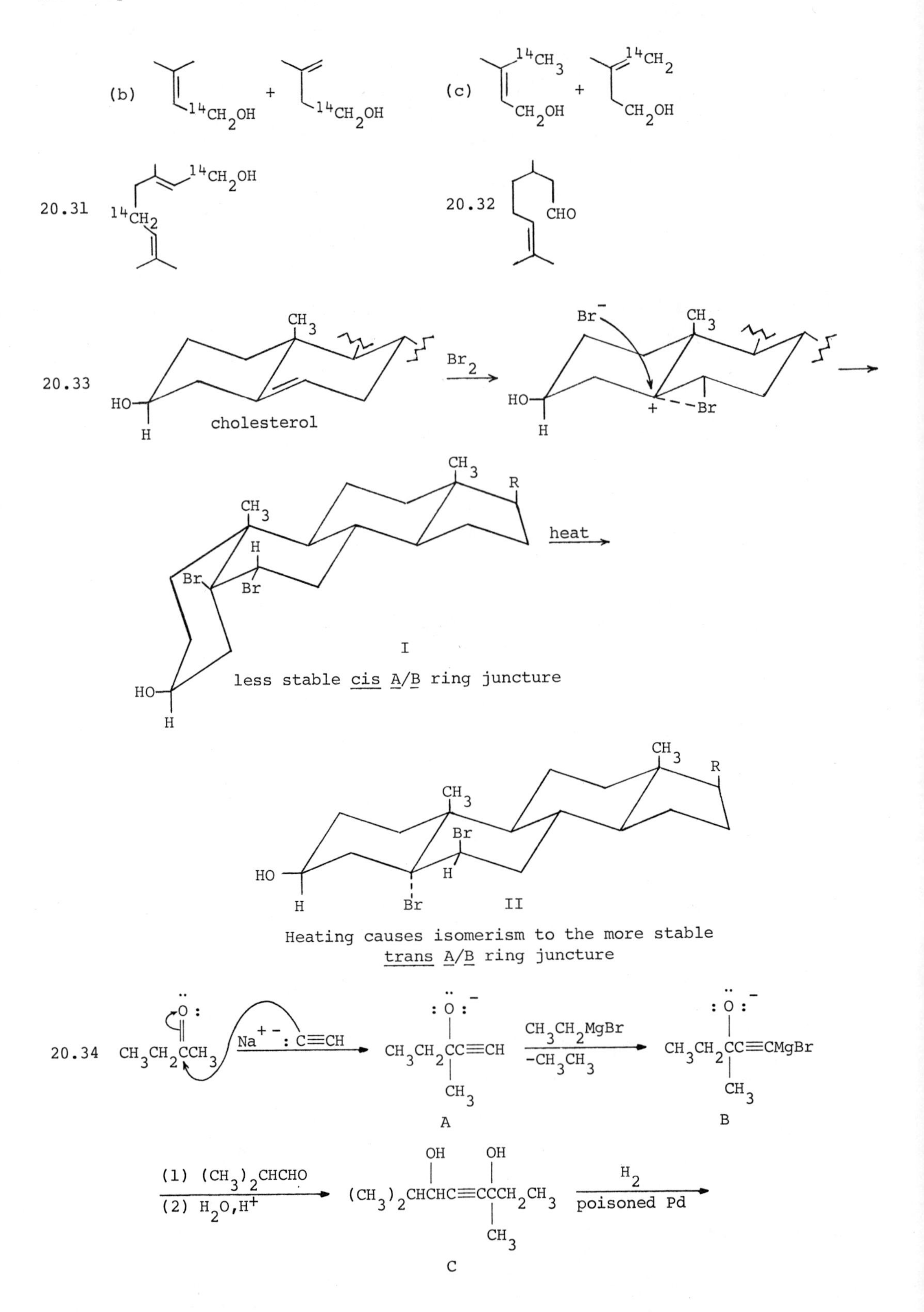
(b)
$^{14}CH_2OH$
+
$^{14}CH_2OH$
(c)
$^{14}CH_3$
CH_2OH
+
$^{14}CH_2$
CH_2OH
20.31
$^{14}CH_2OH$
$^{14}CH_2$
20.32
CHO
20.33
CH_3
HO
H
cholesterol
Br_2
Br^-
CH_3
HO
H
+
Br
CH_3
CH_3
R
H
Br
Br
heat
HO
H
I
less stable cis A/B ring juncture
CH_3
R
CH_3
Br
HO
H
H
Br
II
Heating causes isomerism to the more stable
trans A/B ring juncture
20.34
$CH_3CH_2CCH_3$
O
$Na^+ \; ^-:C\equiv CH$
$CH_3CH_2CC\equiv CH$
CH_3
A
CH_3CH_2MgBr
$-CH_3CH_3$
$CH_3CH_2CC\equiv CMgBr$
CH_3
B
(1) $(CH_3)_2CHCHO$
(2) H_2O,H^+
OH
OH
$(CH_3)_2CHCHC\equiv CCCH_2CH_3$
CH_3
C
H_2
poisoned Pd

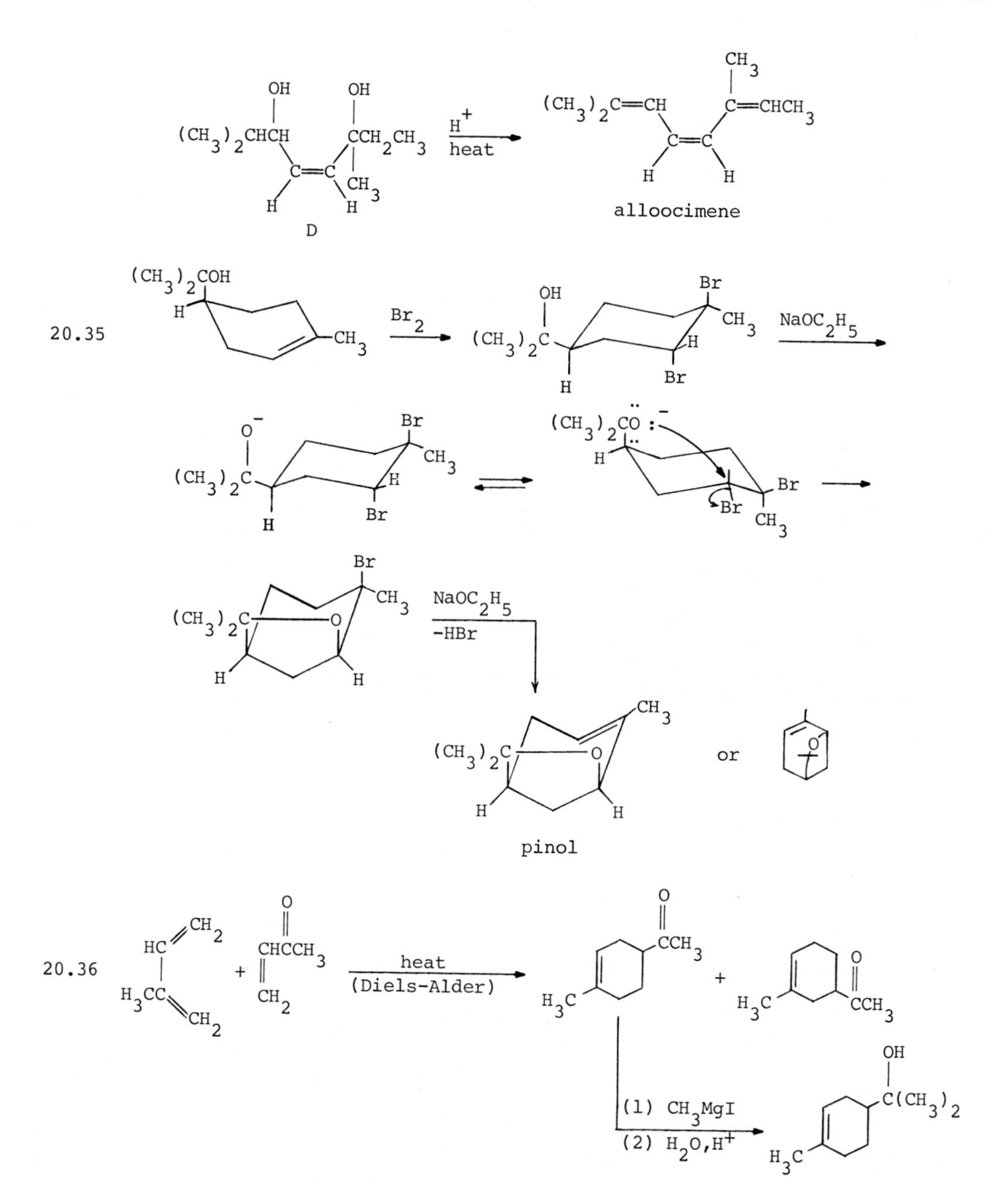

Note that the Diels-Alder reaction would lead to a mixture of isomers, which could be separated before the Grignard reaction. You might have devised a synthetic scheme using a Dieckmann ring closure; however, such a route would require many more steps.

CHAPTER 21

Spectroscopy II: Ultraviolet Spectra, Color and Vision, Mass Spectra

Some Important Features

Ultraviolet and visible spectra arise from the promotion of pi electrons, especially pi electrons that are part of a conjugated system. In general, the larger the conjugated system, the longer is the wavelength of light absorbed. If the conjugated system is sufficiently long, the compound appears colored.

$$\underset{\text{Absorbed light of increasing } \lambda \text{ (decreasing energy)}}{\overrightarrow{C{=}C \qquad C{=}C{-}C{=}C \qquad C{=}C{-}C{=}C{-}C{=}C}}$$

The quantity of radiation absorbed (the molar absorptivity ε) is calculated from the formula $\varepsilon = A/c\,l$ (Section 21.2). Transitions of the $\pi \rightarrow \pi^*$ type generally have a high value for ε compared to $n \rightarrow \sigma^*$ or $n \rightarrow \pi^*$ transitions.

Color usually arises from the absorption of specific wavelengths from the full visible range (white light). Reflection of the remaining, unabsorbed wavelengths to the eye results in color vision.

Some colored organic compounds are discussed in Sections 21.5 and 21.6. Of particular interest are the indicators that change color depending on the pH of the solution. An acid-base reaction of an indicator molecule results in a change in the length of the conjugated system and a change in the wavelength of absorption.

Mass spectra arise from cleavage of molecules into ions and ion-radicals when the molecules are bombarded with high-energy electrons. Loss of a single electron gives rise to the molecular ion, often the farthest peak to the right in a mass spectrum.

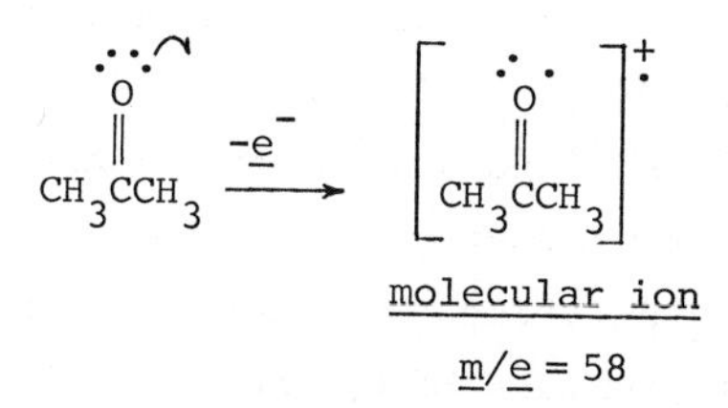

Fission of the molecular ion often occurs at a branch or α to an electronegative atom.

$[R{-}CH(CH_3){-}R]^{+\cdot}$ $\quad$ $[R{-}C(=O){-}R]^{+\cdot}$

Small molecules, such as H_2O, can be lost:

$$[RCH_2CH_2OH]^{+\cdot} \xrightarrow{-H_2O} [RCH{=}CH_2]^{+\cdot}$$

An alkene molecule may be lost in a McLafferty rearrangement (Section 21.11D).

Reminders

In a problem concerning the color change of a compound with a change in pH: (1) look for the most acidic or basic group in the molecule; (2) write the acid-base reaction; (3) write resonance structures for reactant and product.

In mass spectral problems, m/e is the mass of the ion or ion-radical divided by the ionic charge (usually +1).

To determine fragmentation patterns, write the structure of the molecular ion. Mark all likely positions of cleavage, note any small molecules that could be lost, then write the structures of the fragments, keeping track of electrons.

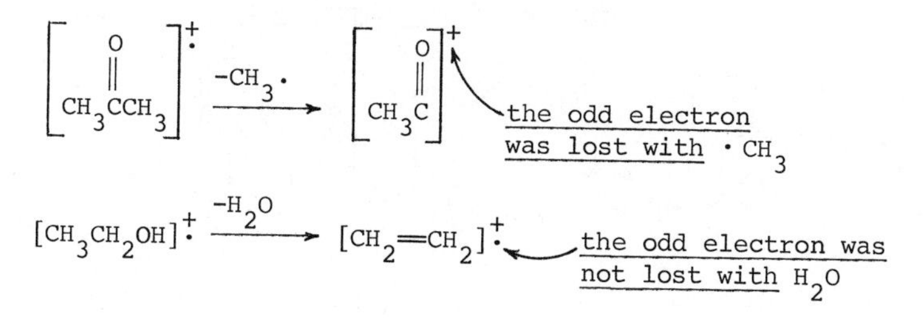

Answers to Problems

21.14 In the uv spectra, the tetraene would show a λ_{max} at a longer wavelength than the triene.

21.15 (a) The two compounds cannot be distinguished with only uv. Both contain the same type of π system (an ester).

(b) The carbonyl group in conjugation with the double bond exhibits different absorption from that of the conjugated diene.

(c) The ketone with the conjugated double bond absorbs at a longer wavelength than the ketone with the nonconjugated double bond.

(d) One structure is a conjugated triene, while the other is a conjugated diene; therefore, these compounds can also be distinguished by their uv spectra.

21.16 (a) $\varepsilon = \dfrac{A}{\underline{c}\ \underline{l}} = \dfrac{1.25}{(9.54 \times 10^{-5}\ \underline{M})(1.0\ \text{cm})} = 1.31 \times 10^{4}$

(b) $\varepsilon = \dfrac{0.75}{(0.038)(1.0)} = 20$

21.17 (a) $\underline{n} \rightarrow \pi^*$ and $\pi \rightarrow \pi^*$ transitions (b) the $\pi \rightarrow \pi^*$ transition at 219 nm

21.18 (a) $\pi \rightarrow \pi^*$ (b) $\pi \rightarrow \pi^*$ and $\underline{n} \rightarrow \pi^*$ (c) $\pi \rightarrow \pi^*$ and $\underline{n} \rightarrow \pi^*$

(d) $\pi \rightarrow \pi^*$ and $\underline{n} \rightarrow \pi^*$

(All could exhibit $\sigma \rightarrow \pi^*$ transitions also.)

21.19 (a), (c), (b), (d), the same order as that of increasing conjugation.

21.20 A conjugated diene, $CH_2{=}CHCH{=}CHCH_3$ or $CH_2{=}C(CH_3)CH{=}CH_2$

21.21 Because of van der Waals repulsions, cis-stilbene cannot be planar; therefore, the degree of conjugation is less and the wavelength of uv absorption is shorter.

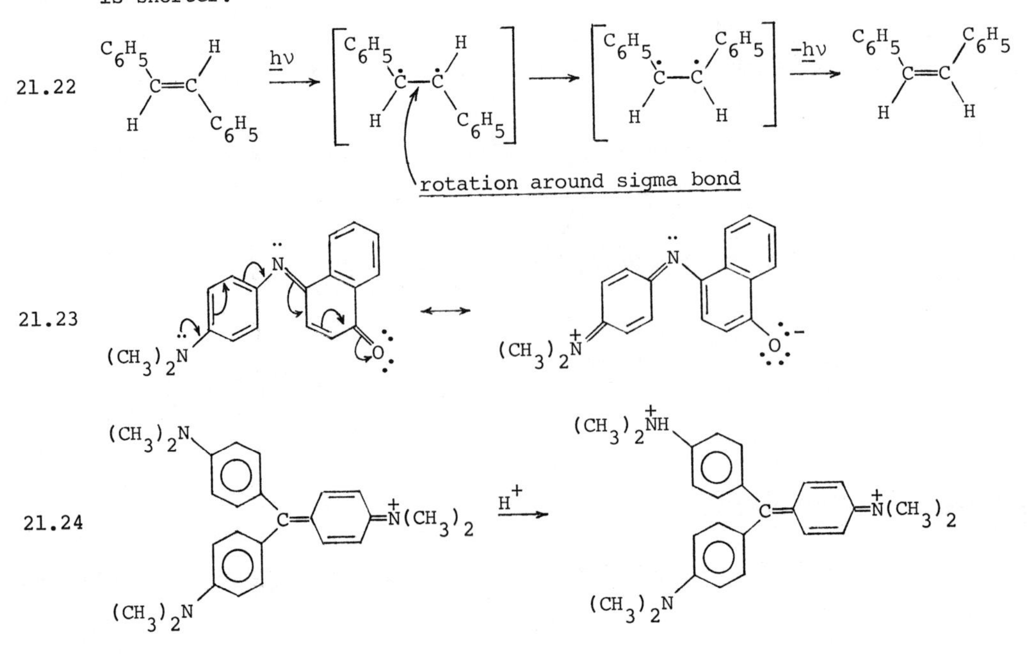

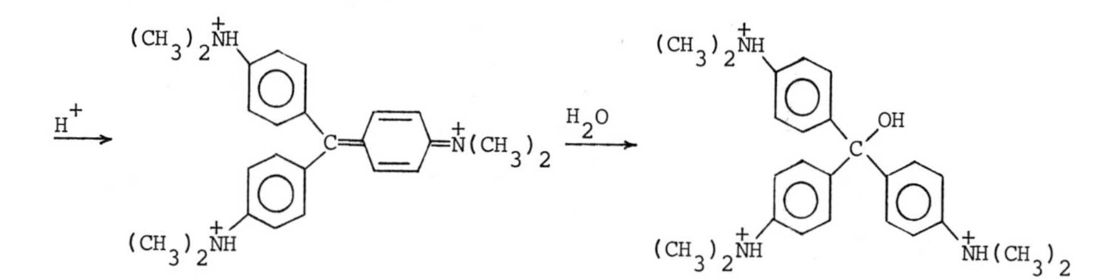

21.25 At pH 6, the structure is as shown in the problem. At pH 9:

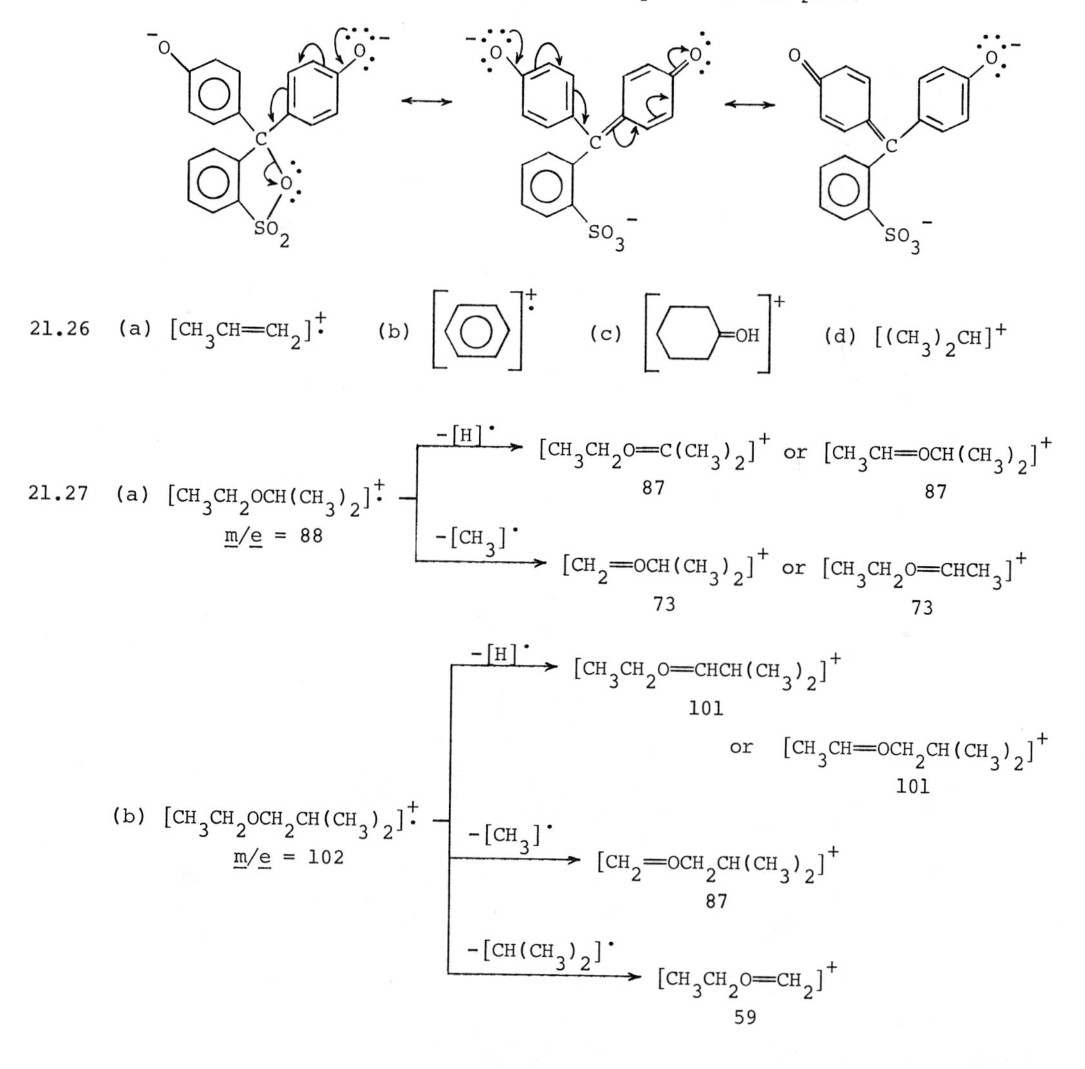

21.26 (a) $[CH_3CH{=}CH_2]^{+\cdot}$ (b)$^{+\cdot}$ (c)$^{+}$ (d) $[(CH_3)_2CH]^{+}$

21.27 (a) $[CH_3CH_2OCH(CH_3)_2]^{+\cdot}$
$\underline{m}/\underline{e} = 88$

$\xrightarrow{-[H]^{\cdot}}$ $[CH_3CH_2O{=}C(CH_3)_2]^{+}$ or $[CH_3CH{=}OCH(CH_3)_2]^{+}$
87 87

$\xrightarrow{-[CH_3]^{\cdot}}$ $[CH_2{=}OCH(CH_3)_2]^{+}$ or $[CH_3CH_2O{=}CHCH_3]^{+}$
73 73

(b) $[CH_3CH_2OCH_2CH(CH_3)_2]^{+\cdot}$
$\underline{m}/\underline{e} = 102$

$\xrightarrow{-[H]^{\cdot}}$ $[CH_3CH_2O{=}CHCH(CH_3)_2]^{+}$
101

or $[CH_3CH{=}OCH_2CH(CH_3)_2]^{+}$
101

$\xrightarrow{-[CH_3]^{\cdot}}$ $[CH_2{=}OCH_2CH(CH_3)_2]^{+}$
87

$\xrightarrow{-[CH(CH_3)_2]^{\cdot}}$ $[CH_3CH_2O{=}CH_2]^{+}$
59

(c) $[(CH_3)_2CHCl]^{+\cdot}$

$M^{+\cdot} = 78$
$M+2 = 80$

$\xrightarrow{-[H]^\cdot} [(CH_3)_2CCl]^+$ 77 (79)

$\xrightarrow{-[CH_3]^\cdot} [CH_3CHCl]^+$ 63 (65)

$\xrightarrow{-[Cl]^\cdot} [(CH_3)_2CH]^+$ 43

$\xrightarrow{-HCl} [CH_2{=}CHCH_3]^{+\cdot}$ 42

(d) $[(CH_3)_2CHCH_2CH_2CH(CH_3)_2]^{+\cdot}$

m/e = 114

$\xrightarrow{-[CH_3]^\cdot} [CH_3CHCH_2CH_2CH(CH_3)_2]^+$ 99

$\xrightarrow{-[(CH_3)_2CHCH_2CH_2]^\cdot} [(CH_3)_2CH]^+$ 43

$\xrightarrow{-[H]^\cdot} [(CH_3)_2CCH_2CH_2CH(CH_3)_2]^+$ 113

(e) $[(CH_3)_2CHOH]^{+\cdot}$

m/e = 60

$\xrightarrow{-[H]^\cdot} [(CH_3)_2COH]^+$ 59

$\xrightarrow{-H_2O} [CH_2{=}CHCH_2]^{+\cdot}$ 42

$\xrightarrow{-[OH]^\cdot} [(CH_3)_2CH]^+$ 43

(f) [cyclopentyl–CH(H)–CH_2–CH_2–CH=O]$^{+\cdot}$ (McLafferty rearrangement, H transfer to O) $\xrightarrow{-\text{cyclopentyl}-CH{=}CH_2}$ $[H_2C{=}CH{-}OH]^{+\cdot}$

m/e = 140 → 44

21.28 (a) $[CH_3CH_2CH_2CH_3]^{+\cdot}$ (58) $\longrightarrow$ $[CH_3CH_2CH_2CH_2]^+$ (57) $\longrightarrow$

$[CH_3CH_2CH_2]^+$ (43) $\longrightarrow$ $[CH_3CH_2]^+$ (29) $\longrightarrow$ $[CH_3]^+$ (15)

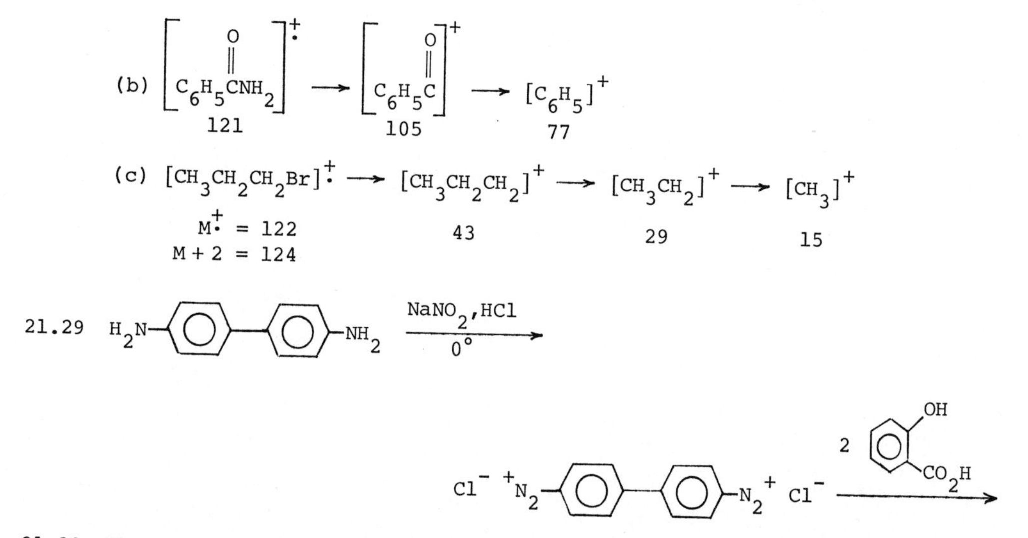

(b) $\left[C_6H_5\overset{\overset{\textstyle O}{\|}}{C}NH_2 \right]^{+\cdot} \longrightarrow \left[C_6H_5\overset{\overset{\textstyle O}{\|}}{C} \right]^{+} \longrightarrow [C_6H_5]^{+}$

$\quad\quad$ 121 $\quad\quad\quad\quad\quad$ 105 $\quad\quad\quad\quad$ 77

(c) $[CH_3CH_2CH_2Br]^{+\cdot} \longrightarrow [CH_3CH_2CH_2]^{+} \longrightarrow [CH_3CH_2]^{+} \longrightarrow [CH_3]^{+}$

$\quad\quad M^{+\cdot} = 122$ $\quad\quad\quad\quad\quad\quad\quad$ 43 $\quad\quad\quad\quad\quad$ 29 $\quad\quad\quad\quad$ 15
$\quad\quad M + 2 = 124$

21.29

21.30 The addition of each alkyl substituent to the conjugated system adds about 5 nm to the λ_{max}.

21.31 (a) 217 + 15 = 232 nm (b) 217 + 30 + 15 = 262 nm

21.32 (a) 229 nm (b) 234 nm (c) 224 + 30 − 5 = 249 nm

21.33 (b), (c), (d). All are methyl ketones and yield acetyl fragments (43), $[C_4H_9]^{+\cdot}$(57), pentanoyl fragments (85), and a molecular ion (100).

21.34 $CH_3\overset{O}{\overset{\|}{C}}CH_3$

21.35 A, $ClCH_2CH_2CH_2OH$ B, CH_3–C_6H_4–CN C, $C_6H_5\overset{O}{\overset{\|}{C}}CH_2CH_2CH_3$ D, C_6H_5I